Pomotherapeutic Insights on Wild Edible Fruits

Parimelazhagan Thangaraj
Editor

Pomotherapeutic Insights on Wild Edible Fruits

Springer

Editor
Parimelazhagan Thangaraj
Department of Botany
Bharathiar University
Coimbatore, Tamil Nadu, India

ISBN 978-981-96-0542-2 ISBN 978-981-96-0543-9 (eBook)
https://doi.org/10.1007/978-981-96-0543-9

This Springer imprint is published by the registered company Springer Nature Singapore Pte Ltd.
The registered company address is: 152 Beach Road, #21-01/04 Gateway East, Singapore 189721, Singapore

To Bioprospecting Research Team

Foreword

Welcome to the fascinating world of wild edible plants and their medicinal treasures. This meticulously researched and thoughtfully organized book invites readers to delve into nature's pharmacy, where every leaf, flower, and root carries a narrative of resilience, adaptation, and healing. Exploring these plants is akin to deciphering a hidden code—a realm of abundance and diversity generously offered by nature. A cornucopia of flavors and nutrients, often overlooked in our modern diets dominated by the convenience of supermarkets. Yet, these plants are far more than culinary curiosities; they harbor profound medicinal properties, revered by cultures worldwide for millennia. They are not merely sustenance; they are ancestral wisdom, passed down through generations, encapsulating the intricate relationship between humans and their environment. This book is, above all, a tribute. A tribute to the planet's botanical richness and the ancestral knowledge that has preserved the understanding of these species' therapeutic potential. Each chapter immerses readers in the unique characteristics and uses of various plants, offering practical insights into identification, harvesting, and preparation. Whether you are an avid forager, an experienced herbalist, or simply curious about the natural world around you, this book serves as a guide to responsibly discover and engage with these gifts of nature. As you navigate the pages of this work, I encourage you to approach them with curiosity and reverence. Every plant you encounter has a story and a purpose—be it as food, medicine, or ecological partner. May this book inspire you to explore, learn, and, above all, deepen your connection with the natural world. Finally, I extend my profound gratitude to the authors and researchers whose dedication has made this project possible. Their passion for wild plants and commitment to sharing knowledge shine through these pages, offering readers a glimpse into a world brimming with abundance and wonder. May we embrace the wisdom of wild edible plants with gratitude and humility, honoring the intricate web of life that sustains us all.

Lucindo J. Quintans Junior
Professor and Director, Department of Physiology
Federal University of Sergipe, Sao Cristovao, Brazil

Preface

Wild edible plants have long been an integral part of traditional diets across various indigenous communities, providing not only essential nutrients but also a wealth of bioactive compounds with potential health benefits. This book aims to bridge the gap between traditional knowledge and modern scientific research by exploring the nutritional, phytochemical, and medicinal properties of wild edible fruits, with a particular focus on those consumed in Tamil Nadu, India, and other ecologically significant regions

The book begins with an ethnobotanical investigation into the nutritional profile of wild edible plants commonly consumed by the Malayali tribal communities in Tamil Nadu (Chapter "Nutritional Analysis on Wild Edible Fruits of Kolli Hills (Eastern Ghats of Tamil Nadu), India"), offering insights into the dietary traditions of indigenous populations. This is followed by an exploration of the nutraceutical potential of edible plants (Chapter "Nutraceutical Potential of Edible Plants for Balancing Dietary Requirements in Human Diet: A Case Study"), emphasizing their role in maintaining a balanced diet. The therapeutic significance of *Muntingia calabura* fruits is reviewed in Chapter "Exploring the Medicinal Potential of *Muntingia calabura Linn.* Fruit: A Comprehensive Overview", highlighting their medicinal benefits. Chapter "Wild Edible Fruits from the Western Ghats: A Rich Source of Bioactive Compounds with Health-Promoting Effects: A Review" presents a comprehensive review of wild edible fruits from the Western Ghats, discussing their bioactive compounds and health-promoting effects. A study on the antioxidant activity of unripe *Syzygium calophyllifolium* fruits using in vitro models is detailed in Chapter "In Silico Analysis of *Syzygium calophyllifolium* Phytocompounds with Type 2 Diabetes Mellitus Targets", providing scientific validation of their pharmacological potential.

Moving forward, Chapter "An Insight to Intensify the Potential Prospects of Underutilized Wild Species of *Piper*" focuses on underutilized wild species of Piper, emphasizing the need to harness their bioactive potential. Chapter "Unveiling the Anti-inflammatory Properties of *Ficus* Fruits: A Promising Frontier for Potential Therapeutic Benefits" investigates the anti-inflammatory properties of *Ficus* fruits, shedding light on their therapeutic promise. The nutritional and phytochemical potential of three lesser-known wild edible fruits is examined in Chapter "Potential Prospects of Underexploited Wild Edible Fruits of Tamil Nadu". In Chapter "Curative Potentialities of *Garcinia gummi-gutta* (L.) Roxb.- a Less Exploited

Tropical Fruit Tree", the book discusses the curative properties of *Garcinia gummi-gutta*, a tropical fruit tree with underexplored medicinal value. A review on anticancer potential of *Psidium guajava* is elaborated in Chapter "*Psidium guajava* L. (Poor Man's Fruit) with Anticancer Potential: A Comprehensive Review". The book also provides region-specific studies, such as an overview of wild edible fruits from the Servarayan and Kalvarayan hill ranges (Chapter "An Overview on Wild Edible Fruits of Servarayan and Kalrayan Hilly Ranges"), followed by an in-depth nutritional and metabolomic analysis of wild edible fruits from the Western Ghats of Tamil Nadu (Chapter "A Review on Nutritional and Metabolic Analysis of Wild Edible Fruits of Western Ghats of Tamil Nadu, India"). The bioactive composition and intestinal anti-inflammatory properties of Brazilian Cerrado fruits are explored in Chapter "Improving the Microbiological, Sensory, and Chemical Constituents of Stored Crystallized "Jenipapo" (*Genipa americana* L.), a Native Brazilian Fruit", extending the discussion to a global context. Chapter "Biomedical Applications and Health Benefits of Wild Noni and Its Fruits" focuses on the biomedical applications and health benefits of wild Noni fruits, a tropical species gaining attention for its therapeutic properties. Chapter "Under-Exploited Wild Edible Fruits of the Western Ghats: A Review on Antioxidant Potential and Phytochemistry" offers a broader exploration of wild edible fruits, while Chapter "Edible Fruit-Yielding Wild Plants of Tamil Nadu: A Preliminary Checklist" presents a state-wide review of the wild edible fruits of Tamil Nadu, consolidating existing knowledge on their diversity and benefits. The book concludes with Chapter "Application of Herbal-Based Edible Coating on Fruits: A Promising Technology to Counter Postharvest Losses", which discusses herbal-based edible coatings for fruits as a promising post-harvest technology, an innovative approach to reducing food wastage and extending shelf life.

We express our sincere gratitude to all the contributors whose expertise and dedication have made this book possible. We also extend our appreciation to the scientific community and indigenous knowledge keepers who have played a crucial role in documenting and preserving the traditional uses of these plants. It is our hope that this book will serve as a catalyst for further research and innovation in the field of wild edible plants and their potential applications in sustainable nutrition and medicine.

Coimbatore, Tamil Nadu, India — Parimelazhagan Thangaraj

Acknowledgments

I sincerely thank the Bharathiar University administration for its continuous encouragement and support. I especially appreciate the Vice Chancellor's committee, the Convenor, and the Registrar for their leadership. Special thanks are extended to the distinguished faculty members of the Department of Botany at Bharathiar University, Coimbatore. The committed researchers Dipshika Motwani N, Benedict Mathews P, Gowtham K, Francis Jegan Raj S, Anish Ruban S, Madhu Bala D, Vetri Velavan S, and Ashanath Fowmitha N. are also greatly appreciated. Their substantial input and careful work were crucial to the thorough compilation of the chapters included in this publication. In addition, I would like to express gratitude in particular to the publishers and their employees for their sincere attempts and dedication in releasing the book on time. Their diligence and professionalism made sure this job was published on time. Without the administration, academics, research team, and publishers' combined encouragement, support, and hard effort, this book would not have been feasible. Their contributions have been important in the effective completion of this project.

Parimelazhagan Thangaraj

Contents

Nutritional Analysis on Wild Edible Fruits of Kolli Hills (Eastern Ghats of Tamil Nadu), India 1
C. Jenipher and M. Ayyanar

Nutraceutical Potential of Edible Plants for Balancing Dietary Requirements in Human Diet: A Case Study 19
Binu Thomas

Exploring the Medicinal Potential of *Muntingia calabura Linn.* Fruit: A Comprehensive Overview 37
Preethi Kathirvel and Vishruthi Sundar Kumar

Wild Edible Fruits from the Western Ghats: A Rich Source of Bioactive Compounds with Health-Promoting Effects: A Review 53
Gowtham Kannan, Benedict Mathews Paul, and Parimelazhagan Thangaraj

In Silico Analysis of *Syzygium calophyllifolium* Phytocompounds with Type 2 Diabetes Mellitus Targets 77
Pavithran Balasubramanian, Saikumar Sathyanarayanan, Rahul Chandran, Heidi Abrahamse, and Parimelazhagan Thangaraj

An Insight to Intensify the Potential Prospects of Underutilized Wild Species of *Piper* 89
M. Pugalenthi, M. A. Deepa, M. Pradheeba, S. Narthanaa, and T. Parimelazhagan

Unveiling the Anti-inflammatory Properties of *Ficus* Fruits: A Promising Frontier for Potential Therapeutic Benefits 115
Madhu Bala Durairajan, Francis Jegan Raj, S. Anish Ruban, and Parimelazhagan Thangaraj

Potential Prospects of Underexploited Wild Edible Fruits of Tamil Nadu 133
M. A. Deepa, M. Pugalenthi, S. Narthanaa, M. Pradheeba, and T. Parimelazhagan

Curative Potentialities of *Garcinia gummi-gutta* (L.) Roxb.: A Less-Exploited Tropical Fruit Tree 163
B. Vishnu, R. M. Deepthy, Parimelazhagan Thangaraj, and E. A. Siril

***Psidium guajava* L. (Poor Man's Fruit) with Anticancer Potential: A Comprehensive Review** 191
Anju Rani George, Kavimani Thangasamy, and Natesan Geetha

An Overview on Wild Edible Fruits of Servarayan and Kalrayan Hilly Ranges 215
V. Sivasankari and P. Viswanathan

A Review on Nutritional and Metabolic Analysis of Wild Edible Fruits of Western Ghats of Tamil Nadu, India 237
Vetri Velavan Sundararajan, Ashanath Fowmitha Nazeer, and Parimelazhagan Thangaraj

Improving the Microbiological, Sensory, and Chemical Constituents of Stored Crystallized "Jenipapo" (*Genipa americana* L.), a Native Brazilian Fruit 263
Raquel P. Campos, Marana Soares, Karuppusamy Arunachalam, Suni Liu, Luciana Miyagusku, and Priscila Aiko Hiane

Biomedical Applications and Health Benefits of Wild Noni and Its Fruits 273
P. Ponmurugan, A. Anitha, and N. Mathivanan

Under-Exploited Wild Edible Fruits of the Western Ghats: A Review on Antioxidant Potential and Phytochemistry 287
J. M. Sasikumar and Sibbala Subramanyam

Edible Fruit-Yielding Wild Plants of Tamil Nadu: A Preliminary Checklist 305
G. Gnanasekaran, E. Angelin Felicia, S. P. Nithya, and W. Arisdason

Application of Herbal-Based Edible Coating on Fruits: A Promising Technology to Counter Postharvest Losses 327
Murugan Rajan, Maria Terezinha Santos Leite Neta, and Narendra Narain

Index 345

Contributors

Heidi Abrahamse Laser Research Centre, Faculty of Health Sciences, University of Johannesburg, Johannesburg, South Africa

E. Angelin Felicia Department of Botany, Madras Christian College (Autonomous), Chennai, Tamil Nadu, India

S. Anish Ruban Bioprospecting Laboratory, Department of Botany, Bharathiar University, Coimbatore, Tamil Nadu, India

A. Anitha Department of Botany, Bharathiar University, Coimbatore, Tamil Nadu, India

W. Arisdason Department of Botany, Madras Christian College (Autonomous), Chennai, Tamil Nadu, India

Karuppusamy Arunachalam Center for Studies in Stem Cells, Cellular Therapy and Toxicological Genetics (CeTroGen), Faculty of Medicine (FAMED), Federal University of Mato Grosso do Sul (UFMS), Campo Grande, Mato Grosso do Sul, Brazil

M. Ayyanar Department of Botany, A.V.V.M. Sri Pushpam College (Autonomous), Poondi, Thanjavur, Tamil Nadu, India

Pavithran Balasubramanian Department of Botany, PSG College of Arts & Science, Coimbatore, Tamil Nadu, India

Raquel P. Campos Faculty of Pharmaceutical Sciences, Food and Nutrition (FACFAN), Federal University of Mato Grosso do Sul Foundation (UFMS). Campo Grande, Mato Grosso do Sul, Brazil

Rahul Chandran Laser Research Centre, Faculty of Health Sciences, University of Johannesburg, Johannesburg, South Africa

M. A. Deepa Department of Botany, Government Arts College (Autonomous), Coimbatore, Tamil Nadu, India

R. M. Deepthy Department of Botany, University of Kerala, Kariavattom, Thiruvananthapuram, Kerala, India

Madhu Bala Durairajan Bioprospecting Laboratory, Department of Botany, Bharathiar University, Coimbatore, Tamil Nadu, India

Natesan Geetha Department of Botany, Bharathiar University, Coimbatore, Tamil Nadu, India

Anju Rani George Department of Botany, Bharathiar University, Coimbatore, Tamil Nadu, India

G. Gnanasekaran Department of Botany, Madras Christian College (Autonomous), Tambaram East, Chennai, Tamil Nadu, India

Priscila Aiko Hiane Faculty of Pharmaceutical Sciences, Food and Nutrition (FACFAN), Federal University of Mato Grosso do Sul Foundation (UFMS). Campo Grande, Mato Grosso do Sul, Brazil

C. Jenipher Department of Botany, A.V.V.M. Sri Pushpam College (Autonomous), Poondi, Thanjavur, Tamil Nadu, India

Gowtham Kannan Bioprospecting Laboratory, Department of Botany, Bharathiar University, Coimbatore, Tamil Nadu, India

Suni Liu Faculty of Pharmaceutical Sciences, Food and Nutrition (FACFAN), Federal University of Mato Grosso do Sul Foundation (UFMS). Campo Grande, Mato Grosso do Sul, Brazil

N. Mathivanan Department of Centre for Advanced Study in Botany, University of Madras, Chennai, Tamil Nadu, India

Luciana Miyagusku Faculty of Pharmaceutical Sciences, Food and Nutrition (FACFAN), Federal University of Mato Grosso do Sul Foundation (UFMS). Campo Grande, Mato Grosso do Sul, Brazil

Murugan Rajan Centre for Research and Postgraduate Studies in Botany, Ayya Nadar Janaki Ammal College (Autonomous), Sivakasi, Tamil Nadu, India

Narendra Narain Laboratory of Flavor and Chromatographic Analysis, Federal University of Sergipe, Sao Cristovao, Brazil

S. Narthanaa Department of Botany, Government Arts College (Autonomous), Coimbatore, Tamil Nadu, India

Ashanath Fowmitha Nazeer Bioprospecting Laboratory, Department of Botany, Bharathiar University, Coimbatore, Tamil Nadu, India

Maria Terezinha Santos Leite Neta Laboratory of Flavor and Chromatographic Analysis, Federal University of Sergipe, Sao Cristovao, Brazil

S. P. Nithya Department of Botany, Madras Christian College (Autonomous), Chennai, Tamil Nadu, India

Benedict Mathews Paul Bioprospecting Laboratory, Department of Botany, Bharathiar University, Coimbatore, Tamil Nadu, India

P. Ponmurugan Department of Botany, Bharathiar University, Coimbatore, Tamil Nadu, India

M. Pradheeba Department of Botany, Government Arts College (Autonomous), Coimbatore, Tamil Nadu, India

Preethi Kathirvel Department of Microbial Biotechnology, Bharathiar University, Coimbatore, Tamil Nadu, India

M. Pugalenthi Department of Botany, Government Arts College (Autonomous), Coimbatore, Tamil Nadu, India

Francis Jegan Raj Bioprospecting Laboratory, Department of Botany, Bharathiar University, Coimbatore, Tamil Nadu, India

J. M. Sasikumar Department of Microbiology, Karpagam Academy of Higher Education (KAHE), Coimbatore, Tamil Nadu, India

Saikumar Sathyanarayanan Department of Botany, PSG College of Arts & Science, Coimbatore, Tamil Nadu, India

E. A. Siril Department of Botany, University of Kerala, Kariavattom, Thiruvananthapuram, Kerala, India

V. Sivasankari Department of Botany, Government Arts College (Autonomous), Coimbatore, Tamil Nadu, India

Marana Soares Faculty of Pharmaceutical Sciences, Food and Nutrition (FACFAN), Federal University of Mato Grosso do Sul Foundation (UFMS). Campo Grande, Mato Grosso do Sul, Brazil

Sibbala Subramanyam Vignans Foundation for Science, Technology and Research, Guntur, India

Vetri Velavan Sundararajan Bioprospecting Laboratory, Department of Botany, Bharathiar University, Coimbatore, Tamil Nadu, India

Parimelazhagan Thangaraj Bioprospecting Laboratory, Department of Botany, Bharathiar University, Coimbatore, Tamil Nadu, India

Kavimani Thangasamy Department of Botany, Bharathiar University, Coimbatore, Tamil Nadu, India

Binu Thomas Department of Botany, St. Joseph's College (Autonomous), Kozhikode, Kerala, India

B. Vishnu Department of Botany, University of Kerala, Kariavattom, Thiruvananthapuram, Kerala, India

Vishruthi Sundar Kumar Department of Microbial Biotechnology, Bharathiar University, Coimbatore, Tamil Nadu, India

P. Viswanathan Department of Botany, Government Arts College (Autonomous), Coimbatore, Tamil Nadu, India

About the Editor

Parimelazhagan Thangaraj is a distinguished academician and researcher, currently serving as Professor and Head of the Department of Botany at Bharathiar University, Coimbatore. He is a graduate of St. Joseph's College, Trichy, a Ph.D. holder from Bharathiar University, and D.Sc. holder from Periyar University, Salem. He brings extensive experience from his tenure as a Scientist with the Defence Research & Development Organization (DRDO), Ministry of Defence, Government of India. With 22 years of research and 13 years of teaching experience, he has also completed post-doctoral research in Hawaii, USA. He has authored ten books in his field of specialization and holds four patents. His prolific research output includes over 132 publications in esteemed international and national peer-reviewed journals with 35 h-index and 4142 citations. In addition to his academic roles, he is the Director of the Bharathiar University's Center for Research and Development and Dean, Faculty of Science. He is an active member of the academic community, serving on the editorial boards of various journals and frequently reviewing scholarly articles. He has guided 16 scholars to receive Ph.D. degree and 13 M.Phil. graduates. He has been recognized for his contributions to science with the National Science Day Medal (2005) and the Laboratory Scientist of the Year award (2005) from DRDO, New Delhi. He oversees multiple research projects funded by Department of Biotechnology (DBT), Department of Science and Technology (DST), University Grants Commission (UGC), Defence Research and Development Organization (DRDO), Tamil Nadu State Council for Higher Education (TANSCHE), and Department of AYUSH, Government of India, with grants totaling over 2.9 Crores. In recognition of his scientific achievements, he has been bestowed with the UGC Mid-Career Award and elected as a Fellow of the Linnaean Society and the Royal Society of Biological Sciences, both located in London. His current work focuses on creating innovative nutraceutical goods and herbal medications to treat common human ailments like cancer, rheumatoid arthritis, and diabetes.

Nutritional Analysis on Wild Edible Fruits of Kolli Hills (Eastern Ghats of Tamil Nadu), India

C. Jenipher and M. Ayyanar

Abstract

Wild edible fruits contribute significance to the local biodiversity and nutrition of the indigenous communities with their rich composition of macro and micronutrients. This study documents a total of 43 wild edible fruit species found in the Kolli Hills region, with a focus on the detailed examination of 12 selected species. The research employs a multidisciplinary approach, combining botanical, ethnobotanical, and nutritional analyses to provide a comprehensive understanding of the identified fruits. Botanical investigations involve taxonomic classification, morphological descriptions, and distribution patterns of the documented fruits. Ethnobotanical surveys are conducted to gather indigenous knowledge, traditional uses, and cultural significance associated with these wild fruits. Additionally, nutritional analyses explore the potential health benefits and dietary value of the selected fruits. The results highlight the ecological importance of these wild edible fruits in the forest ecosystem, serving as a crucial resource for both human sustenance and wildlife. The ethnobotanical findings shed light on the cultural heritage and traditional practices of local communities, emphasizing the need for sustainable conservation efforts. Furthermore, the nutritional analyses provide valuable insights into the potential role of these fruits in promoting health and nutrition, thereby contributing to food security. The findings aim to facilitate conservation initiatives, promote traditional knowledge preservation, and encourage the integration of wild edible fruits into broader discussions on malnutrition, food insecurity, biodiversity conservation, and sustainable development.

C. Jenipher · M. Ayyanar (✉)
Department of Botany, A.V.V.M. Sri Pushpam College (Autonomous), Poondi, Thanjavur, India

P. Thangaraj (ed.), *Pomotherapeutic Insights on Wild Edible Fruits*,
https://doi.org/10.1007/978-981-96-0543-9_1

Keywords

Eastern Ghats · Functional foods · Nutrition · Tribal community · Wild edible fruits

Introduction

Wild edible fruits have a vital role in the food chain of rural communities as they shares a notable part of being fed to ethnic people, mammals, and migratory birds [1]. Among the wild edibles (roots, rhizomes, tubers, leaves, fruits), wild edible fruits are majorly preferred due to their readily availability, aroma, and taste and also because of their therapeutic properties. Apart from being freshly consumed, the fruits are also preserved in different forms (pickles, jams, beverages, marmalade, and compotes). The abundance of dietary fibers, minerals, vitamins, and several secondary metabolites makes them highly nutritional [2]. In addition to providing essential nutrients, the bioactive molecules (organic acids, phenolics, flavonoids) of wild edibles carry therapeutical properties, and this makes them functional foods [3]. The wild edible fruits are also renowned as a vital source of antioxidants.

The nutritional and chemical composition of wild edible fruits is higher than the cultivated fruits that are available in the markets. The increased exposure to biotic interactions in its wild habitat triggers the plant's defense mechanism to produce high content of secondary metabolites [1]. A healthy diet is regarded as the shield in preventing the incidence of several chronic illnesses and disorders. A diet with the required amount of nutrients and minerals can greatly lessen the risk of various chronic diseases like cancer, diabetes, and cardiovascular diseases [4].

About two billion people around the world have mineral deficiencies due to consumption of diets with less or no essential micronutrients, especially iodine, vitamin A, zinc, and iron. Malnutrition is linked to brain development and cognitive impairment [5]. The consumption or ingestion of wild edible fruits is viewed as an effective solution to combat mineral deficiencies as the wild edibles have reports of being an ideal source of macro and trace minerals, namely, boron, copper, iron, manganese, molybdenum, zinc, sodium, potassium, magnesium, calcium, phosphorus, sulfur, and nitrogen [6]. Micronutrients are required by the human body for carrying out essential physiological processes, such as metabolism, muscle contraction, and nerve functioning; it also maintains the pH and osmotic pressure and helps in the synthesis of hormones and enzymes [7].

Non-timber forest produces (NTFPs) include fruits, herbs, mushrooms, lichens, shoots, saps, tubers, rhizomes, etc., and most of the ethnic people rely on these forest produces for their regular diet [8]. Also, NTFPs contribute significantly to the livelihood of tribal inhabitants as an economic uplifter. The wild edible plant species are highly diversified in terms of its habitat, resistance or resilience to biotic factors, and pattern of succession [8]. Most of the potentially edible wild fruits grow in mountainous deciduous forests. More than 75% of the ethnic or indigenous population in developing countries depend on NTFPs for livelihood due to their significant nutritional and therapeutic properties.

According to Siegel et al. [9], the global supply of fruits and vegetables is inadequate to meet or fulfill the nutritional requirements of the world's increasing population. And only 12 food crops and five animal species deliver about 75% of the world's food for mankind [5]. Therefore, it is essential to increase the global production of food crops to meet the rising food demand. An important strategy to meet this rising demand is to employ wild edible food products that are nutritionally higher than the domesticated food crops [10, 11]. Due to the complicated risk of malnutrition and its irrevocable consequences, researchers and health professionals are focusing on nutritional improvement among the people of developing countries through functional foods. About 7039 species of plants have been documented as edible, of which only 417 species are considered as food crops [12]. The modern food system, industrial farming and agriculture, increased reliance on domesticated plants, and genetic modification have led to the disappearance of numerous wild edible plants [5].

Kolli Hills in the Eastern Ghats of Tamil Nadu (India) is renowned for its rich biodiversity and favorable ecological conditions and harbors a diverse array of wild edible fruits, providing both nutritional sustenance and economic opportunities to the indigenous Malayali tribes. This hill is located in Namakkal district of Tamil Nadu, India, with an altitude of 1000–1400 meters above mean sea level (MSL). This hill covers an area of about 280 km^2 with vegetation varying greatly from foothills (scrub jungle) to high ranges (evergreen sholas). Kolli Hills has a pleasing and cold climate throughout the year. The average annual precipitation is around 1440 mm with a temperature range between 13 and 30 °C. The three major rivers emerging from Kolli Hills are Karaipattamur River, Karuvattar River, and Swetandi River. The soil type of Kolli Hills, being red loamy and black soil, is the reason behind the lustrous growth of medicinal plants and wild edible plants [13].

Kolli Hills is also referred as the fertile pocket in the Eastern Ghats, where exotic and tropical fruits also grow in abundant [14]. The ethnic community, residing in mountainous regions, heavily relies on wild edibles for food and medicine, underlining the cultural and practical significance of these resources [15]. Despite their importance and being sold in tribal markets as seasonal fruits, comprehensive studies on the nutritional composition of wild edibles, particularly wild fruits, remain scarce. The present study aims to delve deeper into the nutritional composition and significance of wild edible fruits, particularly in the context of Kolli Hills region, shedding light on their potential contribution to addressing nutritional deficiencies and fostering sustainable food systems.

Ethnic People of the Study Area

Kolli Hills is inhabited by Malayali tribes, who originally belonged to Vellala community of Kanchipuram district and migrated to Kolli Hills about 600 years ago. They called themselves Malayalis, referring to "people who rule the hills," "Malai (mountain) and Ali (rulers)." The main occupations of Malayali tribes are farming, poultry, cattle rearing, carpentry, and edible fruit collection. They involve in the collection of NTFPs, including, medicinal plants, fruits, honey, beeswax, and firewood from the forest. Malayalis are known for their folk knowledge of plants, animals, and agriculture. Their major diet includes seed grains of *Eleusine coracana* (L.) Gaertn. (finger millet), *Setaria italica* (L.) P.Beauv. (foxtail millet), *Echinochloa frumentacea* Link (barnyard millet), and *Oryza sativa* L. (rice). Cash crops, namely, *Manihot esculenta* Crantz (cassava), *Piper nigrum* L. (pepper), *Areca catechu* L. (areca nut), *Elettaria cardamomum* (L.) Maton (cardamom), *Cinnamomum verum* J.Presl (cinnamon), *Coffea arabica* L. (coffee), and *Syzygium aromaticum* (L.) Merr. & L.M.Perry (clove), are cultivated in their farm lands, which improved the economic status of Malayalis.

Wild Edibles of Kolli Hills

From the field visits and literature survey, a total of 43 plant species were documented to be used as wild edible fruits among the Malayali tribes in Kolli Hills, India (Table 1). Previously, Ranjithakani et al. [16] have documented 25 species of wild edible plants (leaves, stems, fruits, rhizomes, and tubers) utilized by Malayali tribes of Kolli Hills; Samydurai et al. [17] documented 38 species of wild edible tubers and rhizomes from Kolli hills; Anand and Deborah [18] recorded 31 species of edible plants whose fruits are widely eaten by the local inhabitants and are sold in local markets as seasonal fruits. The documented fruits were also reported to possess significant therapeutic properties; thus, these fruits were consumed not only for their palatability but also for their wide range of medicinal uses.

The documented 43 wild edible fruits belonged to 35 genera and 24 families. It was observed that the fruits belonging to the family Rutaceae (6 species) were widely used in the study area, which was followed by Rubiaceae (four species), Moraceae (three species), Rhamnaceae (three species), and Solanaceae (three species). The families Annonaceae, Apocynaceae, Ebenaceae, Fabaceae, and Rosaceae were represented with two edible plant species each, and the remaining families with one species. From the documented wild edible fruits, twelve fruits were chosen in the present study based on its abundance and importance in the study area. It includes *Alangium salviifolium* (L.f.) Wangerin, *Carissa carandas* L., *Carissa spinarum* L., *Diospyros melanoxylon* Roxb., *Elaeocarpus serratus* L., *Passiflora subpeltata* Ortega, *Phoenix loureiroi* Kunth, *Rubus ellipticus* Sm., *Rubus niveus* Thunb., *Solanum torvum* Sw., *Toddalia asiatica* (L.) Lam., and *Ziziphus rugosa* Lam. These 12 species were discussed in detail pertaining to their nutritional properties.

Table 1 List of wild edible fruits used by Malayali tribes in Kolli Hills of Tamil Nadu, India

Binomial name	Family	Common names	Vernacular (Tamil) name	Color of mature fruit/ mode of intake
Aegle marmelos (L.) Corrêa	Rutaceae	Bael	Vilvam palam	Yellow/raw
Alangium salviifolium (L.f.) Wangerin	Cornaceae	Sage-leaved alangium	Alingi	Red/raw
Ampelocissus tomentosa (Roth) Planch.	Vitaceae	Hairy wild grape	Kattukodi	Purple-black/ raw
Annona reticulata L.	Annonaceae	Sugar apple, bullock's heart	Ramaseetha	Red/raw
Annona squamosa L.	Annonaceae	Custard apple, sweetsop	Seetha maram	Green/raw
Artocarpus heterophyllus Lam.	Moraceae	Jackfruit	Palamaram	Green/raw
Canthium coromandelicum (Burm.f.) Alston	Rubiaceae	Coromandel boxwood	Sirukeerai	Black/raw
Carissa carandas L.	Apocynaceae	Christ's thorn	Perukala	Black/raw and processed
Carissa spinarum L.	Apocynaceae	Small carissa	Sirukila	Black/raw and processed
Catunaregam spinosa (Thunb.) Tirveng.	Rubiaceae	Emetic nut	Perukarai	Green/roasted
Citrus maxima (Burm.) Merr.	Rutaceae	Pomelo	Pambalimas	Yellow/raw
Clausena anisata (Willd.) Hook.f.	Rutaceae	Horsewood	Yanaikhai	Purple-black/ raw
Coccinia grandis (L.) Voigt	Cucurbitaceae	Ivy gourd	Kovai	Red/raw and cooked
Diospyros ferrea (Willd.) Bakh.	Ebenaceae	Sea ebony	Irumpilli	Yellow-orange/raw
Diospyros melanoxylon Roxb.	Ebenaceae	Coromandel ebony	Karai	Brown/raw
Elaeocarpus serratus L.	Elaeocarpaceae	Ceylon olive	Selamaram	Yellow/raw
Ficus benghalensis L.	Moraceae	Banyan	Aalamaram	Brown/raw
Ficus religiosa L.	Moraceae	Sacred fig	Arasamaram	Brown/raw
Hugonia mystax L.	Linaceae	Climbing flax	Mothirakanni	Red/raw
Lantana camara L.	Verbenaceae	Shrub verbena	Unni chedi	Black/raw
Limonia acidissima L.	Rutaceae	Wood apple	Vilapalam	Brown/raw

(continued)

Table 1 (continued)

Binomial name	Family	Common names	Vernacular (Tamil) name	Color of mature fruit/ mode of intake
Mangifera indica L.	Anacardiaceae	Mango	Maamaram	Yellow/raw and cooked
Mimusops elengi L.	Sapotaceae	Spanish cherry	Makilamaram	Orange-red/ raw
Morinda coreia Buch.-Ham.	Rubiaceae	Indian mulberry	Nuna maram	Black/juice
Murraya koenigii (L.) Spreng.	Rutaceae	Curry leaf tree	Karuveppilai	Black/raw
Opuntia stricta (Haw.) Haw.	Cactaceae	Prickly pear	Chappatikalli	Red/raw
Passiflora subpeltata Ortega	Passifloraceae	White passion	Thatbootpalam	Green-yellow/ raw
Phoenix loureiroi Kunth	Arecaceae	Mountain date palm	Malai eecham	Black/raw
Phyllanthus emblica L.	Phyllanthaceae	Indian gooseberry	Malaineli	Yellow/raw and processed
Physalis peruviana L.	Solanaceae	Cape gooseberry	Kootuthakkali	Orange/raw
Pithecellobium dulce (Roxb.) Benth.	Fabaceae	Madras thorn	Kodukkaapuli	Red/raw
Pogostemon benghalensis Kuntze	Lamiaceae	Bengal shrub	Arissikaai	Brown/raw
Rubus ellipticus Sm.	Rosaceae	Yellow Himalayan raspberry	Yerumaikanni	Yellow/raw
Rubus niveus Thunb.	Rosaceae	Ceylon raspberry	Yerumaikanni	Red-black/raw
Solanum torvum Sw.	Solanaceae	Turkey berry	Chundai	Green/cooked
Solanum trilobatum L.	Solanaceae	Red pea eggplant	Thoothuvalai	Red/raw
Syzygium cumini (L.) Skeels	Myrtaceae	Malabar plum, jambolan	Naval	Purple-black/ raw
Tamarindus indica L.	Fabaceae	Tamarind tree	Puli	Red/cooked
Tarenna asiatica (L.) Kuntze ex K.Schum.	Rubiaceae	Asiatic tarenna	Therani	Black/raw
Toddalia asiatica (L.) Lam.	Rutaceae	Orange climber	Milai karanai	Black/raw
Ziziphus mauritiana Lam.	Rhamnaceae	Indian jujube	Illanthai	Brown/raw
Ziziphus oenopolia (L.) Mill.	Rhamnaceae	Jackal jujube	Soorai mullu	Black/raw
Ziziphus rugosa Lam.	Rhamnaceae	Zunna berry	Kattuelanthai	Brown/raw

Table 2 Proximate composition of documented wild edible fruits

Binomial name/ parameters	Energy (Kcal)	Carbohydrates (g/100 g)	Proteins (g/100 g)	Lipids (g/100 g)	References
A. salviifolium	50.00–55.15	11.67	2.07–3.89	0.12–60.00	[20–23]
C. carandas	745.00–753.00	0.51–0.94	0.39–6.20	0.0029–7.50	[24–26]
C. spinarum	167.50	12.43–25.92	1.24–3.64	1.30–4.94	[27–30]
D. melanoxylon	491.75	26.70–26.93	0.57–8.64	0.20–0.71	[26, 31, 32]
E. serratus	–	–	4.92	1.10	[33]
P. subpeltata	–	0.02	0.88	–	[34]
P. loureiroi	–	1.28	3.63	3.00	[35]
R. ellipticus	372.12–374.06	0.08–86.40	1.40–8.83	0.015–2.86	[36–40]
R. niveus	364.42	85.35	3.28	1.10	[39, 40]
S. torvum	421.00	0.33–71.42	0.31–16.49	0.70–8.81	[41–45]
T. asiatica	–	–	1.15	–	[28]
Z. rugosa	–	15.00–39.11	0.58–1.12	1.35–5.00	[28, 46, 47]

Alangium salviifolium (L.f.) Wangerin

Alangium salviifolium is native to Bangladesh, India, Comoros, Nepal, Kenya, Sri Lanka, Pakistan, West Himalaya, and Tanzania [19]. The ripe fruits (red in color) of this thorny tree are eaten by the Malayali tribes of Kolli Hills. This tree starts to flower in the month of March and the fruits mature by the month of June. The fruits of *A. salviifolium* contained energy of 50.00–55.15 Kcal [20, 21]. The proximate composition (Table 2) of its fruits revealed high contents of carbohydrates (11.67 g/100 g), proteins (2.07, 2.96, and 3.89 g/100 g), and lipids (0.12, 0.23, and 60.00 g/100 g) [20–23]. The macro and micro-minerals, namely, potassium, sodium, calcium, magnesium, iron, zinc, manganese, copper, and vitamin C, were present in the *A. salviifolium* fruits with notable quantity (Table 3). The high content of potassium (168.00–179.65 mg/100 g), vitamin C (62.46–117.80 mg/100 g), calcium (12.00 mg/100 g), sodium (5.14 mg/100 g), iron (0.01–1.37 mg/100 g), and lipids (0.71 mg/100 g), were ascertained from this fruit [20, 21, 23]. Apart from being an edible fruit, it was also reported to treat diabetes, skin diseases, rheumatism, and hemorrhages by the Malayali tribes of Kolli Hills [18].

Carissa carandas L.

Carissa carandas is a woody shrub that originated in India and was later introduced to Bangladesh and Sri Lanka [19]. The ripe fruits of *C. carandas* are black in color, known for its usage as a condiment. This plant starts to flower in the month of August and the fruits mature during October. The fruits are considered to be a rich source of potassium (85.00 mg/100 g), phosphorus (10.5–66 mg/100 g), calcium (29.70–115.00 mg/100 g), magnesium (89.60 mg/100 g), sodium (57.00 mg/100 g),

Table 3 Mineral composition (mg/100 g) of documented wild edible fruits

Binomial name/ parameters	Sodium	Potassium	Phosphorus	Calcium	Magnesium	Iron	Zinc	Manganese	Copper	Vitamin C	References
A. salviifolium	5.14	168.00–179.65	–	12.00	–	0.01–1.37	–	–	–	62.46–117.80	[20, 21]
C. carandas	57.00	85.00	10.5–66	29.7–115	89.60	12.0–150	3.94	1.60	1.22	1.64	[24, 25]
C. spinarum	–	1281.50	32.10	29.00	–	2.95–3.45	1.25	0.92	–	17.80	[27, 29]
D. melanoxylon	0.09–87.11	123.00–305.52	20.00–167.30	58.42–90.86	10.80–56.29	0.26–0.66	0.12–0.67	1.26	0.004	2.00	[26, 31, 32]
E. serratus	–	–	–	–	–	–	–	–	–	5.93	[33]
P. subpeltata	2.00	181.00	10.00	5067.00	2.00	173.00	19.00	6.00	–	–	[34]
P. loureiroi	126.00	83.00	19.00	40.00	37.00	8.00	38.00	–	–	150.00	[35]
R. ellipticus	89.43	1.80–680.16	0.05–1.26	0.95–450.00	5.60–118.72	4.24	12.77	1.94	0.02	1.05–19.80	[36–39, 48]
R. niveus	56.30	720.00	1.48	390.00	179.00	3.26	8.13	2.43	1.07	10.20	[39, 49]
S. torvum	2.18–155.20	526.00 –1975.00	8.70–66.00	2.26–329.00	29.00–49.60	1.85–60.10	1.00–14.65	1.95–10.83	0.62–1.49	2.44–2.68	[41–44, 50]
T. asiatica	27.47	599.09	–	66.48	5.84	3.36	1.35	5.84	0.17	22.02	[28]
Z. rugosa	1.80–52.00	168.00–836.00	16.50–149.00	17.10–910.00	22.0–83.00	1.87	3.88	4.64	–	21.26	[28, 46, 47]

iron (12.00–150.00 mg/100 g), zinc (3.94 mg/100 g), manganese (1.6 mg/100 g), copper (1.22 mg/100 g), and vitamin C (1.64 mg/100 g) [24, 25]. The amount of carbohydrates, proteins, and lipids were reported as 0.51–0.94 g/100 g, 0.39–6.20 g/100 g, and 0.0029–7.50 g/100 g, respectively [24–26]. The energy provided by *C. carandas* is 745–753 Kcal [24]. Malayali tribes of Kolli Hills use these fruits in treating rheumatoid arthritis, atherosclerosis, ulcer, and asthma [18].

Carissa spinarum L.

Carissa spinarum is native to India, Sri Lanka, parts of Africa, Australia, and China [19]. The flowering and fruiting season of *C. spinarum* is similar to that of *C. carandas*, and the ripe fruits are purple to black in color. This fruit provides energy of 167.50 Kcal [27]. Comparing to the nutritional composition of *C. carandas*, the nutritional composition of *C. spinarum* is higher with 12.43–25.92 g/100 g of carbohydrates, 1.24–3.64 g/100 g of proteins, and 1.30–4.94 g/100 g of lipid [27–30]. The mineral composition was also high in this fruit in terms of potassium (1281.50 mg/100 g), phosphorus (32.10 mg/100 g), calcium (29.00 mg/100 g), and vitamin C (17.80 mg/100 g) [27, 29]. The fruits are eaten to expel parasitic worms from the body and to treat scurvy and gastric distress [18].

Diospyros melanoxylon Roxb.

Diospyros melanoxylon is native to India and Sri Lanka and introduced to Bangladesh [19]. The flowering and fruiting season of this tree falls between the months of March and July. The energy provided by the fruits of *D. melanoxylon* is 491.75 Kcal [31]. Carbohydrates in the range between 26.70 and 26.93 (g/100 g) and proteins in the range between 0.57 and 8.64 (g/100 g) were present in the fruits [26, 31, 32]. Nearly 0.092–87.11 (mg/100 g) of sodium, 123.00–305.52 (mg/100 g) of potassium, 20.00–167.30 (mg/100 g) of phosphorus, 58.42–90.86 (mg/100 g) of calcium, 10.80–56.29 (mg/100 g) of magnesium, 0.26–0.66 (mg/100 g) of iron, 0.12–0.67 (mg/100 g) of zinc, 1.26 (mg/100 g) of manganese, 0.004 (mg/100 g) of copper, and 2.00 (mg/100 g) of vitamin C were contained by the fruits of *D. melanoxylon* [31, 32]. Fruits are used to treat diarrhea, nervous and mental disorders, and cardiovascular disorders [18].

Elaeocarpus serratus L.

Elaeocarpus serratus is native to India, Sri Lanka, and Bangladesh and was introduced to the Caroline Islands and Mauritius [19]. This fruit-bearing small tree starts to flower in the month of November and the fruits mature in the month of February. de Lima et al. [33] have evaluated the contents of proteins (4.92 g/100 g), lipids (1.10 g/100 g), and vitamin C (5.93 mg/100 g). The rest of the other macro and micronutrients were not estimated in this fruit. The ripe fruits are eaten as such by the Malayali tribes of Kolli Hills.

Passiflora subpeltata Ortega

Passiflora subpeltata is native to Colombia, Guatemala, El Salvador, Central and Gulf Mexico, Northeast, Southeast and Southwest Mexico, Panamá, and Venezuela. This plant was later introduced into Kenya, Madeira, Madagascar, India, Hawaii, Haiti, Cuba, Bangladesh, Philippines, Sri Lanka, Tanzania, Tonga, etc. [19]. This climbing plant starts to flower in the month of December and the fruits mature in March. The small fleshy fruits of *P. subpeltata* are reported to have high content of calcium (5067.00 mg/100 g) than the other minerals [34]. The proximate composition of *P. subpeltata* (Table 2) is less with a content of 0.02 g/100 g of carbohydrate and 0.88 g/100 g of protein [34]. The other minerals estimated were potassium (181.00 mg/100 g), phosphorus (10.00 mg/100 g), magnesium (2.00 mg/100 g), sodium (2.00 mg/100 g), iron (173.00 mg/100 g), zinc (19.00 mg/100 g), and manganese (6.00 mg/100 g) [34]. The micro-mineral composition was superior than the major nutrients in this fruit.

Phoenix loureiroi Kunth

Phoenix loureiroi is native to Bangladesh, India, the East and West Himalayas, Myanmar, Nepal, Cambodia, Vietnam, Taiwan, Philippines, Pakistan, Thailand, Hainan, Southeast China, and Laos [19]. The fruits of *P. loureiroi* are small in size and black in color when ripen. The flowerings start at the month of January and the fruits mature by April. The fruits are eaten to enhance digestion and to treat fever, constipation, cough, and asthma. The fruits are also believed to nurture and protect the heart, eyes, hair, and mental health [18]. The proximates and the nutrients were recorded as 1.28 g/100 g of carbohydrates, 3.63 g/100 g of proteins, 3.00 g/100 g of lipids, 126.00 mg/100 g of sodium, 83.00 mg/100 g of potassium, 19.00 mg/100 g of phosphorus, 40.00 mg/100 g of calcium, 37.00 mg/100 g of magnesium, 38.00 mg/100 g of zinc, 8.00 mg/100 g of iron, and 150.00 mg/100 g of vitamin C [35].

Rubus ellipticus Sm.

Rubus ellipticus is widely distributed in China, India, Nepal, Pakistan, Sri Lanka, Philippines, Vietnam, Tibet, Tanzania, Thailand, Hawaii, Jamaica, Malawi, Mozambique, Jawa, Ecuador, South Wales, and Queensland [19]. The flowering and fruiting period of this shrub is between the months of November and June. The ripe fruits of *R. ellipticus* are eaten by the Malayali tribes of Kolli Hills [18]. The ripe fruits of this plant can provide energy of 372.12–374.06 Kcal, and the carbohydrate content ranged between 0.08 and 86.4 g/100 g. Bhusal et al. [36] have recorded less protein content of 1.40 g/100 g, whereas Rana et al. [37] have recorded 8.83 g/100 g of proteins. The lipid content did not show much variation (0.015–2.86 g/100 g) [37–39, 48]. High content of sodium (89.43 mg/100 g), potassium (680.00 mg/100 g), phosphorus (1.26 mg/100 g), calcium (450.00 mg/100 g), magnesium (119.00 mg/100 g), iron (4.24 mg/100 g), zinc (12.8 mg/100 g), manganese (1.94 mg/100 g), copper (0.020 mg/100 g), and vitamin C (19.80 mg/100 g) were present in the fruits [37–39, 48].

***Rubus niveus* Thunb.**

Rubus niveus is native to India and parts of Africa, South America, and China [19]. Unlike other wild edible plants, *R. niveus* bears its flowers and fruits throughout the year. The energy represented by the fruits of *R. niveus* was 364.42 Kcal [39, 40]. The rich profile of potassium, sodium, calcium, magnesium, zinc, and vitamin C displayed by these fruits reveals their potential as edible and nutritious fruit. Each 100 g of *R. niveus* fruits contained 85.35 g of carbohydrates, 3.30 g of proteins, 1.10 g of lipids, 56.30 mg of sodium, 720.00 mg of potassium, 1.48 mg of phosphorus, 390.00 mg of calcium, 179.00 mg of magnesium, 3.26 mg of iron, 8.13 mg of zinc, 2.43 mg of manganese, 1.07 mg of copper, and 10.20 mg of vitamin C [39, 40, 49].

***Solanum torvum* Sw.**

Solanum torvum is native to parts of Northern and Southern America and was introduced to the regions of India, Australia, Africa, etc. [19]. The flowering and fruiting season of *S. torvum* is over the year. The plant is widely used by the Malayali tribes. The ripe fruits are eaten, while the unripe fruits are made as curries or used as preserved. The fruits of this plant can provide energy of 421 Kcal [41]. Hundred grams of fruit powder was found to have 71.42 g of carbohydrate, 16.49 g of protein, 8.81 g of lipid, 155.20 mg of sodium, 1840.00 mg of potassium, 66.00 mg of phosphorus, 329.00 mg of calcium, 49.60 mg of magnesium, 60.10 mg of iron, 14.65 mg of zinc, 10.83 mg of manganese, 1.49 mg of copper, and 2.68 mg of vitamin C [41–45, 50].

***Toddalia asiatica* (L.) Lam.**

Toddalia asiatica is widely distributed in India and Africa [19]. The ripe and mature fruits of *T. asiatica* is black in color, whose flowering begins in the month of July and fruits mature in the month of December. The fruits of this plant are used in many forms rather than consuming it. Its fruits are used in treating malarial fever, cough, sprains, fever, epilepsy, indigestion, and neuralgia. About 1.15 g of proteins were determined to be present in 100 g of fruit powder. The fruits of this plant also displayed a decent nutritive profile, 100 g of fruit powder was proven to contain, 27.47 mg of sodium, 599.09 mg of potassium, 66.48 mg of calcium, 5.84 mg of magnesium, 3.36 mg of iron, 1.35 mg of zinc, 5.84 mg of manganese, 0.17 mg of copper, and 22.02 mg of vitamin C [28].

Ziziphus rugosa Lam.

Ziziphus rugosa is native to India and the regions surrounding it [19]. The flowers of this shrub form in the month of November and the fruits mature by the month of May. The fruits of *Z. rugosa* contain good composition of proximates and nutrients (Tables 2 and 3). Hundred grams of this fruit contains, 39.11 g of carbohydrates, 1.12 g of proteins, 5.00 g of lipids, 52.00 mg of sodium, 836.00 mg of potassium, 1490.00 mg of phosphorus, 910.00 mg of calcium, 830.00 mg of magnesium, 1.87 mg of iron, 3.88 mg of zinc, 4.64 mg of manganese, and 21.26 mg of vitamin C [28, 46, 47].

Based on the results of the nutritional analysis carried out by various researchers on the wild edible fruits and the functioning of different nutrients and minerals in the human body, it is known that the documented wild edible fruits can help the human body in combating malnutrition and associated diseases. The nutritional composition of wild edible plants examined by various researchers has shown much variation among the nutrients and minerals determined. The variation in the nutritional profile is be due to the genetic factor, habitat characteristics (water availability, altitudinal gradient, soil condition), and biotic interactions experienced by the wild edible plant that grows in different regions [8, 51]. The variation in nutrient profile can also due to the harvesting stage or period and technique adopted for study. The growth forms of wild edible plants include shrubs, trees, herbs, and climbers, the diversity in growth form makes them vary in terms of its habitat requirements and its responses to forest or habitat succession pattern and disturbances [52]. The variation exhibited by the wild edible fruits can also be due to the degradation of nutrients and minerals while processing. However, the results showed that the documented wild edible fruits possess high content of potassium, phosphorus, magnesium, sodium, calcium, zinc, iron, and vitamin C, therefore by incorporating these fruits in daily diet, the mineral deficiencies, and associated diseases can be alleviated.

Wild Edible Fruits as Functional Foods

Functional foods are those which provides health benefits (therapeutical) to mankind in addition to nutrition. Functional foods are regarded as the first line of defense against a number of illnesses and disorders. Hence, the documented wild edible fruits can be considered as functional foods as they deliver high content of macro and micronutrients required for the normal functioning of the body.

The wild edible fruits of *Carissa carandas, Diospyros melanoxylon, Rubus ellipticus*, and *Solanum torvum* were recorded with significantly higher protein content (Table 2). An important nutrient supplement required by the human body is proteins. The dietary proteins comprise amino acid that repairs the body tissues, antibody synthesis, and functioning of bodily chemicals and hormones [4, 53]. The edible fruits having high crude lipids or fats are viewed as the basis for fat soluble vitamins that are crucial for free radical scavenging property [2]. And the body's ability in retaining the oil or fat-soluble vitamins, namely, vitamin A and carotenes, depends on lipid content present in the diet [4]. The fruits of *Alangium salviifolium* (60.00 g/100 g) and *Solanum torvum* (8.81 g/100 g) had notable amount of crude lipids [22, 44].

The micronutrient calcium is crucial for improving the bone strength and provides rigidity. Calcium is also involved in the functioning of nerves and muscles, coagulation of blood, vasodilation, nerve transmission, vascular contraction, muscle and intracellular functioning, cell porosity, and secretion of hormones [4, 6, 7]. The wild edible fruits such as *Carissa carandas, Carissa spinarum, Diospyros melanoxylon, Toddalia asiatica, Passiflora subpeltata, Rubus ellipticus, Rubus niveus, Solanum torvum,* and *Ziziphus rugosa* were recorded with remarkably high calcium content (Table 3). Potassium, being an intracellular cation, plays a vital role in the functioning of the human body. This helps in the secretion of hormones and energy transduction and regulates the synthesis of glycogen and protein [7]. Consumption of diet with required amount of potassium aids in reducing the risk of increased blood pressure and cardiovascular diseases [6]. Therefore, the fruits, namely, *Carissa spinarum, Toddalia asiatica, Rubus ellipticus, Rubus niveus, Solanum torvum,* and *Ziziphus rugosa* with high potassium content can be incorporated in our daily diet.

The minerals, zinc, and magnesium are the cofactor for about 300 enzymes in our body. Being an essential component of enzymes, zinc is involved in the formation and degradation of nucleic acids, lipids, proteins, and carbohydrates and in the metabolism of microminerals [53]. Moreover, magnesium regulates the biochemical reactions, including signal transduction; protein, RNA, and DNA synthesis; nerve and muscle function; neuro and muscular conductions; and blood pressure and glucose regulation in human body [6]. Among the discussed 12 wild edible fruits, the fruits of *Phoenix loureiroi, Passiflora subpeltata, Rubus ellipticus, Rubus niveus,* and *Solanum torvum* revealed high content of zinc. And the fruits of *Carissa carandas, Diospyros melanoxylon, Phoenix loureiroi, Rubus ellipticus, Rubus niveus, Solanum torvum,* and *Ziziphus rugosa* are revealed to be the major sources of magnesium.

Sodium is known for controlling blood pressure, and it is advised to consume <2.00 g of sodium per day to avoid encountering heart diseases caused by increased blood pressure. The high ratio of potassium (K) to sodium (Na) in the diet prevents arteriosclerosis and hypertension; also, this combination is believed to protect the human body from many ailments [4, 7]. All the discussed 12 wild edible fruits contained less sodium and potassium content. Therefore, blood pressure can be decreased or controlled upon consumption of these edible fruits. The micromineral copper influences iron in binding with red blood cells, thereby preventing major diseases [4]. Another essential micronutrient, iron, is essential in energy production as it is the vital part in cytochromes involved in the electron transport functionality. Also, iron is the functional constituent of myoglobin and hemoglobin [6, 53]. Iron content was predominant in the fruits of *Carissa carandas, Passiflora subpeltata,* and *Solanum torvum* (Table 3).

The production of energy, hemoglobin structure, and immune system performance are all significantly regulated by manganese. In addition, it lowers and regulates blood pressure, maintains vitamin K levels, and inhibits blood clotting [7]. The fruits of *Toddalia asiatica, Passiflora subpeltata,* and *Solanum torvum* contained a higher amount of manganese than the other wild edible fruits (Table 3). Vitamin C is a natural antioxidant that protects the skin and helps in iron absorption [54]. This water-soluble vitamin C acts as the first line of defense mechanism against free radicals [2]. The vitamin C also helps the body to counteract bacterial and viral infections and scurvy [4]. The high content of vitamin C was recorded in the fruits of *Alangium salviifolium, Carissa spinarum,* and *Phoenix loureiroi* (Table 3). Therefore, these fruits can help in skin protection and iron absorption and can help the body to combat deadly diseases.

According to the Food Safety and Standards Authority of India (FSSAI), the recommended dietary allowance (RDA) of nutrients and minerals for Indian men and women are 46–54.00 g/d of protein, 1000.00 mg/d of calcium and phosphorus (each), 370.00–440.00 mg/d of magnesium, 19.00–29.00 mg/d of iron, 13.2–17.00 mg/d of zinc, 65.00–80.00 mg/d of vitamin C, 2000.00 mg/d of sodium, 3500.00 mg/d of potassium, 1.7 mg/d of copper, and 4.00 mg of manganese [55]. Following this RDA through healthy diet, malnutrition can be reversed, and the diseases and disorders associated with this can be effectively managed. From Tables 2 and 3, it is obvious that the documented wild edible fruits can meet the dietary requirements of humans. Though the mineral components are required in minimal quantity, they have a vital role in the formation and regulation of cells, tissues, and chemical reactions that occur in the human body [3]. Globally, the wild edibles have proved its efficiency as an income generator and as functional food. In spite of this, the wild edibles are regarded as underutilized and untapped resources, which can have a valuable role in the rural livelihood development, economic empowerment, and food security among the ethnic communities through scientific intervention [56].

Conclusions

The present study underscores the invaluable role of wild edible fruits in Kolli Hills, not only as a source of biodiversity but also as a repository of indigenous knowledge and cultural heritage. Through a comprehensive multidisciplinary approach, encompassing botanical, ethnobotanical, and nutritional analyses, we have documented 43 wild edible fruit species, with a focused examination of 12 selected species. With this insight, nutritional analyses have highlighted the significant potential of these fruits in promoting health and nutrition, thereby contributing to food security and well-being in the region. By recognizing the ecological, cultural, and nutritional importance of wild edible fruits, this study aims to inform and inspire conservation initiatives that prioritize the sustainable management and utilization of these invaluable resources.

Acknowledgement The authors thank the Science for Equity, Empowerment, and Development (DST-SEED), Department of Science and Technology, Government of India, New Delhi (Grant No. DST/SEED/TSP/STI/2020/333) for financial support.

References

1. Popović Z, Matić R, Stefanović M, Vidaković V, Bojović S. Biodiversity of wild fruits with medicinal potential in Serbia. In: Biodiversity and biomedicine. Academic Press; 2020. p. 161–88.
2. Bayang JP, Laya A, Kolla MC, Koubala BB. Variation of physical properties, nutritional value and bioactive nutrients in dry and fresh wild edible fruits of twenty-three species from Far North region of Cameroon. J Agric Food Res. 2021;4:100146.
3. Ali F, Khan N, Ali K, Amin M, Khan ME, Jones DA. Assessment of variability in nutritional quality of wild edible fruit of *Monotheca buxifolia* (Falc.) A. DC. Along the altitudinal gradient in Pakistan. Saudi J Biol Sci. 2023;30(1):103489.
4. Seal T, Chaudhuri K, Pillai B. Nutritional and toxicological aspects of selected wild edible plants and significance for this society. S Afr J Bot. 2023;159:219–30.
5. Mokria M, Gebretsadik Y, Birhane E, McMullin S, Ngethe E, Hadgu KM, Hagazi N, Tewolde-Berhan S. Nutritional and ecoclimatic importance of indigenous and naturalized wild edible plant species in Ethiopia. Food Chem Mol Sci. 2022;4:100084.
6. Nyero A, Achaye I, Anywar GU, Malinga GM. Inorganic nutrients and heavy metals in some wild edible plants consumed by rural communities in northern Uganda: implications for human health. Heliyon. 2023;9(8):e18999.
7. Jalali M, Fakhri R. Evaluation of macro and trace elements content of wild edible Iranian plants and their contribution to dietary reference intakes. J Food Compos Anal. 2021;102:104049.
8. Pohjanmies T, Jašková A, Hotanen JP, Manninen O, Salemaa M, Tolvanen A, Merilä P. Abundance and diversity of edible wild plants in managed boreal forests. For Ecol Manag. 2021;491:119151.
9. Siegel KR, Ali MK, Srinivasiah A, Nugent RA, Narayan KV. Do we produce enough fruits and vegetables to meet global health need? PLoS One. 2014;9(8):e104059.
10. Santhi VP, Sriramavaratharajan V, Murugan R, Masilamani P, Gurav SS, Sarasu VP, Parthiban S, Ayyanar M. Edible fruit extracts and fruit juices as potential source of antiviral agents: a review. J Food Meas Charact. 2021;15(6):5181–90.
11. Santhi VP, Masilamani P, Sriramavaratharajan V, Murugan R, Gurav SS, Sarasu VP, Parthiban S, Ayyanar M. Therapeutic potential of phytoconstituents of edible fruits in combating emerging viral infections. J Food Biochem. 2021;45(8):e13851.
12. Royal Botanic Gardens, Kew. State of the World's plants and fungi. 2020.
13. Lakshmi G. Ecological studies on the vegetation of Kolli hills, Salem district, Tamilnadu. Ph.D. thesis submitted to Bharathiar University, Coimbatore. 1995.
14. Arunachalam G, Karunanithi M, Subramanian N, Ravichandran V, Selvamuthukumar S. Ethno medicines of Kolli Hills at Namakkal District in Tamilnadu and its significance in Indian Systems of Medicine. J Pharm Sci Res. 2009;1(1):1.
15. Pathak R, Pant V, Negi VS, Bhatt ID, Belwal T. Introduction to Himalayan region and wild edible diversity. In: Himalayan fruits and berries. Academic Press; 2023. p. 1–12.
16. Ranjithakani P, Geetha S, Lakshmi G, Murugan S. Preliminary survey of wild edibles of Kolli Hills of Salem. Anc Sci Life. 1992;11(3–4):133.

17. Samydurai P, Thangapandian V, Aravinthan V. Wild habits of Kolli Hills being staple food of inhabitant tribes of Eastern Ghats, Tamil Nadu, India. Indian J Nat Prod Resour. 2012;3(3):432–7.
18. Anand SP, Deborah S. Enumeration of wild edible fruits from Boda hills and Kolli hills. Int J Appl Biol Pharm Technol. 2016;7:96–102.
19. International Plant Names Index. The Royal Botanic Gardens, Kew, Harvard University Herbaria & Libraries and Australian National Herbarium (Retrieved 4 March 2024).
20. Kumar A, Gupta AK. Importance and uses of underutilized edible fruit Ankol. Indian Hortic. 2020;65(2): https://epubs.icar.org.in/index.php/IndHort/article/view/105325.
21. Ghosh S, Sarkar T, Chakraborty R. Ankol plant (*Alangium salvifolium*)-The treasure trove of bioactives and medicinal potential. Food Biosci. 2023;51:102230.
22. Tanwer BS, Vijayvergia RE. Phytochemical evaluation and quantification of primary metabolites of *Alangium salviifolium*. Int J Pharm Bio Sci. 2010;1:1–6.
23. Sreekumar VB, Sreejith KA, Hareesh VS, Sanil MS. An overview of wild edible fruits of Western Ghats, India. Genet Resour Crop Evol. 2020;67:1659–93.
24. Arif M, Kamal M, Jawaid T, Khalid M, Saini KS, Kumar A, Ahmad M. *Carissa carandas* Linn. (Karonda): an exotic minor plant fruit with immense value in nutraceutical and pharmaceutical industries. Asian J Biomed Pharm Sci. 2016;6(58):14–9.
25. Mahajan M, Bons HK, Dhillon GK, Sachdeva PA. Unlocking the impact of drying methods on quality attributes of an unexploited fruit, karonda (*Carissa carandas* L.): a step towards food and nutritional security. S Afr J Bot. 2022;145:473–80.
26. Misra S, Misra MK. Ethnobotanical and nutritional evaluation of some edible fruit plants of southern Odisha, India. Int J Adv Agric Sci Technol. 2016;3(1):1–30.
27. Siyum ZH, Meresa TA. Physicochemical properties and nutritional values of *Carissa Spinarum* L./ "Agam" fruit. Int J Fruit Sci. 2021;21(1):826–34.
28. Mahapatra AK, Mishra S, Basak UC, Panda PC. Nutrient analysis of some selected wild edible fruits of deciduous forests of India: an explorative study towards non conventional bio-nutrition. Adv J Food Sci Technol. 2012;4(1):15–21.
29. Chauhan AM, Tanwar B, Arneja IN. Influence of processing on physiochemical, nutritional and phytochemical composition of *Carissa spinarum* (karonda) fruit. Asian J Pharm Clin Res. 2015;8(6):254–9.
30. Dhatwalia J, Kumari A, Verma R, Upadhyay N, Guleria I, Lal S, Thakur S, Gudeta K, Kumar V, Chao JC, Sharma S. Phytochemistry, pharmacology, and nutraceutical profile of *Carissa* species: an updated review. Molecules. 2021;26(22):7010.
31. Hmar BZ, Mishra S, Pradhan RC. Physico-chemical, mechanical and antioxidant properties of kendu (*Diospyros melanoxylon* Roxb.). Curr Res Nutr Food Sci J. 2017;5(3):214–22.
32. Jamil Z, Mohite AM, Sharma N. Phyto-chemical and nutritional profiling of tendu fruit (*Diospyros melanoxylon* roxb.) and evaluation of its shelf stability. Plant Arch. 2021;21(1):656–64.
33. de Lima FF, Breda CA, Cardoso CA, Duarte MC, Sanjinez-ArgandoÃ±a EJ. Evaluation of nutritional composition, bioactive compounds and antimicrobial activity of *Elaeocarpus serratus* fruit extract. Afr J Food Sci. 2019;13(1):30–7.
34. Shanmugam S, Gomes IA, Denadai M, dos Santos LB, de Souza Araújo AA, Narain N, Neta MT, Serafini MR, Quintans-Júnior LJ, Thangaraj P. UHPLC-QqQ-MS/MS identification, quantification of polyphenols from *Passiflora subpeltata* fruit pulp and determination of nutritional, antioxidant, α-amylase and α-glucosidase key enzymes inhibition properties. Food Res Int. 2018;108:611–20.
35. Rajan M, Shanmugam S, Guedes TJ, Thangaraj P, Barbosa PF, Narain N. Profiles of nutritional, bioactive compounds and cytotoxic activity of Dwarf date palm (*Phoenix loureiroi*) fruit. J Food Meas Charact. 2020;15:1611–25.
36. Bhusal S, Pant DR, Joshi GP, Adhikari M, Raut JK, Pandey MR, Bhatt LR. Antioxidant activity and nutraceutical potential of selected Nepalese wild edible fruits. Sci World. 2020;13(13):8–13.

37. Rana YS, Tiwari OP, Krishan R, Sharma CM. Determination of nutritional potential of five important wild edible fruits traditionally used in Western Himalaya. Int J Life Sci. 2018;6(1):79–86.
38. Saklani S, Chandra S, Badoni PP, Dogra S. Antimicrobial activity, nutritional profile and phytochemical screening of wild edible fruit of *Rubus ellipticus*. Int J Med Aromat Plants. 2012;2(2):269–74.
39. Ahmad M, Masood S, Sultana S, Hadda TB, Bader A, Zafar M. Antioxidant and nutraceutical value of wild medicinal *Rubus* berries. Pak J Pharm Sci. 2015;28(1):241–7.
40. Bhatt SC, Naik B, Kumar V, Gupta AK, Kumar S, Preet MS, Sharma N, Rustagi S. Untapped potential of non-conventional *Rubus* species: bioactivity, nutrition, and livelihood opportunities. Plant Methods. 2023;19(1):114.
41. Melila M, Dossou BR, Etse KD, Firmin S. Biochemical study and evaluation of the nutritional value of *Solanum torvum* (Swartz) fruits used as fruiting vegetables in Togo. J Food Nutr Res. 2021;9(11):579–84.
42. Akoto O, Borquaye LS, Howard AS, Konwuruk N. Nutritional and mineral composition of the fruits of *Solanum torvum* from Ghana. Int J Chem Biomol Sci. 2015;1(4):222–6.
43. Nadeeshani H, Samarasinghe G, Wimalasiri S, Silva R, Hunter D, Madhujith T. Comparative analysis of the nutritional profiles of selected *Solanum* species grown in Sri Lanka. J Food Compos Anal. 2021;99:103847.
44. Abraham JD, Sekyere EK, Gyamerah I. Effect of boiling on the nutrient composition of *Solanum torvum*. Int J Food Sci. 2022; Royal Botanic Gardens, Kew. State of the World's Plants and Fungi.
45. Balkrishna A, Shankar R, Joshi RA, Joshi M, Prajapati UB, Srivastava A, Arya V. Nutraceutical studies of *Solanum torvum* Swartz. Curr Sci. 2023;125(7):784.
46. Vinayaka KS, Kumaraswamy HG. Nutritive composition of fruit of *Zizyphus rugosa* lam. Res Rev Biomed Biotechnol. 2011;2:20–4.
47. Krishnamurthy SR, Sarala P. Determination of nutritive value of *Ziziphus rugosa* Lamk.: a famine edible fruit and medicinal plant of Western Ghats. Indian J Nat Prod Resour. 2012;3(1):20–7.
48. Lamichhane A, Lamichhane G, Devkota HP. Yellow Himalayan raspberry (*Rubus ellipticus* Sm.): ethnomedicinal, nutraceutical, and pharmacological aspects. Molecules. 2023;28(16):6071.
49. Schulz M, Chim JF. Nutritional and bioactive value of *Rubus* berries. Food Biosci. 2019;31:100438.
50. Otu PN, Sarpong F, Gidah JE, Labanan AM, Anim D. Characterization of Turkey berry (*Solanum torvum*)-fresh, dry & powder. Afr J Food Integr Agric. 2017;21:9–14.
51. Angami T, Wangchu L, Debnath P, Sarma P, Singh B, Singh AK, Hazarika BN, Singh MC, Touthang L, Lungmuana S, Ayyanar M. Exploring the nutritional potential and anti-nutritional components of wild edible fruits of the Eastern Himalayas. J Food Meas Charact. 2024;18(1):150–67.
52. Sureshkumar J, Jenipher C, Gurav SS, Amiri-Ardekani E, Kalaskar MG, Ayyanar M. Local knowledge on wild edible and economically important pteridophytes of India: a review. Hum Ecol. 2023;51(5):937–77.
53. Chaudhary MK, Tripathi D, Misra A, Singh SP, Srivastava PK, Gupta V, Acharya R, Srivastava S. Nutritional characteristics of *Stereospermum chelonoides* (Lf) DC., an underutilized edible wild fruit of dietary interest. Heliyon. 2024;10(2):e24193.
54. Yimer A, Forsido SF, Addis G, Ayelign A. Phytochemical profile and antioxidant capacity of some wild edible plants consumed in Southwest Ethiopia. Heliyon. 2023;9(4):e15331.
55. Food safety and standards authority of India. Recommended dietary allowances and estimated average requirements: nutrient requirements for Indian – 2020, Science and Standards Division, FDA Bhawan, New Delhi – 110 002, p. 1–4.
56. Krupa J, Murugan R, Gangapriya P, Amalraj S, Gurav S, Sam Arulraj M, Ayyanar M. *Moringa concanensis* Nimmo. seed extracts as a potential source of bioactive molecules, antioxidants and enzyme inhibitors. J Food Meas Charact. 2022;16(5):3699–711.

Nutraceutical Potential of Edible Plants for Balancing Dietary Requirements in Human Diet: A Case Study

Binu Thomas

Abstract

The present study on edible plants, which are distributed in Kozhikode district, documented a total of 50 species of plants belonging to 38 genera and 25 families. Out of these 25 families represented, Cucurbitaceae and Solanaceae are the first dominant families with six species each, followed by Moraceae with four species, and then Amaranthaceae, Annonaceae, and Myrtaceae with three species; each and other families are represented with two and single species, respectively. They used different plant parts like fruits (37 Nos.), leaves (9 Nos.), seed (1 No.), tuber (1 No.), rhizome/corm (2 Nos.), etc. Are used for edible purposes. Some of the documented edible plants also have various medicinal properties too.

Keywords

Edible plants · Dietary requirements · Human diet · Kerala

Introduction

Most of the people commonly use some wild plants in their regular dish. They consume these plants or plant parts after boiling, frying, or preparing curry or chutneys or as raw. The various anti-nutritional elements, which are present in the edible plant parts, were removed during cooking process. Moreover, the edibles, which are eaten as raw, are comparatively negligible percentage of anti-nutritional elements. Various plant parts, of course, are consumed following various processes, which vary according to species or parts of the plant as well as according to the people who

B. Thomas (✉)
Department of Botany, St. Joseph's College (Autonomous),
Devagiri, Kozhikode, Kerala, India

P. Thangaraj (ed.), *Pomotherapeutic Insights on Wild Edible Fruits*,
https://doi.org/10.1007/978-981-96-0543-9_2

use the items. Tribal communities widely use wild edibles for balancing dietary equilibrium [1].

The tribal as well as other common people of various parts of the country commonly use some wild plants in their regular dish. They consume their plants or plant parts after boiling or frying or preparing curry or chutney or as raw [2]. The consumption of these edibles varies with respect to the species or parts of the plant as well as according to the tribes and communities that use these items. The wild edible plant parts, like leaves, fruits, tubers, rhizomes, etc., may provide an adequate amount of vitamins and minerals for humans. These are excellent sources of carotenes, ascorbic acid, riboflavin, folic acid, and essential minerals such as calcium, iron, and phosphorus [3].

The tribal people fully depended on both wild and cultivated plants for their food, medicine, shelter, and other uses. The community obtained knowledge of plants from their ancestors through oral tradition. The local tribal communities cultivate cereals in forest lands, foothills, and plain dry lands, and their associated naturally growing plants, popularly known as "weeds," are consumed seasonally. They usually used a variety of leaves, fruits, tubers, rhizomes, flowers, etc., for their dietary purposes. However, these tribal families also depend on minor forest products for their daily needs [4].

The studies revealed that people living in many regions collected edible wild plants from their surroundings for their dietary equilibrium. The traditional knowledge regarding various wild edibles was transferred orally to the next generation. Moreover, many ethnobotanical studies conducted by various authors in the countries also documented variety of wild edibles, which are utilized by tribal and local inhabitants. Their food habit also linked with their culture [5]. Wild edible plants are the precious gift of nature and most of the ethnic communities are strongly dependent on them for their day-to-day life. Food plants are not only supplement to food quality but also important during starvation for survival and thus make significant contributions to their nutrition throughout the year [6].

Leafy vegetables, also called potherbs, green, vegetable greens, leafy greens, or salad greens, are plant leaves eaten as vegetables, sometimes accompanied by tender pedicles and shoots. Leafy vegetables most often come from short-lived herbaceous plants such as Amaranthus and spinach. Woody plants, whose leaves can be eaten as leaf vegetables, include *Adanosonia, Bauhinia, Moringa,* and so on. The leaves of many fodder crops are also edible by humans but usually only eaten under famine condition [7].

The earth is filled with overwhelming plant diversity. The efforts of the man are to categorize them based on their structural and functional features. Among the functional features, the health benefits of edible plant parts and their bioactive compounds and their incorporation in food have simulated wider interest and demand across the world [8]. People have been trying to understand and utilize biodiversity from time immemorial mainly for food. When they were hunter-gatherers, they ate many of wild plants and their products. The trial-and-error practices of tribal and local inhabitants helped them understand which edible plant species and functional parts they could find in their local vegetation. When man began to settle in life, the demand for food plants increased day by day to fulfill their needs [9]. The present study mainly focused on documenting edible plants in the study area with their other attributes.

Study Area

Kozhikode District

Kozhikode, also known as *Calicut*, is a city in the state of Kerala in Southern India on the Malabar Coast and was formed on January 1, 1957. Kozhikode is the largest urban area in the state and 195TH largest urban area in the world. The city of Kozhikode is 410 kilometers North of the State capital Thiruvananthapuram. It is located at approximately 11.25°N 75.77°E. It has an elevation of 1 meter along the coast with the city's eastern edges rising to at least 15 meters. The rainy season is during the southwest monsoon, which sets in the first week of June and extends up to September. The northeast monsoon extends from the second half of October through November. The average annual rainfall is 3266 mm, and the best weather is found toward the end of the year, in December and January [10] (Plate 1).

Materials and Methods

The present study was based on an extensive survey and field observations during the year December 2022 to March 2023. In this study, attempts were made to document edible plants from Kozhikode district. The documentation was mainly based on field observation, discussions with local peoples, as well as scrutinizing the literature review. During the field visits, the plant specimens were collected at different reproductive stages to prepare herbarium specimens. The collected specimens were identified taxonomically with the help of available floras and literature [11–14]. The nomenclature of each species has been brought up to data as per the rules given in the International Code of Nomenclature (ICN). The specimens were processed for the preparation of the Herbarium by standard methods [15]. The voucher specimens were deposited in the Herbaria of PG & Research Department of Botany, St. Joseph's College, Calicut (DEV), for future reference.

Results and Discussion

Edible Plants and Their Edible Potentiality

The present study on edible plants, of which are distributed in Ramanattukara, Kozhikode district, reveals that there are about 50 species of plants belonging to 38 genera and 25 families. The present documentation is mainly based on filed observation, discussion with local inhabitants, as well as scrutinizing literature. The

A) Map of India Showing Kerala State

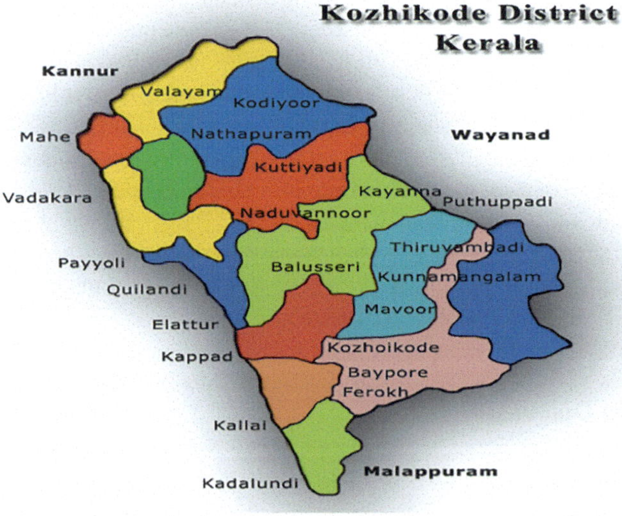

B) Map of Kozhikode district

Plate 1 Map illustrating the study area

various plant parts and their mode of consumption are also discussed (Table 1). When any plant or their parts are edible means, those plants have a higher percentage of nutritional elements when compared to anti-nutritional elements. Similarly, when any plants are said to be nonedible or poisonous, that means those plants are having more percentage of anti-nutritional elements when compared to nutritional elements.

Studies on wild edible plants, used by the *Kattunaikka, Paniya,* and *Kuruma* tribes of Wayanad District, Kerala, India, were done by Narayanan et al. [16]. According to them, there are about 165 edible plants used by these tribes for daily nutritional requirements. Moreover, they also conserved such potential wild edibles through sustainable utilization.

Analysis of Dominant Families and Genera of Edible Plants

The present documentation reveals that 50 species of edible plants are distributed in the 25 families. Out of these 25 families, Cucurbitaceae and Solanaceae are the first dominant families with six species each, followed by Moraceae with four species, Amaranthaceae, Annonaceae, and Myrtaceae with three species each, and other families are represented two and single species, respectively (Fig. 1).

Similarly, the analysis of dominant genera among the total of 38 genera represented shows that genera like *Annona*, *Amaranthus,* and *Artocarpus* are the dominant ones with three species each, followed by *Averrhoa*, *Citrus*, *Syzygium*, *Passiflora*, *Capsicum,* and *Solanum* with two species each, and all other families are represented with single species, respectively.

There are ethnobotanical studies on wild edible plants of *Gond, Halba,* and *Kawar* tribes in Salekasa Taluk of Gondia district, Maharashtra, by Patale et al. [17]. They documented 80 wild edible plant species, under 69 genera and 38 families, respectively. Among the 38 families represented, five belong to monocots and 33 are of dicots. The analysis dominant families revealed that most widely utilized species belong to Amaranthaceae (seven species) and Araceae (seven species), followed by Caesalpiniaceae (six species), Fabaceae (five species), and Solanaceae (four species), respectively.

Analysis of Life Forms/Growth Forms of Edible Plants

Analysis of various life/growth forms of edible plants in the study area consists of dominant tress (23 Nos.), followed by herbs (15 Nos.), climbers (10 Nos.), and shrubs (2 Nos.) (Fig. 2).

Studies on leafy vegetable plants of South Odisha were done by Misra and Misra [7]. According to them, a total of 106 leafy vegetable plant species under 88 genera and 49 families were reported from the study area. Various ethnic groups and rural people consume around 78 wild plants as unconventional sources of food, while 28

Table 1 List edible plants and their edible potentiality

Sl. no.	Botanical name	Family	Edible potentiality
1.	*Annona muricata* L. (Pl.2A)	Annonaceae	Ripe fruits are eaten
2.	*Annona reticulata* L.	Annonaceae	Ripe fruits are eaten
3.	*Annona squamosa* L. (Pl.2B)	Annonaceae	Ripe fruits are eaten
4.	*Garcinia gummi-gutta* (L.) Robs.	Clusiaceae	Young fruits are used to prepare pickles, dried fruits used for flavoring for curries
5.	*Abelmoschus esculentus* (L.) Moench.	Malvaceae	Young fruits are used as vegetables
6.	*Averrhoa bilimbi* L.	Oxalidaceae	Fruits used for making pickles
7.	*Averrhoa carambola* L. (Pl.2C)	Oxalidaceae	Ripe fruits are eaten as raw
8.	*Citrus limon* (L.) Burm. F.	Rutaceae	Fruits are used to prepare pickle; fruit juice is rejuvenating
9.	*Citrus maxima* (Burm.f.) Merr. (Pl.2D)	Rutaceae	Ripe fruits are eaten as raw
10.	*Nephelium lappaceum* L. (Pl.2E)	Sapindaceae	Ripe fruits eaten as raw
11.	*Mangifera indica* L.	Anacardiaceae	Young fruits used to prepare pickles also for the preparation of various dishes. Ripe fruits eaten as raw
12.	*Spondias pinnata* (L. F.) Kurz. (Pl.2F)	Anacardiaceae	Young fruits used to prepare pickles
13.	*Moringa pterygosperma* Gaertn.	Moringaceae	Leaves and fruits used as vegetable
14.	*Phaseolus vulgaris* L.	Fabaceae	Pods and young leaves used as vegetables
15.	*Senna tora* (L.) Roxb.	Caesalpiniaceae	Young leaves used as vegetables
16.	*Tamarindus indica* L.	Mimosaceae	Ripe fruits eaten as raw, also for the preparation of various dishes
17.	*Terminalia catappa* L.	Combretaceae	Seeds eaten as raw
18.	*Psidium guajava* L.	Myrtaceae	Ripe fruits eaten as raw
19.	*Syzygium aqueum* (Burm.f.) Alston (Pl.3A)	Myrtaceae	Ripe fruits eaten as raw
20.	*Syzygium cumini* (L.) Skeels (Pl.3B)	Myrtaceae	Ripe fruits eaten as raw
21.	*Passiflora edulis* Sims (Pl.3C)	Passifloraceae	Ripe fruits eaten as raw
22.	*Passiflora foetida* L. (Pl.3D)	Passifloraceae	Ripe fruits eaten as raw

(continued)

Table 1 (continued)

Sl. no.	Botanical name	Family	Edible potentiality
23.	*Carica papaya* L.	Caricaceae	Young fruits used as vegetables; ripened fruits also eaten as raw
24.	*Citrullus lanatus* (Thunb.) Mats.	Cucurbitacae	Fruit juice is rejuvenating
25.	*Coccinia grandis* (L.) Voight (Pl.3E)	Cucurbitacae	Young leaves and fruits used as vegetable
26.	*Cucumis sativus* L.	Cucurbitacae	Fruits used as vegetable also eaten in raw form
27.	*Cucurbita maxima* Duch.	Cucurbitacae	Fruits used as vegetable
28.	*Momordica charantia* L.	Cucurbitacae	Fruits used as vegetable
29.	*Trichosanthes anguina* L.	Cucurbitacae	Leaves and fruits used as vegetable
30.	*Centella asiatica* (L.) Urban (Pl.3F)	Apiaceae	Leaves used as vegetable
31.	*Manilkara zapota* (L.) P.Royen	Sapotaceae	Ripe fruits eaten as raw
32.	*Capsicum annuum* L.	Solanaceae	Fruits used for the preparation of various dishes
33.	*Capsicum frutescens* L.	Solanaceae	Fruits used for the preparation of various dishes
34.	*Lycopersicon esculentum* Mill.	Solanaceae	Fruits used as vegetable
35.	*Physalis angulata* L.	Solanaceae	Fruits eaten as raw
36.	*Solanum melongena* L.	Solanaceae	Fruits used as vegetable
37.	*Solanum torvum* Sw.	Solanaceae	Fruits used as vegetable
38.	*Boerhavia diffusa* L. (Pl.4A)	Nyctaginaceae	Young leafy twigs used as vegetable
39.	*Amaranthus hybridus* L.	Amaranthaceae	Leaves used as vegetables
40.	*Amaranthus tricolor* L. (Pl.4B)	Amaranthaceae	Leaves used as vegetables
41.	*Amaranthus viridis* L.	Amaranthaceae	Leaves used as vegetables
42.	*Phyllanthus emblica* L.	Euphobiaceae	Fruits used to prepare pickles. Fruit juice is also used to control diabetes
43.	*Sauropus androgynus* (L.) Merr. (Pl.4C)	Euphobiaceae	Young leaves used as vegetables
44.	*Artocarpus heterophyllus* Lam.	Moraceae	Young fruits and seeds are cooked before eating, while ripe fruits are eaten as raw
45.	*Artocarpus hirsutus* Lam.	Moraceae	Ripe fruits are eaten as raw
46.	*Artocarpus incisus* (Thunb.) L.f. (Pl.4D)	Moraceae	Fruits are used as vegetable
47.	*Morus alba* L. (Pl.4E)	Moraceae	Young fruits used to prepare pickles. Ripe fruits also eaten as raw
48.	*Dioscorea alata* L. (Pl.4F)	Dioscoreaceae	Tubers cooked before eating
49.	*Colocasia esculenta* (L.) Schott.	Araceae	Rhizome cooked before eating
50.	*Amorphophallus paeoniifolius* (Dennst.) Nicolson	Araceae	Corn is cooked before eating

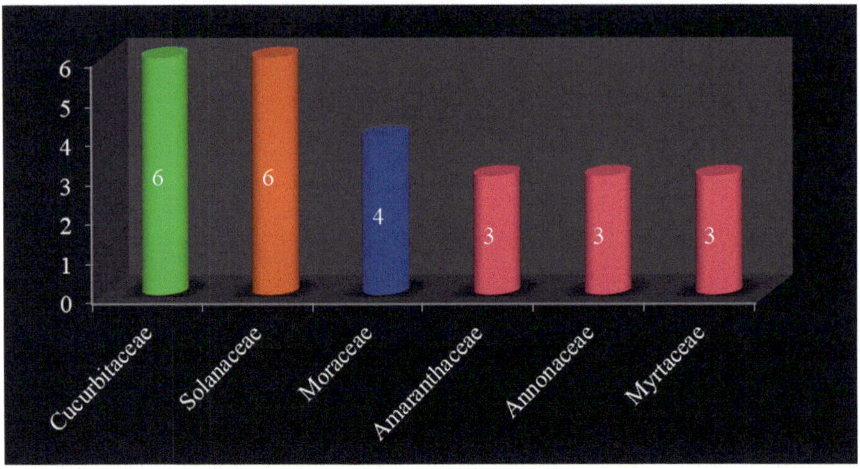

Fig. 1 Analysis of dominant families of edible plants

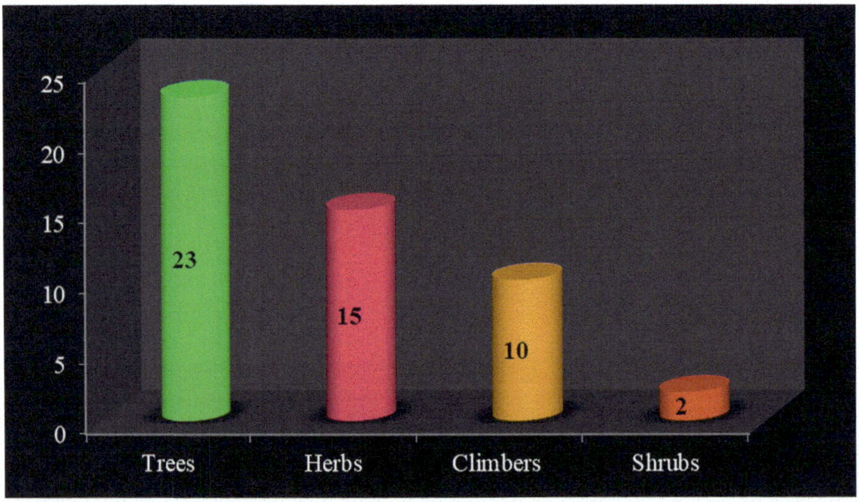

Fig. 2 Life form/growth form analysis of edible plants

plants are under cultivation. The life form analysis of documented plants reveals that 79 are herbs, 16 are shrubs, and 11 are tree species, respectively. They also recommend for the cultivation of such wild leafy vegetables around habitats of both tribal's as well as local inhabitants of an area.

Analysis of Edible Plant Parts

The present study on edible plants reveals that there are diverse kinds of plant parts that are used for edible purposes. Some edible plant parts are eaten raw, while some are cooked before eating. During the cooking process, some amount of anti-nutritional elements in the particular edible part of a plant becomes eliminated. Similarly, the juice prepared from some fruits is also rejuvenating. The different parts/products like fruits (37 Nos.), leaves (9 Nos.), seed (1 No.), tuber [1], rhizome/corm (2 Nos.), etc. are used for edible purposes (Fig. 3)

Studies on indigenous plant foods, which are commonly consumed by the tribal communities in Dumbriguda area of Viskhapatanam district, Andhra Pradesh, India, was done by Rao et al. [18]. According to them, the present study encompasses the in-depth investigation on uncultivated vegetables in the study area. The results of the study also revealed that a total of 55 indigenous food plants were documented, in which their various useful parts include tubers, rhizome, roots, young leaves, buds, bulbs, inflorescence, unripe/ripe fruits, and seeds.

Plant Species Which Are Showing Medico-Potentialities

The present documentation also highlighted medicinal importance of some edible plants. The present results reveal that there are about 14 species of edible plants, which belong to 11 families that also possess potential medicinal properties to cure various ailments, like hypertension, diabetes, and cholesterol, and enhance memory power, immunity power, etc., There are various plant parts like fruits (8 Nos.), leaves (3 Nos.), whole plant (1 No.), bark (1 No.), and Seed (1 No.), respectively.

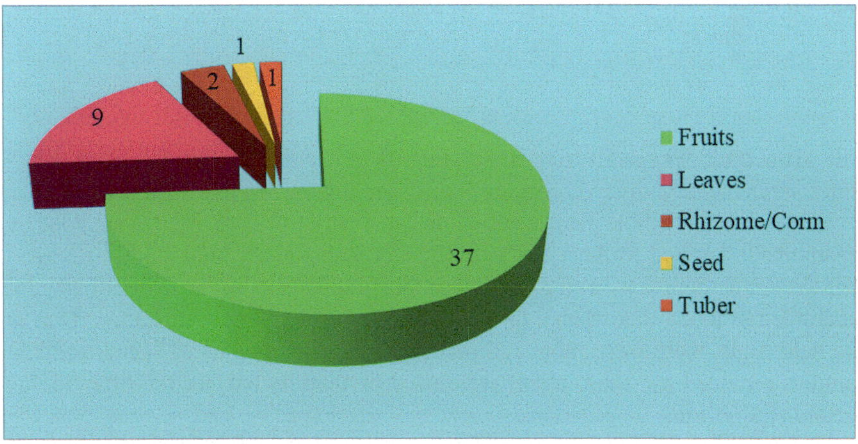

Fig. 3 Analysis of plant parts/products for edible purposes

Table 2 List of plant species that shows medico-potentialities

Sl. no.	Botanical name	Family	Edible potentiality
1.	*Abelmoschus esculentus* (L.) Moench.	Malvaceae	Fruits are cut into small pieces, and it is kept in the water overnight. The consumption of such water after removing fruits from it in the next day morning is used to control diabetes
2.	*Annona muricata* L.	Annonaceae	The fruits and seeds are reported that high antioxidant potential; moreover, they also have anticancerous properties
3.	*Averrhoa bilimbi* L.	Oxalidaceae	Fruit juice is reported to reduce hypertension
4.	*Averrhoa carambola* L.	Oxalidaceae	Fruit juice is reported to reduce cholesterol
5.	*Capsicum frutescens* L.	Solanaceae	Fruits are also used to control cholesterol as well as hypertension
6.	*Carica papaya* L.	Caricaceae	Leaf juice is used to improve immunity power
7.	*Centella asiatica* (L.) Urban	Apiaceae	Whole plant juice also enhances the memory power
8.	*Coccinia grandis* (L.) Voight	Cucurbitacae	Eating raw fruits also reduces diabetes
9.	*Momordica charantia* L.	Cucurbitacae	Juice from leaves is also used to control diabetes
10.	*Garcinia gummi-gutta* (L.) Robs.	Clusiaceae	Fruit juice is also reported to reduce cholesterol
11.	*Psidium guajava* L.	Myrtaceae	Young leaves and fruits are reported to control diabetes
12.	*Syzygium cumini* (L.) Skeels	Myrtaceae	Teaspoon full of dried seed powder is mixed in a glass of water and taken it in an empty stomach to control diabetes
13.	*Phyllanthus emblica* L.	Euphobiaceae	Fruit juice is also used to control diabetes
14.	*Moringa pterygosperma* Gaertn.	Moringaceae	Crushed bark is pasted in the cotton cloth and tied around the fractured bone, for immediate cure

The extract/juice prepared from these plant parts is either directly taken or sometimes applied externally to cure/control different diseases (Table 2 and Fig. 4).

Studies on wild edible plants of Koch Bihar, District, West Bengal, were done by Bandyopadhyay and Mukherjee [19]. Their ethnobotanical field studies revealed that the ethnic people as well as the other local inhabitants of the district had considerable traditional knowledge on wild edible plants and their utilization. During the field study, there were about 125 plant species belonging to 102 genera and 54 families recorded and these plants were used by them to balance their day-to-day dietary equilibrium.

Studies on wild edible plants traditionally used by *Kadar* tribes of Vazhachal forest Division, Thrissure, Kerala, were done by Chaithanya et al. [20]. According

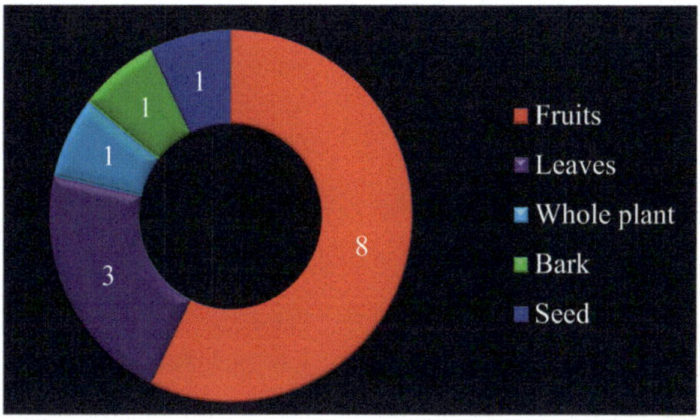

Fig. 4 Analysis of plant parts for medicinal purposes

to them, a total of 55 species of edible plants were documented from the study area, of which 23 species were consumed as fruits, while 12 are tubers and rhizomes, 12 are leafy vegetables, and the seeds of 8 species were used for edible purposes. They also noticed that *Kadar* tribes have very good traditional knowledge regarding potential edible plants. They still use a variety of such edible plants for nutritional requirements in their daily diet.

Nativity of Some Introduced Edible Plants

Out of 50 edible plant species documented, there are about 22 plants introduced here from different countries, like Central America, Malaysia, Indonesia, South America, Tropical Africa, Tropical America, South America, Pacific Islands, Afghanistan, etc. These introduced plants are now naturalized and well-grown in our habitats. In spite of these, some species of introduced plants are sometimes a threat to our native flora and fauna; they are called invasive or exotic weeds. The present documentation highlights the edible or economic potentiality of some introduced plants, which have become major constituents in our daily diet (Table 3).

Distribution Analysis of Edible Plants

The study documents about 50 species of edible plants from the study area. These plants were distributed variously in the study area. Based on their distribution pattern and occurrence in the study area, they are categorized as common (32 Nos.), uncommon (9 Nos.), and rare (9 Nos.) (Table 4 and Fig. 5) (Plates 2, 3, and 4).

Table 3 List of plant species that are introduced from other countries

Sl no.	Botanical name	Family	Nativity
1.	*Annona muricata* L.	Annonaceae	Central America
2.	*Annona reticulata* L.	Annonaceae	Central America
3.	*Annona squamosa* L.	Annonaceae	Central America
4.	*Averrhoa bilimbi* L.	Oxalidaceae	Malaysia
5.	*Averrhoa carambola* L.	Oxalidaceae	Indonesia
6.	*Nephelium lappaceum* L.	Sapindaceae	Malaysia
7.	*Phaseolus vulgaris* L.	Fabaceae	Central America
8.	*Senna tora* (L.) Roxb.	Caesalpiniaceae	South America
9.	*Tamarindus indica* L.	Mimosaceae	Tropical Africa
10.	*Terminalia catappa* L.	Combretaceae	Malaysia
11.	*Psidium guajava* L.	Myrtaceae	Tropical America
12.	*Syzygium aqueum* (Burm.f.) Alston	Myrtaceae	Malaysia
13.	*Passiflora edulis* Sims	Passifloraceae	America
14.	*Passiflora foetida* L.	Passifloraceae	America
15.	*Carica papaya* L.	Caricaceae	Tropical America
16.	*Citrullus lanatus* (Thunb.) Mat.	Cucurbitacae	Africa
17.	*Manilkara zapota* (L.) P.Royen	Sapotaceae	South America
18.	*Capsicum annuum* L.	Solanaceae	West Indies
19.	*Capsicum frutescens* L.	Solanaceae	Tropical America
20.	*Lycopersicon esculentum* Mill.	Solanaceae	South America
21.	*Artocarpus incisus* Thunb.	Moraceae	Pacific Islands
22.	*Morus alba* L.	Moraceae	Afghanistan

Table 4 Distribution of edible plants in the study area

Sl. no.	Botanical name	Family	Common	Uncommon	Rare
1.	*Annona muricata* L.	Annonaceae			√
2.	*Annona reticulata* L.	Annonaceae	√		
3.	*Annona squamosa* L.	Annonaceae			√
4.	*Garcinia gummi-gutta* (L.) Robs.	Clusiaceae	√		
5.	*Abelmoschus esculentus* (L.) Moench.	Malvaceae	√		
6.	*Averrhoa bilimbi* L.	Oxalidaceae		√	
7.	*Averrhoa carambola* L.	Oxalidaceae			√
8.	*Citrus limon* (L.) Burm. F.	Rutaceae	√		
9.	*Citrus maxima* (Burm.f.) Merr.	Rutaceae			√
10.	*Nephelium lappaceum* L.	Sapindaceae			√
11.	*Mangifera indica* L.	Anacardiaceae	√		
12.	*Spondias pinnata* (L. F.) Kurz.	Anacardiaceae			√
13.	*Moringa pterygosperma* Gaertn.	Moringaceae	√		
14.	*Phaseolus vulgaris* L.	Fabaceae	√		

(continued)

Table 4 (continued)

Sl. no.	Botanical name	Family	Common	Uncommon	Rare
15.	*Senna tora* (L.) Roxb.	Caesalpiniaceae	√		
16.	*Tamarindus indica* L.	Mimosaceae	√		
17.	*Terminalia catappa* L.	Combretaceae	√		
18.	*Psidium guajava* L.	Myrtaceae	√		
19.	*Syzygium aqueum* (Burm.f.) Alston	Myrtaceae	√		
20.	*Syzygium cumini* (L.) Skeels	Myrtaceae		√	
21.	*Passiflora edulis* Sims	Passifloraceae		√	
22.	*Passiflora foetida* L.	Passifloraceae			√
23.	*Carica papaya* L.	Caricaceae	√		
24.	*Citrullus lanatus* (Thunb.) Mats.	Cucurbitacae		√	
25.	*Coccinia grandis* (L.) Voight	Cucurbitacae	√		
26.	*Cucumis sativus* L.	Cucurbitacae	√		
27.	*Cucurbita maxima* Duch.	Cucurbitacae	√		
28.	*Momordica charantia* L.	Cucurbitacae	√		
29.	*Trichosanthes anguina* L.	Cucurbitacae	√		
30.	*Centella asiatica* (L.) Urban	Apiaceae		√	
31.	*Manilkara zapota* (L.) P.Royen	Sapotaceae		√	
32.	*Capsicum annuum* L.	Solanaceae	√		
33.	*Capsicum frutescens* L.	Solanaceae	√		
34.	*Lycopersicon esculentum* Mill.	Solanaceae	√		
35.	*Physalis angulata* L.	Solanaceae			√
36.	*Solanum melongena* L.	Solanaceae	√		
37.	*Solanum torvum* Sw.	Solanaceae	√		
38.	*Boerhavia diffusa* L.	Nyctaginaceae	√		
39.	*Amaranthus hybridus* L.	Amaranthaceae	√		
40.	*Amaranthus tricolor* L.	Amaranthaceae	√		
41.	*Amaranthus viridis* L.	Amaranthaceae	√		
42.	*Phyllanthus emblica* L.	Euphobiaceae		√	
43.	*Sauropus androgynus* (L.) Merr.	Euphobiaceae	√		
44.	*Artocarpus heterophyllus* Lam.	Moraceae	√		
45.	*Artocarpus hirsutus* Lam.	Moraceae			√
46.	*Artocarpus incisus* (Thunb.) L.f.	Moraceae		√	
47.	*Morus alba* L.	Moraceae		√	
48.	*Dioscorea alata* L.	Dioscoreaceae	√		
49.	*Colocasia esculenta* (L.) Schott.	Araceae	√		
50.	*Amorphophallus paeoniifolius* (Dennst.) Nicolson	Araceae	√		

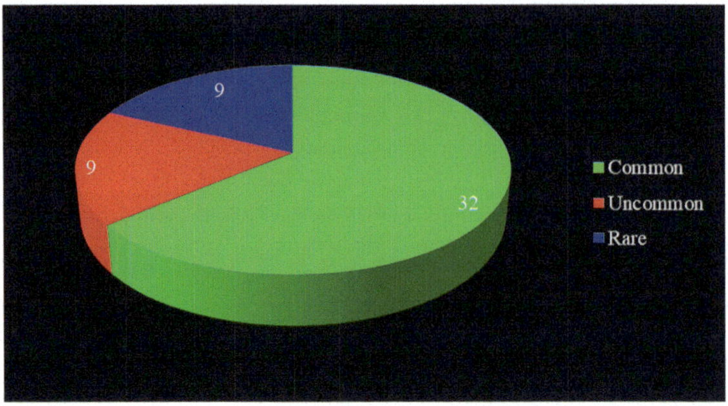

Fig. 5 Analysis of the distribution pattern of edible plants in the study area

Conclusion

The present study on edible plants, which are distributed in Kozhikode district, reveals that there are about 50 species of plants belonging to 38 genera and 25 families. These plants were distributed variously in the study area. Based on their distribution pattern and occurrence in the study area, they are categorized as common (32 Nos.), uncommon (9 Nos.), and rare (9 Nos.). Out of these 25 families, Cucurbitaceae and Solanaceae are the first dominant families with six species each, followed by Moraceae with four species; Amaranthaceae, Annonaceae, and Myrtaceae with three species each; and other families, represented two and single species, respectively. The different plant parts, like fruits (37 Nos.), leaves (9 Nos.), seeds (1 No.), tuber [1 No.], rhizome/corm (2 Nos.), etc., are used for edible purposes. The present documentation also highlighted medicinal importance of some edible plants. It consists of 14 species of edible plants, which belong to 11 families that possess potential medicinal properties to cure various ailments. Moreover, out of 50 edible plant species documented, there are about 22 plants introduced here from different countries, like Central America, Malaysia, Indonesia, South America, Tropical Africa, Tropical America, South America, Pacific Islands, Afghanistan, etc.

The present study is a preliminary approach in this field. A more detailed study on this aspect is required to disclose the edible potentiality of many less-known plant species in forests as well as other unexplored areas of the country. Tribal people of the state are more conscious about the edible plant wealth of their surroundings. However, native inhabitants are not that much conscious of it. In this context, the present study also suggests conducting some more explorative approach to various uncultivated areas to reveal the edible/economic potential of our prospective plants; each individual has an equal responsibility to conserve these valuable plants through sustainable utilization.

A). *Annona muricata* L.

B). *Annona squamosa* L.

C). *Averrhoa carambola* L.

D). *Citrus maxima* (Burm.f.) Merr.

E). *Nephelium lappaceum*

F). *Spondias pinnata* (L. F.) Kurz

Plate 2 Common edible plants of the study area

A). ***Syzygium aqueum*** (Burm.f.) Alston

B). ***Syzygium cumini*** (L.) Skeels

C). ***Passiflora edulis*** Sims

D). ***Passiflora foetida*** L.

E). ***Coccinia grandis*** (L.) Voight

F). ***Centella asiatica*** (L.) Urba

Plate 3 Common edible plants of the study area (Cont...)

A). ***Boerhavia diffusa*** L.

B). ***Amaranthus tricolor*** L.

C). ***Sauropus androgynus*** (L.) Merr.

D). ***Artocarpus incisus*** (Thunb.) L.f.

E). ***Morus alba*** L.

F). ***Dioscorea alata*** L.

Plate 4 Common edible plants of the study area (Cont...)

References

1. Nath N. Wild edible vegetables from Western Assam. Acad J Biosci. 2015;3(12):1044–50.
2. Jose DK, Sasidharan N. Checklist of wld edible plants of Ardom Wildlife Sanctuary, Kerala, India, South Indian. J Boil Sci. 2016;2(1):141–4.
3. Sheela L, Nath KG, Vijayalakshmi D, Geetha M, Roopa B. Proximate composition of under-utilized green leafy vegetables in southern Karnataka. Hum Ecal. 2004;15(3):227–9.
4. Sanjasi Rao ML, Yesudas S, Sakkrri K. Indigenous plant foods which are commonly consumed by the tribal communities in Dumbariguda area of Vishakapatanam District, Andhra Pradesh, India. Biolife. 2014;2(3):866–75.
5. Chen B, Qui Z. Consumer attitude towards edible wild plants, a case study. Int J For Res. 2012;12:1–16.
6. Reyes-Garcia V, Vadez T, Huanca WL, Wilkie D. Knowledge and consumtion of wild plants; a comparative study in two two Tsimane villages in the Bolivian Amazon. Ethnobot Res Appl. 2005;3:201–7.
7. Misra S, Misra MK. Leafy vegetable plants of South Odisha, India. Int J Agric Food Sci. 2013;3(4):131–7.
8. Nemli K, Tanyolac T. Green leafy vegetables occupy an important place among the food crops. J Nat Sci. 2015;12:14–9.
9. Bhatt ID, Rawat S, Badhani A, Rawal RS. Nutraceutical potential of selected wild edible fruits of the Indian Himalayan region. Food Chem. 2016;215:84–91.
10. And CM, Thomas B. Traditional worshiping plants from selected sacred groves of Kozhikode District, Kerala, India. Res J Recent Sci. 2017;6(4):7–13.
11. Hooker JD. The Flora of British India. I–VIILondon: Reeve & Co; 1872–1897.
12. Gamble JS, Fischer CEC. The Flora of the Presidency of Madras. London: Adlard & Son Ltd; 1915–1936.
13. Manilal KS, VV Sivarajan. Flora of Calicut. Bishen Singh Mahendrapal Singh, Dehra Dun; 1982
14. Sasidharan N. Biodiversity documentation for Kerala. Part-6: flowering plants. Peechi, Thrissur: Kerala Forest Research Institute; 2004
15. Santapau H, Hentry AN. A dictionary of the flowering plants in India. New Delhi: Council of Scientific & Industrial Research; 1973.
16. Narayanan MK, Anilkumar R, Balakrishnan N, Sivadasan V. Wild edible plants used by the Kattunaikka, Paniya and Kuruma tribes of Wayanad District, Kerala, India. J Med Plants Res. 2011;5(5):3520–9.
17. Patale CK, Praveen ND, Sushama N. Ethnobotanical studies on wild edible plants of Gond, Halba and Kawar tribes of Salekasa Taluk, Gondia district, Maharashtra state, India. Int Res J Pharm. 2015;6(8):512–8.
18. Rao MS, Yesudas S, Sarkkari K. Indigenous plant foods which are commonly consumed by the tribal communities in Dumbriguda area of Viskhapatanam district, Andhra Pradesh, India. Biolife. 2014;2(3):866–75.
19. Bandyopathy S, Mukherjee SK. Wild edible plants of Koch Bihar district West Bengal. Nat Prod Rad. 2009;8(1):64–72.
20. Chaithanya VV, Sugiya N, Aneesh KS. Wild edible plants traditionally used by *Kadar* tribes of Vazhachal forest Division, Thrissure, Kerala. J Global Bio Sci. 2015;4(9):344–9.

Exploring the Medicinal Potential of *Muntingia calabura Linn.* Fruit: A Comprehensive Overview

Preethi Kathirvel and Vishruthi Sundar Kumar

Abstract

The only member of the genus *Muntingia*, *Muntingia calabura*, is a flowering plant in the Elaeocarpaceae family. This is a fast-growing fruit tree. This pioneer species is resistant to drought, alkaline and acidic soil, and harsh environmental conditions. The crimson fruits are commonly known as Jamaican cherries. The flowers are used to relieve cramps and spasms in the abdomen and as an antiseptic. It is effective in treating headaches and colds. The fruit extract from *M. calabura* L. was tested for its antibacterial properties against the gram-positive strains (*S. aureus, S. epidermidis,* and *S. mutans*) and gram-negative (*C. violaceum* and *E. coli*) bacteria that are relevant to human infections. The fruit extract exhibited notable antibacterial activity against the microorganisms under investigation. The fruits of *Muntingia calabura* are antioxidant-rich due to the presence of phytochemical compounds like phenolic, flavonoids, and tannins. The investigated methanolic *Muntingia calabura* fruit extract is a chemotherapeutic drug to stop the proliferation of cancerous cells, which proves to treat cancer cells. Consequently, it is evident from these pharmacological findings that *M. calabura* fruits and leaves have a historic folkloric use in the management of pain-inflammatory illnesses, which proves to be an anti-inflammatory agent too. Hence, these findings could help to identify the active ingredients in wild edible fruits by analyzing their antimicrobial, antioxidant, anti-inflammatory, and cytotoxicity activities in more detail.

Keywords

Muntingia calabura Linn. fruit · Elaeocarpaceae · Antioxidant · Anti-inflammatory · Antibacterial · Cytotoxicity activities

P. Kathirvel (✉) · V. Sundar Kumar
Department of Microbial Biotechnology, Bharathiar University, Coimbatore, India

P. Thangaraj (ed.), *Pomotherapeutic Insights on Wild Edible Fruits*,
https://doi.org/10.1007/978-981-96-0543-9_3

Graphical Abstract

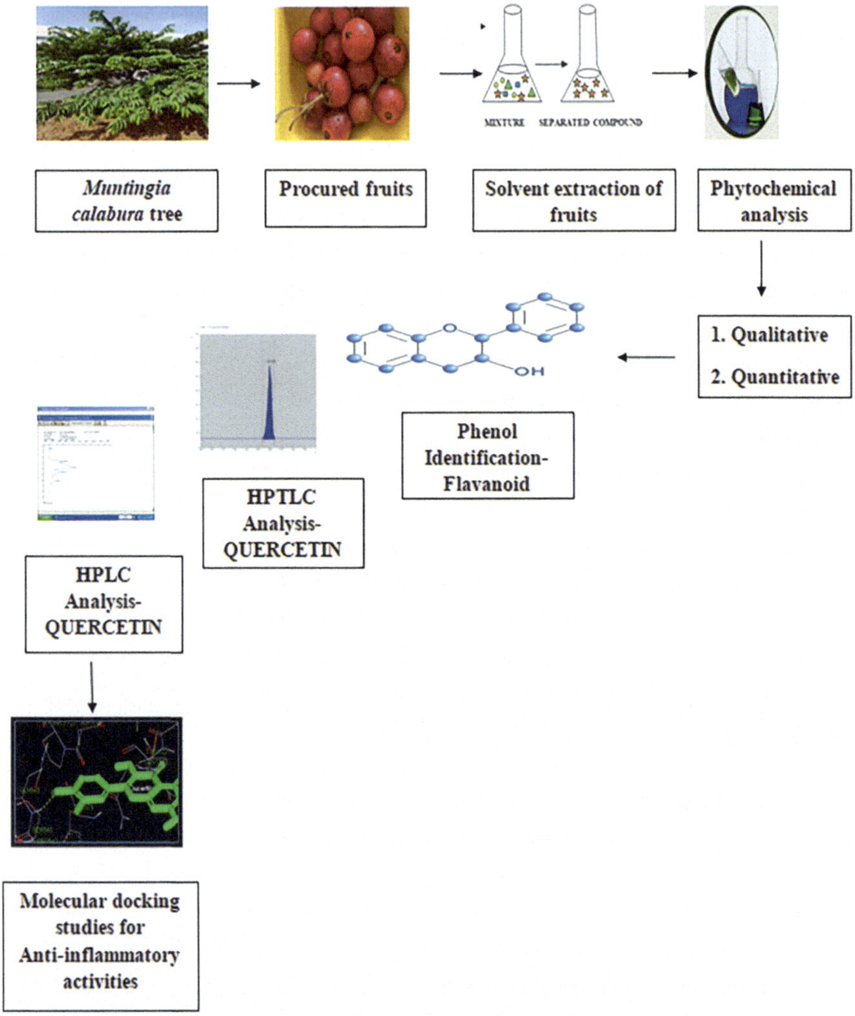

Introduction

Before the discovery of microbes, it was already established that some plants possessed medicinal properties. Plants have been used as medicines for common infectious diseases since ancient times [1]. The earliest form of medicine was herbal, with roots in traditional medicine that has been practiced across various countries for thousands of years. Natural products are a major source of therapeutic compounds. Certain contemporary medications utilized in pharmacotherapy have historical roots in herbal medicine. Apart from their bioactive components, many plant species that have been used historically also have biological properties [2].

Numerous ailments are still being treated with some of these age-old remedies today. Research has demonstrated that when it comes to treating skin, respiratory, biliary, gastrointestinal, and urinary infections, essential oils derived from plants are superior to their extracts [1].

Since ancient times, Africans have treated inflammatory illnesses with medicinal herbs. Of the more than 5000 species of plants found in Zimbabwe, 10% have possible medical applications. There are reportedly more than thirty thousand blooming plants in South Africa, some of which are utilized to treat and manage conditions involving inflammatory pain in both human and animal subjects. It has been suggested that natural anti-inflammatory compounds found in Egyptian plants may have few adverse effects [3].

Approximately 80% of the global population, particularly for primary healthcare requirements, mostly uses plant-based traditional medicine, according to a 1993 report by the Director of WHO Traditional Medicine [4]. One of the countries with the most diverse cultural and medical practices is India. Ayurveda, Unani, and Siddha are the three major traditional medical systems. Eighty percent of people in underdeveloped nations are dependent on plant-based phytomedicines, according to a 2003 estimate by the World Health Organization. Numerous unique bioactive phytocompounds from plants continue to have therapeutic benefits like anti-inflammatory, anticancer, antibacterial, and other qualities [5].

Medicinal Properties

Herbal Medicines

Herbal medicines and their derivatives have been used for thousands of years in Asia. During the decoction process, boiling water is used to extract all herbal medicines found in oriental herbal medicine preparations, regardless of whether the herbs are sold as singles or in composite formulas. This could be the primary reason why it is harder to maintain the quality control of Eastern herbal drugs than it is for Western medications [6].

Some conventional medical systems commonly use minerals, particularly heavy metals, for specific therapeutic applications [7]. Plants contain chemical components that can positively impact the human body, making them valuable for therapeutic purposes. The study of these organic compounds is known as phytochemistry by pharmacists, focusing on medicinal plants' biochemistry. It is crucial to consider using plant extracts and phytochemicals that have antibacterial properties essential in medicinal processes. These could consist of one particular component or a combination of several. Secondary metabolites are substances that are metabolic byproducts that result from the primary compounds that the plant does not use. They are stored in plant cells instead. The effectiveness of these components depends on whether they have therapeutic properties. While proteins, sugars, and carbohydrates are inactive plant parts, they still have medicinal applications. In contrast, alkaloids, glycosides, volatile oils, tannins, and other secondary metabolites are examples of active ingredients [8].

According to Carmona and Pereira [9] herbal medicine began to be viewed by doctors as an "alternative," untrained, archaic, and nonscientific form of treatment for people who could not afford "real" medical care. The majority of traditional medical practices, however, involve the use of plant extracts or their active ingredients, and between 70% and 95% of the world's population receives their main medical treatment from them, according to the World Health Organization (WHO).

India has one of the largest plant medicine traditions in the world. The rural populations of India have approximately 25,000 powerful plant-based medicines used in traditional medicine. More than 1.5 million practitioners of traditional medicine systems employ medicinal plants for preventative, therapeutic, and promotion purposes. In India, 7800 pharmaceutical manufacturing companies use an estimated 2000 tons of herbs annually [10].

How far an herb's traditional use guarantees the safety of a related herbal medication. According to Moreira et al. [11], two categories of herbal pharmaceuticals are designated for registration by the Brazilian National Health Surveillance Agency: "traditional phytotherapeutic products" (TPTP) and "phytotherapeutic medicines." Recently, the proposed rule was made available for public review and comment. A product is deemed to have a traditional use if it has been demonstrated to be safe, acknowledged in the literature, or backed by studies on ethnopharmacology and/or ethnobotany. Safety and efficacy evidence from preclinical and/or clinical research is no longer necessary to receive approval for commercialization once the traditional usage has been established.

Synthetic Medicines

Synthetic medicines can pose a risk to one's health even though they are effective in treating several illnesses. Plant-based pharmaceutical alternatives could be a better option. Because they are less likely to have adverse effects and are cheaper than synthetic inhibitors, natural inhibitors, such as those derived from medicinal herbs, can be useful replacements. According to scientific research, the pharmacological effects of the various *M. calabura* L. sections differ. An effective method for treating chronic degenerative disorders is enzyme inhibition, with a focus on the α-glucosidase and α-amylase enzymes responsible for converting glucose. When these enzymes are blocked, less glucose is absorbed, which can lead to decreased blood glucose levels after meals. Another stage in treating gouty arthritis and hyperuricemia is to inhibit the xanthine oxidase pathway. The leaves of *M. calabura* L. have been found to possess anti-inflammatory, antibacterial, anticarcinogenic, and antioxidant properties. The roots of this plant also contain flavonoids that exhibit anticancer effects. Though there have been numerous studies on the pharmacological properties of *M. calabura* L. leaves, research on the fruit is limited, with only a few publications highlighting its anti-inflammatory and antioxidant abilities [12].

According to Karimi et al. [13], a large number of people who look for medicinal plants have already used medication. As agonists or potentiators of other medications, medicinal plants can function. Herbal treatments usually promote the body's natural healing process, while synthetic medications target specific disease symptoms as determined by scientific pathology.

These medications, known as synthetic pharmaceuticals, are created in laboratories using various techniques and are not present in nature. Even though herbal remedies sometimes have less potency than synthetic medications, they are nonetheless thought to be less poisonous or cause fewer negative effects. Numerous instances concerning the adverse consequences of synthetic medications are documented in the literature, such as the well-known antipyretic medication paracetamol, which has a serious adverse effect on liver damage [14]. The biological roles of just a small percentage of the estimated 250,000–500,000 plant species have been thoroughly studied. Despite this, during the past 50 years, natural products and the chemicals derived from them have continued to be highly significant parts of pharmacopeia. This is because combinatorial chemistry and high-throughput screening techniques are used to create novel medication [15].

Ayurvedic Medicines

Numerous investigations into herbal remedies have had unfavorable outcomes, yet certain plants have demonstrated promising outcomes in in-vivo animal models and small-scale clinical trials. Herbal remedies sold as dietary supplements are governed under the Dietary Supplement Health and Education Act (DSHEA) because the Drugs and Cosmetics Act of 1940 does not demand evidence of efficacy or safety [16].

Ayurveda seeks to return biological systems to normal using a range of techniques, including internal herbal remedies, surgery, purifying therapies, and food and exercise guidelines. Ayurveda treats patients by identifying their ailments and then employing internal remedies, dietary restrictions, and behavioral changes to treat them. Plant-based therapy is a fundamental component of Ayurvedic treatment. Most traditional Ayurvedic treatments are polyherbal, made up of 3–30 distinct herbs. To make a steady, well-balanced mixture, these ingredients are carefully blended. When two plants are paired together, one or two become dominant, and the other plants take on a supporting role. With differing results, the auxiliary herbs will act as catalysts to support proper absorption, transportation, and toxicity reduction [17].

Muntingia calabura Linn

Hundreds to thousands of distinct chemical compounds, each with a unique biological function, are produced by higher plants that can perform photosynthesis. These substances play a significant ecological role [18]. One of the Philippine Islands' medicinal plants is *Muntingia calabura* L, a member of the Elaeocarpaceae family. It was brought to southern Taiwan, where it was grown for edible and decorative uses in gardens and at the sides of roads [19]. Some common names for *Muntingia calabura* L. include Jamaican cherry, strawberry cherry tree, jam tree, Chinese cherry, Madras pea pumpkin, and cotton candy berry. Indian Japanese cherry, Singapore cherry, bird's cherry, calabura, poor man's cherry, Panama berry, Pazham, and cherry chettu are some of the other names for this fruit [5].

Muntingia calabura L. is frequently referred to as the "Jamaican cherry." Especially among the Malay people, *Muntingia calabura* L. is frequently referred to as "kerukup siam" in Malaysia. It is indigenous to southern Mexico, tropical South America, Central America, the Greater Antilles, Trinidad, and St. Vincent. Moreover, *M. calabura* L. is widely planted as a wayside tree in Malaysia [20].

Morphology

Growing to a maximum height of 7–12 meters in just 2 years, the tree's two main characteristics are its quick growth and fruit-bearing ability. The tall, angular, long, pointed leaves have tiny, sticky hairs covering them. The vegetation is evergreen. Its tiny flowers have five apparent green sepals, five white petals, and a significant quantity of yellow stamens. They also have a faint odor. Flowers are fully pollinated and open in January. The leaves are evergreen at all stages. Fruits are round, ranging in width from 1.25 to 1.25 cm, and have smooth, thin, fragile skin that is red or occasionally yellow. The pulp is soft and juicy, with a light brown color and a wonderfully sweet, musky flavor that resembles figs. Hundreds of tiny, yellowish seeds, too little to be discernible when eating, are found inside fruits. All through the year, the tree bears fruit. The growth of the plant is facilitated by soil that has a minimum pH of 5.6–8.5, full sun exposure, and a temperature that is cold and hardy [5].

This plant is a slender, quickly-growing tree with almost horizontally extending branches that can develop to heights of 7.5–12 meters. Evergreen alternating leaves of which *M. calabura* L. are 5–12.5 cm long, lanceolate or oblong, long pointed at the apex, oblique at the base, dark green in color, minutely hairy on the upper surface, and hairy in texture on the underside in gray or brown. The fruits are many, round, and 1.25–1.25 cm wide. They have a light-brown, soft, juicy pulp with a flavor similar to figs and are covered in exceedingly small, yellowish seeds. They have thin, smooth, tender skin, either red or yellow, and thin, tender flesh. The blooms, which have a diameter of 1.22–2.25 cm, are produced singly, in pairs, or in groups of three in the axils of the leaves [20].

Fig. 1 Leaves and flower of *Muntingia calabura* L. fruit

Scientific Classification [5] (Figs. 1 and 2)

Kingdom: Plantae
(Unranked): Magnoliophyta
Subphylum: Euphyllophytina
Infraphylum: Radiatopes
(Unranked): Magnoliopsida
Subclass: Magnoliidae
Superorder: Rosanae
Order: Malvales
Family: Muntingiaceae
Genus: Muntingia
Species: calabura

Phytochemistry of *Muntingia calabura* Linn Fruit

According to Chaudhari et al. [21], screening was performed to ascertain whether or not the extract included several active components, including phenolic compounds, carbohydrates, flavonoids, glycosides, saponins, alkaloids, lipids or fixed oils, proteins and amino acids, and tannins.

To determine whether alkaloids, flavonoids, glycosides, tannins, and phlobatannins, which may be responsible for the antibacterial properties, were present or absent in *M. calabura* L. extracts (n-hexane, ethyl acetate, and methanol) for both ripe and unripe fruits, a qualitative phytochemical screening was carried out first. All of the extracts from ripe and unripe fruits contained flavonoids [22]. The presence of polar solvents, including water, methanol, and ethanol, has led to a higher concentration of flavonoids, saponins, and tannins. Nonpolar solvents, however,

Fig. 2 *Muntingia calabura* L. fruits

have lower concentrations. The findings show that although the amounts of the bioactive components are varied, this plant contains saponin, tannin, and flavonoids. Flavonoids' biological impacts, such as their antibacterial and anticancer properties, are well-documented. According to the study's findings, the phytochemical substances that were discovered may be the bioactive components that give the plants' leaves their therapeutic properties. Genistein and epigallocatechin gallate (EGCG) are the bioactive components identified in *Muntingia calabura* L. leaves. The catechin is partially composed of both of those ingredients. Tea contains potent antioxidants called catechins, which are thought to be responsible for many of these health advantages. The leaves of *Muntingia calabura* L. have been extracted, and this shows that the plant has significant bioactive chemicals that may be used to produce plant species with antibacterial activity [18] (Table 1).

Traditional Application

In the field of complementary and alternative medicine, in particular, the fruit, i.e., *M. calabura* L., has recently piqued the curiosity of researchers worldwide. Many phenolic compounds, including phenolic acids, cytotoxic flavonoids, and chalcones, are present in high concentrations in the organic extract of leaves and stems of *M. calabura* L. These compounds are assumed to be the source of the plant's medicinal properties [33].

Table 1 Parts, types of extract, and activities of *Muntingia calabura* L.

Parts of *Muntingia calabura* L.	Extract	Activity	References
Leaves	Ethanol	Acute toxicity	[23]
Leaves	Methanol	Antiulcer	[24]
Fruits	Petroleum ether, chloroform, ethyl acetate, butanol, methanol, butyl hydroxy toluene	Antioxidant	[25]
Fruits	Methanol	Anti-inflammation	[26]
Fruits	Methanol	Anticancer	[27]
Leaves	Ethanol	Antibacterial	[28]
Leaves	Methanol	Antibacterial and cytotoxic	[29]
Leaves	Aqueous	Antinociceptive, anti-inflammatory, and antipyretic	[30]
Fruits	Ethanol	Enzyme inhibitory, antioxidant, and antibacterial	[12]
Root	Methanol, chloroform, and petroleum ether	Antifungal	[31]
Leaves, flowers and fruits	Water, ethanol, hydroethanol, and petroleum ether	Antityrosinase and antioxidant	[32]
Stem bark	Methanol	Cytotoxic	[19]
Leaves	n-Hexane, ethyl acetate, and methanol	Antioxidant and antimicrobial	[21]
Fruits	n-Hexane, ethyl acetate, and methanol	Antibacterial	[22]

Muntingia calabura L. is widely cultivated in hotter areas of Southeast Asia and India, where it is frequently planted next to roads to denote a person's tolerance for air pollution. Different portions of *M. calabura* L. have been used to treat a range of illnesses. The flowers are used topically to alleviate edema and serve as an antibacterial on the lower extremities. By boiling or steeping the leaves in water, they can be used to treat swelling of the prostate gland, colds, and headaches. Headaches, the beginnings of colds, spasms, and dyspepsia are all treated with flowers. They can also be used as tonics and sedatives. Vietnam, however, also employs the plant's roots as emmenagogues. The fruits are frequently made into tarts or jam, and the leaf infusion is offered as a beverage similar to tea [20]. Particularly in Malaysian and Peruvian communities, infusions made from its leaves, stems, bark, and flowers have long been used in folk medicine to treat infections, colds, headaches, and stomach ulcers [33].

It is commonly known that flavonoids have biological effects, such as antibacterial and anticancer properties. They are potent antibacterial and anticancer agents in vitro, effectively combating a variety of microbes and cancer cells, respectively.

The antibacterial and cytotoxic properties of the pure components extracted from *Muntingia calabura* L. leaves were greater than those of the crude extract. Synergism between the extract's constituent components may be the cause of this behavior [29]. Especially flavonoids, which can be combined with other ingredients to create natural antioxidants in oral dosage forms like vitamins and topicals for skin care products, are plants that are effective as antioxidants because they include carotenoids and polyphenols [34].

Antioxidant Activity

The antioxidant capacity of the *M. calabura* L. fruit's ethanol extract was evaluated using a variety of assays, including phosphomolybdenum, ferric thiocyanate, CUPRAC, DPPH radical scavenging, and ferrous ion chelating assays. By giving a proton to DPPH radicals, the antioxidant chemicals in the extract can scavenge free radicals and quench them into a more stable state. The DPPH test measures this process. The present investigation included CUPRAC, ferric thiocyanate, and phosphomolybdenum tests to quantify the reductive potential of the *M. calabura* L. extract. By donating an electron, reductants found in the extract led to the reduction of metal ions into their reduced form [12].

Muntingia calabura L. leaf extracts were screened for phytochemicals and found to contain flavonoids, phenols, steroids, triterpenes, tannins, reducing sugars, and saponins. According to [34], it was suggested that this plant might be useful as a natural antioxidant source because of the significant TPC value and the strong antioxidant activity (DPPH) of the leaf methanol extract.

Gallic acid equivalents (GE) were used in this work to express the total phenolic contents in the various fractions of *M. calabura* L. that were evaluated using the Folin-Ciocalteau technique. The MeOH extract produced the maximum phenolic content, followed by the EA extract, BuOH extract, and CHCL fraction, according to the assessment of TP among the various fruit fractions. The extract from PE has the least amount of phenolics. The study's conclusion suggested that the MeOH extract contained total phenolics [25].

Under varied extraction conditions, the antioxidant activity of the fruit evaluated was *Muntingia calabura* L. peel using the DPPH radical, ORAC, ABTS cation radical, FRAP, and total phenolic contents assays. Additionally, a method for phenolic component analysis in Brazilian calabura peel samples was devised. It employs methanol and water as the solvents, magnetic stirring as the extraction method, and HPLC-DAD and UPLC-ESI-MS/MS analysis. The polar solvents were the most successful in eliciting antioxidant activity from the calabura peel, evident in all extraction techniques and assays. The HPLC-DAD method accurately detected phenolic compounds (excluding chlorogenic acid), with recoveries ranging from 72% to 107% and precision values of 4% [35].

The ORAC assay of calabura fractions evaluated antioxidant activity differently from TPC and CT, even though the DPPH and ABTS antioxidant assays demonstrated a tendency similar to TPC. The glycosylated fraction was the most antioxidant fraction, according to the ORAC experiment, followed by the free, esterified, and insoluble-bound fractions. The calabura fruit-free fraction exhibited the highest DPPH and ABTS antioxidant activity levels. According to DPPH, ABTS, and ORAC assays, the antioxidant activity was comparable to that of popular berry fruits [36].

Muntingia calabura L. was evaluated using 1,1-diphenyl-2-picrylhydrazyl (DPPH) [32], while Singleton and Rossi used Folin-Ciocalteu (FC) to estimate total phenolic content, with gallic acid and tannic acid serving as the calibration standards. Significant antioxidant activity and substantial reducing power were also demonstrated by *Muntingia calabura* L. leaf extract [37].

It used several techniques with various processes to examine the efficacy of extracts and fractions from Kersen fruit (*Muntingia calabura* L.). The purpose of this study is to evaluate Kersen fruit as an exogenous antioxidant agent made of natural components. This study employed a range of antioxidant testing techniques, including hydroxyl radicals, nitric oxide, bleaching of β-carotene, and iron-chelating agents. Based on the IC_{50} values, the antioxidant capacity of Kersen fruits was ascertained. The linear regression equation of the sample concentration yields the IC_{50} value; the lower the IC_{50} value, the higher the activity, and vice versa. Extracts and fractions derived from Kersen fruit exhibit antioxidant properties that help prevent Parkinson's disease, Alzheimer's disease, and other degenerative diseases caused by free radicals. Hydroxyl radicals are one of the most reactive components of ROS (reactive oxygen species). The antioxidant properties of phenolic and flavonoid components found in Kersen fruit's EE (ethanol extract), EF (aqueous-ethanol fraction), and EAF (ethyl acetate fraction) can counteract free radicals through the HAT and SET mechanisms. Based on the suppression of the β-carotene breakdown rate in the emulsion medium, the β-carotene bleaching method was used to investigate the antioxidant activity of Kersen fruit extracts and fractions. Antioxidant activity was measured using the iron-chelating method to determine a sample's capacity to chelate iron ions from a reagent [38].

Anti-Inflammatory Activity

Similar to the effects of allopathic medicines, numerous plant extracts of botanical medicinal plants have been demonstrated to alleviate disease symptoms. Moreover, chemical therapies frequently have serious side effects attached to them. So, it's necessary to create natural product ingredients that are safer and have fewer adverse effects. Here, new findings on the anti-inflammatory properties of *Muntingia calabura* L. fruit extracts in carrageenan-induced acute inflammation of the hind paw have been observed for the first time. But the majority of the time, the extracts had notable, dose-dependent effects. *Muntingia calabura* L. fruit extracts may have anti-inflammatory properties because of the biologically active chemicals they contain. One well-established paradigm for assessing the anti-inflammatory properties of pharmaceuticals is the suppression of Carrageenan-induced inflammation in rats, which has been widely employed to evaluate natural products for their anti-edematous impact [26].

The membrane stabilization assay was utilized to investigate the anti-inflammatory properties of the root proteins. Since the membranes of HRBCs and lysosomes are identical, the ability to inhibit hypotonicity-induced lysis of the membranes is considered a marker of anti-inflammatory action. Crude proteins from the roots of *Muntingia calabura* L. have the highest level of anti-inflammatory activity at 1000 μg/ml, which is equivalent to the 200 μg/ml concentration of the common medication diclofenac sodium. The crude protein's ability to reduce inflammation depended on its concentration. The proteins demonstrated a stabilizing impact on the membrane by preventing the lysis of the erythrocyte membrane caused by hypotonicity. The stability of the erythrocyte membrane suggests that the proteins found in the roots of *Muntingia Calabura* L. may be able to stabilize the lysosomal membrane, as the two membranes are similar. Comparing *Muntingia Calabura* L. root boiling water extract to other antioxidants, the proteins show promise as both anti-inflammatory and strong antioxidants [39].

In RAW264.7 murine macrophage cell lines, an ethanol extract of the fruit *Muntingia calabura* L. inhibits the production of pro-inflammatory mediators by lipopolysaccharide. The fruit extract was first subjected to a phytochemical analysis by the researchers, and the findings showed that it contained four flavonoids and eleven phenolic acids, of which gallic acid was present in significant amounts. The production of nitrate, PGE2, and pro-inflammatory cytokines such as tumor necrosis factor (TNF), interleukin-1ß (IL-1ß), and interleukin-6 (IL-6) in lipopolysaccharide (LPsac)-stimulated RAW264.7 cells were later found to have greatly decreased after pretreatment with the test extract [40].

Saponins and alkaloids were not found in the petroleum ether fraction (PEF) or ethyl acetate fraction (EAF), but phytochemical analysis revealed the presence of steroids, condensed tannins, and flavonoids. When comparing EAF to PEF, flavonoids were found in greater amounts. Furthermore, it was discovered that PEF contained triterpenes while EAF did not. Alkaloids, steroids, flavonoids, and triterpenes were absent from aqueous fractions (AQF), which, on the other hand, revealed the existence of only condensed tannins and saponins [24].

According to Zakaria et al. [30], the *Muntingia calabura* L. extract exhibited concentration-independent anti-inflammatory activity; the carrageenan-induced rat paw edema test is a commonly employed method to evaluate the anti-inflammatory efficacy of extracts as well as medications. When compared to the other concentrations of *Muntingia calabura* L. aqueous extract (MCAE), the highest concentration of MCAE produced less activity. Edema is formed by prostaglandins and other pro-inflammatory chemicals produced by polymorphonuclear leucocytes and kinins. It was indicated that the medicine had anti-inflammatory qualities when the MCAE group's paw edema thickness was much lower than that of the control group. The conventional use of the *M. calabura* L. leaves in helping to alleviate pain associated with stomach ulcers through the central mechanism has been scientifically validated by this discovery.

Antibacterial Activity

The fruit extract from *M. calabura* L. was tested for its antibacterial properties against the gram-positive strains (*S. aureus, S. epidermidis,* and *S. mutans*) and gram-negative (*C. violaceum* and *E. coli*) bacteria that are relevant to human infections. The fruit extract exhibited notable antibacterial activity against the microorganisms under investigation. The investigations revealed that *C. violaceum* was the most sensitive, as observed from the lowest MIC and MBC values. On the other hand, *S. epidermidis* and *E. coli* exhibited the least degree of activity [12].

The Kirby-Bauer method was used to assess each fraction's antibacterial activity to determine which fraction was most effective at inhibiting the growth of *S aureus*. Ethanol extract, 96% ethanol as a negative control, and vancomycin antimicrobial susceptibility disks as a positive control were evaluated in comparison. When compared to the inhibition zone of the related fraction, the results indicated that the ethyl acetate fraction was the most active and was linked to the creation of the biggest inhibition zone diameter [41].

The antimicrobial compounds from the root of *M. calabura* L. were extracted using three different solvents: polar methanol, medium-polar chloroform, and petroleum ether (which is the least polar). Compared to other solvent extracts, the methanol extract of *Muntingia calabura* L. demonstrated a broad spectrum of antimicrobial activity, according to the results of the studies on antimicrobial activity against fungal infections. Increased inhibition of *A. solani, Fusarium oxysporum f.*sp. *lycopersici, Pythium* sp., *Phytophthora* sp., *Rhizoctonia solani, Aspergillus niger,* and *Colletotrichum* sp. was observed in the methanol extract of *M. calabura* L. root. The *M. calabura* L. root methanol extract was found to have a stronger suppressive effect on the pathogens compared to the extracts made of petroleum ether and chloroform [31].

Cytotoxic Effects

The study's findings demonstrated that the methanol extract of *M. calabura* L. fruit inhibited the proliferation of human cancer cell lines, such as the NIN 3 T3 cell line for mouse embryonic carcinoma, the HeLa cell line for cervical cancer, and the Hep-2 cell line for laryngeal epithelial carcinoma. The extract demonstrated dose-dependent and cell-specific antiproliferative cytotoxicity against the tested cell lines. The results suggest that the *Muntingia calabura* L. fruit extract succeeded in inhibiting cell growth in vitro. The human cancer cell line HeLa was found to be maximally cytotoxic to methanolic fruit extract. Notable morphological alterations were seen when HeLa cells treated with fruit extract were compared to the control. Eventually, the degenerative morphological modification caused the cells to perish. In a concentration-dependent manner, the number of colonies dropped drastically, suggesting that the *M. calabura* L. fruit extract effectively inhibited the capacity of cancer cells to multiply and reduced their malignancy. The results of the study showed how effective the fruit extract was against the cell lines that were examined [27].

M. calabura's cytotoxic potential was determined using the brine shrimp lethality test. This assay was designed to evaluate extracts' bioactivity, which is typically linked to their cytotoxic and antitumor effects. Plotting the percentage of dead shrimp nauplii versus extract concentrations of *M. calabura* L. revealed the species' LC_{50} value of 455.93 μg/ml. To the best of our understanding, this is the first investigation into the cytotoxicity test of *M. calabura* L. The LC_{50} value was determined to be less than 1000 μg/ml based on the bioactivity assessment, showing cytotoxicity. Even though the fruit is edible, the LC_{50} value of *M. calabura* L. indicated the presence of cytotoxic compounds, which are responsible for the observed toxicological activity. The reported lethality of the *M. calabura* L. extracts against brine shrimps could demonstrate the presence of anticancer components [42].

Conclusion

The investigation of the medicinal qualities of *Muntingia calabura* L. fruit concludes by highlighting its possible importance in the field of complementary therapy. It is clear from the investigation of its different constituents—antioxidants, and phytochemicals—that this fruit has significant health benefits. As a medicinal agent, *Muntingia calabura* L. shows promise due to its anti-inflammatory and antibacterial qualities and its possible function in managing chronic diseases, including diabetes and cardiovascular problems. It assists in preventing Alzheimer's disease, Parkinson's disease, and other degenerative illnesses brought on by free radical damage. It is also a good option for places where access to traditional healthcare facilities is limited because of its affordability and ease. To completely understand its methods of action and maximize its therapeutic benefits, further investigation is required. The research conducted on *Muntingia calabura* L. fruit emphasizes the value of investigating traditional medicines for their potential benefits to contemporary medicine. It also emphasizes the importance of further research into the medicinal qualities of natural compounds.

References

1. Rios JL, Recio MC. Medicinal plants and antimicrobial activity. J Ethnopharmacol. 2005;100:80–4.
2. Marrelli M. Medicinal Plants. Plan Theory. 2021;10:1355.
3. Oguntibeju O. Medicinal plants with anti-inflammatory activities from selected countries and regions of Africa. J Inflamm Res. 2018;11:307–17.
4. Haq I. Safety of medicinal plants. Pakistan J Med Res. 2004;43:203–10.
5. Nirmala M, Priya R, Shankar T, Malarvizhi A. Medicinal values of Muntingia calabura leaves. In: Pharmacological benefits of natural products; 2018. p. 238–53.
6. Liang YZ, Xie P, Chan K. Quality control of herbal medicines. J Chromatogr B. 2004;812:53–70.

7. Zhang J, Wider B, Shang H, Li X, Ernst E. Quality of herbal medicines: challenges and solutions. Complement Ther Med. 2012;20:100–6.
8. Upadhye M, Kuchekar M, Pujari R, Kadam S, Gunjal P. Muntingia calabura: a comprehensive review. J Pharm Biol Sci. 2021;9:81–7.
9. Carmona F, Pereira AM. Herbal medicines: old and new concepts, truths and misunderstandings. Rev Bras Farmacogn. 2013;23:379–85.
10. Verma S, Singh SP. Current and future status of herbal medicines. Vet World. 2008;11:347–50.
11. Moreira DD, Teixeira SS, Monteiro MH, De-Oliveira AC, Paumgartten FJ. Traditional use and safety of herbal medicines. Rev Bras Farmacogn. 2014;24:248–57.
12. Simamora A, Santoso AW, Rahayu I, Timotius KH. Enzyme inhibitory, antioxidant, and antibacterial activities of ethanol fruit extract of Muntingia calabura Linn. J Herbmed Pharmacol. 2020;9:346–54.
13. Karimi A, Majlesi M, Rafieian-Kopaei M. Herbal versus synthetic drugs; beliefs and facts. J Nephropharmacol. 2015;4:27–30.
14. Nisar B, Sultan A, Rubab SL. Comparison of medicinally important natural products versus synthetic drugs-a short commentary. Nat Prod Chem Res. 2018;6:308.
15. Yuan H, Ma Q, Ye L, Piao G. The traditional medicine and modern medicine from natural products. Molecules. 2016;21:559.
16. Sharma AK, Kumar R, Mishra A, Gupta R. Problems associated with clinical trials of Ayurvedic medicines. Rev Bras Farmacogn. 2010;20:276–81.
17. Kumar S, Dobos GJ, Rampp T. The significance of ayurvedic medicinal plants. J Evid Based Complementary Altern Med. 2017;22:494–501.
18. Surjowardojo P, Sarwiyono IT, Ridhowi A. Quantitative and qualitative phytochemicals analysis of Muntingia calabura. J Biol Agric Healthc. 2014;4:84–8.
19. Chen JJ, Lin RW, Duh CY, Huang HY, Chen IS. Flavones and cytotoxic constituents from the stem bark of Muntingia Calabura. J Chin Chem Soc. 2004;51:665–70.
20. Mahmood ND, Nasir NL, Rofiee MS, Tohid SM, Ching SM, Teh LK, Salleh MZ, Zakaria ZA. Muntingia calabura: a review of its traditional uses, chemical properties, and pharmacological observations. Pharm Biol. 2014;52:1598–623.
21. Chaudhari RN, Jain AK, Chatap VK. Phytochemical screening, antioxidant and antimicrobial potential of leaves extract of Muntingia calabura. J Adv Sci Res. 2020;11:218–24.
22. Ariffin S, Zulkefli J, Saleh AM. Phytochemical analysis of Muntingia calabura Linn. and its antibacterial properties via in vitro evaluation. Malays J Anal Sci. 2022;26:766–73.
23. Ibrahim IA, Abdulla MA, Abdelwahab SI, Al-Bayaty F, Majid NA. Leaves extract of Muntingia calabura protects against gastric ulcer induced by ethanol in Sprague-dawley rats. Clin Exp Pharmacol. 2012;5:2161–1459.
24. Balan T, Sani MHM, Mumtaz Ahmad SH, Suppaiah V, Mohtarrudin N, Zakaria ZA. Antioxidant and anti-inflammatory activities contribute to the prophylactic effect of semi-purified fractions obtained from the crude methanol extract of Muntingia calabura leaves against gastric ulceration in rats. J Ethnopharmacol. 2015;164:1–15.
25. Preethi K, Vijayalakshmi N, Shamna R, Sasikumar JM. In vitro antioxidant activity of extracts from fruits of Muntingia calabura Linn. from India. Pharmacogn J. 2010;2:11–8.
26. Preethi K, Premasudha P, Keerthana K. Anti-inflammatory activity of Muntingia calabura fruits. Pharm J. 2012;4:51–6.
27. Preethi K, Sasikumar JM. In vitro cytotoxic effects of Muntingia Calabura Linn. Fruits against human cancer cell lines. Int J Biotechnol Biochem. 2011;7:379–84.
28. Gurning K, Simanjuntak HA, Purba H, Situmorang RF, Barus L, Silaban S. Determination of total tannins and antibacterial activities ethanol extraction seri (Muntingia calabura L.) leaves. J Phys. 2021;1811:012121.
29. Sufian AS, Ramasamy K, Ahmat N, Zakaria ZA, Yusof MI. Isolation and identification of antibacterial and cytotoxic compounds from the leaves of Muntingia calabura L. J Ethnopharmacol. 2013;146:198–204.

30. Zakaria ZA, Hazalin NAMN, Zaid SNHM, Ghani MA, Hassan MH, Gopalan HK, Sulaiman MR. Antinociceptive, anti-inflammatory and antipyretic effects of Muntingia calabura aqueous extract in animal models. J Nat Med. 2007;61:443–8.
31. Ramasamy R, Nanjundan J, Ponnusamy M. Solvent extraction and evaluation of antifungal activity of Muntingia calabura root against fungal phytopathogens. Int J Curr Microbiol Appl Sci. 2017;6:77–83.
32. Narayanaswamy N, Duraisamy A, Balakrishnan K. Tyrosinase inhibition and anti-oxidant properties of Muntingia calabura extracts: in vitro studies. Int J Pharm Bio Sci. 2011;2:294–303.
33. Seixas DP, Palermo FH, Rodrigues TM. Leaf and stem anatomical traits of Muntingia calabura L. (Muntingiaceae) emphasizing the production sites of bioactive compounds. Flora. 2021;278:151802.
34. Pungot NH. Potential of Malaysian cherry leaves (Muntingia calabura) as an antioxidant agent. Sci Lett. 2020;14:103–9.
35. Rotta EM, Haminiuk CW, Maldaner L, Visentainer JV. Determination of antioxidant activity and phenolic compounds of Muntingia calabura Linn. peel by HPLC-DAD and UPLC-ESI-MS/MS. Int J Food Sci Technol. 2017;52:954–63.
36. Pereira GA, Arruda HS, de Morais DR, Eberlin MN, Pastore GM. Carbohydrates, volatile and phenolic compounds composition, and antioxidant activity of calabura (Muntingia calabura L.)fruit. Food Res Int. 2018;108:264–73.
37. Ayesha Siddiqua AS, Premakumari KB, Roukiya Sultana RS, Vithya V, Savitha S. Antioxidant activity and estimation of total phenolic content of Muntingia calabura by colorimetry. Int J Chem Tech Res. 2010;2:205–8.
38. Nur S, Aswad M, Yulianti R, Sami FJ, Burhan A, Fadri A, Khairi N. Antioxidant activity profile of extract and fraction of kersen (Muntingia calabura L.) fruits prepared by different methods. In: IOP conference series: earth and environmental science. IOP Publishing; 2022. p. 012066.
39. Khan MA, Dinesha Ramadas DR, Mundasada SC, Kumar SN, Kashyap HR, Chikkanna D. In vitro anti-inflammatory activity of proteins isolated from Muntingia Calabura plant root. World J Pharm Res. 2015;4:2766–73.
40. Narendra R, Jyothi Y. Pharmacological activities of Muntingia Calabura: an overview of the last lustrum. Int J Pharm Sci Res. 2020;11:6020–7.
41. Rizky HM, Nanik S, Nurkhasanah N, Ricke D. Antibacterial activity and tlc-bioautography analysis of the active fractions of Muntingia calabura L. leaves against Staphylococcus aureus. Jurnal Farmasi Sains dan Komunitas. 2020;17:69–75.
42. Muniyappan G, Adenam NM, MY MY, NH MH, Kernain D, Karimah H. Phytochemical screening of Muntingia Calabura fruit for antioxidant and cytotoxic activities. J Adv Res Appl Sci Eng Technol. 2022;28:116–25.

Wild Edible Fruits from the Western Ghats: A Rich Source of Bioactive Compounds with Health-Promoting Effects: A Review

Gowtham Kannan, Benedict Mathews Paul,
and Parimelazhagan Thangaraj

Abstract

The Western Ghats of India are renowned for their rich biodiversity, including a plethora of wild edible fruits that have been integral to the diet and traditional medicine of indigenous communities for centuries. These fruits, besides their culinary appeal, harbor a diverse array of bioactive compounds with significant health-promoting effects. This review aims to explore the nutritional and medicinal properties of select wild edible fruits native to the Western Ghats, highlighting their bioactive compounds and potential health benefits. Drawing upon scientific literature and indigenous knowledge, this paper provides a comprehensive overview of the phytochemical composition, antioxidant activity, anti-inflammatory properties, and other bioactivities associated with these fruits. Additionally, the review discusses the cultural significance and sustainable harvesting practices of these fruits, emphasizing the importance of conserving both the biodiversity of the Western Ghats and the traditional knowledge associated with these valuable resources. Overall, this review underscores the potential of wild edible fruits from the Western Ghats as functional foods and sources of bioactive compounds for promoting human health and well-being.

Keywords

Western Ghats · Wild edible fruits · Health benefits · Traditional knowledge · Biodiversity conservation

G. Kannan · B. M. Paul · P. Thangaraj (✉)
Bioprospecting Laboratory, Department of Botany, Bharathiar University, Coimbatore, Tamil Nadu, India

P. Thangaraj (ed.), *Pomotherapeutic Insights on Wild Edible Fruits*,
https://doi.org/10.1007/978-981-96-0543-9_4

Introduction

Tropical regions, such as the Western Ghats, are known for their rich biodiversity and abundance of exotic fruit-yielding plants. These fruits not only provide a unique taste and sensory experience, but they are also packed with bioactive compounds that have health-promoting effects. These bioactive compounds found in wild edible fruits from the Western Ghats have gained popularity due to their potential to enhance overall health and well-being. By exploring the potential of underutilized fruit crops in the region, we can not only fulfill the basic requirements of health-conscious consumers but also contribute to the local community's self-sufficiency and economic security. This can be achieved through promoting the consumption of these wild edible fruits and highlighting their nutritional and medicinal benefits [1]. The Western Ghats region is a treasure trove of unique and exotic fruits that not only tantalize the taste buds but also offer a plethora of health benefits. In this review, we will delve into the diverse range of bioactive compounds found in these wild edible fruits and their potential in promoting overall health and well-being. Moreover, we will explore the significance of underutilized fruit crops in the region and their role in meeting the demands of health-conscious consumers while uplifting the local community. Through this exploration, we aim to shed light on the nutritional and medicinal advantages of these wild edible fruits from the Western Ghats, emphasizing their value in bolstering both individual health and local economies [2, 3]. By promoting the consumption of these wild edible fruits, we can tap into their unique bioactive compounds and harness their potential in improving health outcomes. These fruits not only provide a delicious and unique taste experience but also offer numerous health benefits. Some of the key health-promoting effects of these wild edible fruits include their antioxidant properties, which help in scavenging free radicals and reducing oxidative stress in the body. Additionally, these fruits are rich in vitamins, minerals, and dietary fiber, making them valuable for maintaining a healthy digestive system and supporting overall immune function. The bioactive compounds present in these fruits, such as polyphenols, flavonoids, and carotenoids, have been linked to a range of health benefits, including anti-inflammatory, anticancer, and cardiovascular-protective effects [4].

Importance of Wild Edible Fruits in Local Diets

Wild edible fruits play a significant role in local diets, especially in regions like the Western Ghats of Tamil Nadu in India. These fruits are not only a source of nutrition but also a rich reservoir of bioactive compounds with health-promoting effects. Consuming wild edible fruits can provide various benefits to individuals, including the intake of essential vitamins, minerals, and antioxidants. Furthermore, these fruits are often traditional foods that have been consumed by local communities for generations, carrying cultural and historical importance [5].

Furthermore, wild edible fruits offer a sustainable and environmentally friendly food source. They grow naturally in the local ecosystem and do not require extensive cultivation or chemical inputs. This makes them a valuable resource for communities living in the Western Ghats, where sustainable and locally sourced food options are crucial for both individual well-being and the overall ecological balance [6]. By incorporating wild edible fruits into their diets, local communities can diversify their nutrient intake and promote food security. Additionally, the consumption of wild edible fruits can also contribute to the preservation and conservation of biodiversity. By cultivating and protecting these wild edible fruit plants, communities can ensure their availability for future generations while also preserving the local indigenous knowledge systems associated with their collection and use [7].

Significance of Bioactive Compounds in Promoting Health

There is a growing recognition of the importance of bioactive compounds in promoting health. These compounds, found in various fruits and plants, have been shown to possess a wide range of beneficial effects on human health. Experimental evidences suggest that fruits contain various bioactive constituents including flavonoids, polyphenols, carotenoids, and micronutrients such as vitamins and minerals [8]. These bioactive compounds have been associated with antioxidant, anti-inflammatory, antimicrobial, cardioprotective, and neuroprotective properties. Furthermore, studies have shown that bioactive compounds can play a role in preventing and treating chronic health complications such as diabetes, obesity, cancer, cardiovascular disease, immune system decline, and certain neurological disorders. Therefore, incorporating these bioactive compounds into our diets through the consumption of fruits and other plant-based foods can be an effective strategy for maintaining and improving our health. Recent research has also revealed the potential of bioactive compounds in promoting gut health. The consumption of these compounds has been linked to the modulation of the gut microbiota, which plays a crucial role in overall health and well-being. Additionally, bioactive compounds have been shown to have a positive impact on metabolic health, including the regulation of blood sugar levels and improvement of lipid profiles [9].

Incorporating a diverse range of fruits, vegetables, nuts, seeds, and whole grains into the diet can provide a rich source of bioactive compounds. Additionally, the synergistic effects of consuming these compounds in whole foods, rather than isolated supplements, may offer enhanced health benefits due to the complex interactions of different nutrients and phytochemicals. Thus, it is evident that bioactive compounds found in fruits and plants play a significant role in promoting health and preventing chronic diseases. Therefore, it is important to include a variety of fruits and plant-based foods in our diet to ensure an adequate intake of these bioactive compounds for optimal health and well-being [10].

Given the wide-ranging health benefits associated with bioactive compounds, further exploration and understanding of their mechanisms of action could offer promising avenues for the development of preventive and therapeutic interventions for various health conditions [11].

Diversity of Wild Edible Fruits in the Western Ghats

The Western Ghats region in India is known for its rich biodiversity, including a diverse range of wild edible fruits. These fruits are not only an important source of nutrition for the local communities but also have potential economic value. The Western Ghats is home to a wide variety of wild edible fruits that have sustained indigenous communities for centuries. These fruits are not only valued for their nutritional benefits but also for their cultural significance. The diverse range of wild edible fruits in the Western Ghats includes the juicy kokum, tangy bilimbi, succulent forest apples, sweet wild mangoes, and various types of berries [12].

Many of these wild fruits are not only consumed locally but also have the potential for commercial exploitation. The rich flavors and unique characteristics of these fruits make them a valuable resource for the food and beverage industry. Additionally, the conservation and sustainable management of these wild edible fruits can contribute to the preservation of the region's rich biodiversity [13].

Exploring the diverse array of wild edible fruits in the Western Ghats can provide valuable insights into traditional knowledge systems, biodiversity conservation, and opportunities for sustainable development. This unique ecosystem holds immense potential for further research and exploration, offering new avenues for promoting food security and economic growth for local communities [14]. Furthermore, understanding the nutritional and economic value of these wild edible fruits is crucial in order to promote their conservation and cultivation on a larger scale [13]. By conserving and cultivating these wild edible fruits, we can not only uplift the economic status of local communities but also promote sustainable practices that benefit both people and the environment.

Description of the Western Ghats Biodiversity Hotspot

The Western Ghats biodiversity hotspot is a region of immense ecological and biological importance located along the western coast of India. The Western Ghats biodiversity hotspot covers an area of approximately 160,000 square kilometers and is known for its incredible biodiversity, with thousands of species of flora and fauna found nowhere else on Earth. The region's topography ranges from high mountain peaks to dense forests and lush grasslands, creating a diverse range of habitats for its unique wildlife [15].

The Western Ghats is home to several endemic and endangered species, including the Malabar giant squirrel, the Nilgiri tahr, and the lion-tailed macaque. The region's rich biodiversity also extends to its plant life, with a variety of rare and medicinal plants found in its forests. These plants are often utilized by local communities for traditional medicine and have attracted the attention of researchers for their potential in pharmaceutical development [16].

The hotspot's importance is not only ecological but also cultural, as it plays a crucial role in the lives of the indigenous communities that have inhabited the area for centuries. The Western Ghats is not only a remarkable natural treasure but also a living heritage that holds significant cultural and spiritual value for the people who call it home. Considering the importance of preserving this unique hotspot, conservation efforts have been implemented to protect and sustain its biodiversity [17].

Conservation efforts in the Western Ghats biodiversity hotspot have focused on establishing protected areas, promoting sustainable land use practices, and engaging local communities in conservation initiatives. These efforts aim to mitigate human-wildlife conflict, preserve critical habitats, and safeguard the livelihoods of indigenous communities that depend on the region';s natural resources [18].

Furthermore, the Western Ghats has been recognized as a UNESCO World Heritage Site, underscoring its global significance and the need for its protection. The designation has brought attention to the importance of the biodiversity hotspot on an international level and has led to increased support for conservation and sustainable development in the region.

Despite these efforts, the Western Ghats still faces numerous threats, including deforestation, habitat fragmentation, and poaching. Addressing these challenges requires continued collaboration between government bodies, non-governmental organizations, local communities, and other stakeholders to ensure the long-term preservation of this invaluable natural treasure. By working together, it is possible to safeguard the Western Ghats biodiversity hotspot for future generations while respecting its cultural and ecological significance [19].

Types of Wild Edible Fruits Found in the Western Ghats Region

The Western Ghats region in India is known for its rich biodiversity and diverse flora. In this region, there are numerous types of wild edible fruits that can be found [13]. Some of the wild edible fruits found in the Western Ghats region include: jackfruit, mango, guava, passion fruit, wild berries (such as mulberries and raspberries), custard apple, bamboo shoots, fig fruits, wild grapes, wild bananas, tamarind, hog plum, and gooseberry. Additionally, several varieties of wild berries, including strawberries and blackberries, can also be found in the Western Ghats region. These fruits are not only consumed by the tribes living in the region but also hold great potential for their nutritional value and medicinal properties [15]. The wild edible fruits found in the Western Ghats region are not only a source of food for the local tribes but also have significant nutritional and medicinal value for a wider population. However, it is important to note that proper identification and knowledge of these wild edible fruits is crucial to ensure their safe consumption. It is also important to mention that the conservation and cultivation of these wild edible fruits can provide economic opportunities for the local communities in the Western Ghats region, contributing to their upliftment and economic well-being in the future. The 20 wild edible fruits listed in the study conducted with the Badagas of Nilgiri District are essential for their conservation and cultivation purposes [20]. Therefore, efforts should be made to conserve these wild edible fruits and promote their cultivation on a larger scale, as this can not only uplift the economic status but also contribute to the overall food security and sustainability of the region. The unique ecosystem of the Western Ghats region provides a diverse range of wild edible fruits that have sustained the local tribes for generations. In addition to their nutritional value, many of these fruits also possess medicinal properties that have been traditionally used for various health purposes. Proper identification and knowledge of these wild edible fruits are essential to ensure their safe consumption and to fully benefit from their potential [4].

Conserving and cultivating these wild edible fruits can not only provide economic opportunities for the local communities but also contribute to the overall food security and sustainability of the region. Efforts to promote their cultivation on a larger scale are crucial for uplifting the economic status of the region and ensuring the preservation of these valuable resources for future generations. The Western Ghats region is not only a treasure trove of biodiversity but also a source of traditional knowledge and sustainable living practices. By conserving and promoting the cultivation of these wild edible fruits, the region can continue to thrive while preserving its natural resources [7]. In addition to their economic and nutritional significance, the cultivation of wild edible fruits can also contribute to the conservation of the unique ecosystem of the Western Ghats. These fruits are adapted to thrive in the specific environmental conditions of the region, and their cultivation can help maintain the ecological balance and prevent the loss of valuable plant species [21]. Furthermore, the promotion of sustainable harvesting practices and the development of value-added products from these wild fruits can create market opportunities and potential sources of income for the local communities. Whether it's processing wild berries into jams and preserves or developing innovative uses for fruits like jackfruit and wild grapes, there is potential for entrepreneurship and economic growth centered around these natural resources [22]. As we look toward the future, it is essential to recognize the importance of preserving the rich biodiversity of the Western Ghats and supporting the traditional knowledge systems associated with wild edible fruits. Initiatives aimed at conservation, cultivation, and value addition can not only safeguard the region's natural heritage but also foster sustainable development and prosperity for the communities that call the Western Ghats home.

Importance of Preserving Indigenous Fruit Species

Preserving indigenous fruit species is crucial for maintaining biodiversity and sustaining local ecosystems. These fruits have adapted to specific environmental conditions over centuries, making them resilient and valuable genetic resources. Additionally, they often play significant cultural and traditional roles within local communities, being used in rituals, celebrations, and traditional medicine. Furthermore, these fruits can provide important food sources, particularly during times of climate change and environmental stress. Preserving indigenous fruit species is not only important for maintaining biodiversity but also for safeguarding valuable cultural and traditional knowledge while promoting food security and resilience in the face of environmental challenges [4]. Preserving indigenous fruit species also contributes to the economic development of local communities. These fruits have the potential to be commercialized, providing sustainable livelihoods for indigenous farmers and contributing to the local economy. By promoting the cultivation and consumption of indigenous fruits, it is possible to create new market opportunities and value chains, thus empowering local communities and enhancing their economic independence [23]. Moreover, the preservation of indigenous fruit species supports the preservation of traditional agricultural practices and knowledge. By continuing to grow and utilize these fruits, communities can maintain and pass down traditional farming techniques, promoting sustainable and environmentally friendly agricultural methods [24].

In order to achieve these goals, it is essential to work in partnership with local communities, providing them with the resources and support needed to preserve and cultivate indigenous fruit species. This can involve initiatives such as supporting community-based seed banks, providing training in sustainable farming practices, and creating market linkages for the sale of indigenous fruits. In conclusion, the preservation of indigenous fruit species is imperative for maintaining biodiversity, supporting local economies, preserving traditional knowledge, and promoting sustainable agriculture. It is crucial to recognize the value of these fruits and to work toward their conservation and promotion within local and global contexts.

Bioactive Compounds in Wild Edible Fruits in Western Ghats

Wild edible fruits are an important source of nutrition and contribute significantly to the nutritional security of people around the world. The Western Ghats region is particularly rich in biodiversity, and many of the wild edible fruits found there are also known for their bioactive compounds. These bioactive compounds, such as antioxidants, flavonoids, and polyphenols, have been shown to have various health benefits, including anti-inflammatory and anticancer properties. Research has indicated that the bioactive compounds in wild edible fruits from the Western Ghats may play a role in preventing chronic diseases and promoting overall health. By further exploring the specific bioactive compounds present in these fruits, we can gain a better understanding of their potential nutritional and medicinal value [25].

In addition to their nutritional significance, the preservation and sustainable harvest of wild edible fruits in the Western Ghats are crucial for maintaining ecological balance and supporting local communities. Efforts to conserve these valuable resources while also harnessing their potential for nutrition and health could lead to a more sustainable and resilient food system in the region. Wild edible fruits in the Western Ghats are not only a source of essential nutrients but also a treasure trove of bioactive compounds that offer numerous health benefits. These compounds, including antioxidants, flavonoids, and polyphenols, play a significant role in preventing chronic diseases and promoting overall well-being. Research has shown that these bioactive compounds exhibit anti-inflammatory and anticancer properties, making the consumption of these fruits even more crucial for human health [26].

Furthermore, a deeper exploration into the specific bioactive compounds present in these wild edible fruits can provide valuable insights into their potential medicinal and nutritional value. Understanding the unique composition of these compounds can pave the way for harnessing their health-promoting properties and incorporating them into various health and nutrition initiatives. The preservation and sustainable harvest of these wild edible fruits are not only important for human consumption but also crucial for maintaining the ecological balance in the Western Ghats. Efforts to conserve these diverse fruits while also leveraging their nutritional and medicinal properties have the potential to contribute to a more sustainable and resilient food system in the region, benefiting both the local communities and the environment.

Overview of the Bioactive Compounds Present in Wild Edible Fruits

Wild edible fruits are a valuable source of bioactive compounds that have potential health benefits for humans. Wild edible fruits are not only a valuable source of bioactive compounds, but they also offer a diverse range of flavors, textures, and colors. These fruits have been consumed by indigenous cultures for centuries, not only for their nutritional value but also for their medicinal properties. The bioactive compounds present in wild edible fruits include antioxidants, vitamins, minerals, and phytochemicals, all of which play a crucial role in promoting human health and preventing various chronic diseases. Additionally, the consumption of wild edible fruits contributes to the preservation of biodiversity and the sustainability of natural ecosystems. Understanding the specific bioactive compounds found in different wild edible fruits can provide valuable insights into their potential health benefits and contribute to the development of functional foods and nutraceuticals.

Health-Promoting Properties of These Bioactive Compounds

Bioactive compounds found in plants have been recognized for their potential health-promoting properties [27]. Research has shown that bioactive compounds found in plants, such as flavonoids, carotenoids, and polyphenols, have a wide range of health-promoting properties. These compounds have been linked to antioxidant effects, anti-inflammatory properties, and potential anticancer activity. Additionally, they have been associated with cardiovascular benefits, including improved heart health and reduced risk of certain chronic diseases. The diverse array of bioactive compounds found in plants underscores their potential to contribute to overall well-being and disease prevention. It is evident that the health-promoting properties of bioactive compounds found in plants are vast and diverse. These compounds not only provide antioxidant and anti-inflammatory effects but also show promise in potential anticancer activity. Furthermore, their association with cardiovascular benefits highlights their significance in improving heart health and reducing the risk of chronic diseases. The diverse array of bioactive compounds found in plants underscores their potential to contribute to overall well-being and disease prevention. Moreover, ongoing research continues to unveil new health-promoting properties of these compounds, further emphasizing their importance in promoting a healthy lifestyle. Researchers have identified a wide range of health-promoting properties associated with bioactive compounds found in plants [11].

Nutritional and Medicinal Value of Wild Edible Fruits in Western Ghats

Wild edible fruits play a vital role in the diet and overall nutrition of rural communities, particularly those residing in hilly and mountainous regions. These fruits serve as a valuable source of nutrition, especially in areas where access to cultivated fruits may be limited. Furthermore, these wild fruits have not only nutritional value but also possess medicinal properties that contribute to the traditional healthcare practices of local communities [28]. The Western Ghats region in India is renowned for its rich biodiversity and abundant wild edible fruits that have sustained local communities for generations. These fruits are not only a source of sustenance but also hold significant medicinal value, often serving as remedies for common health ailments. The traditional knowledge associated with the use of these wild fruits for medicinal purposes is a testament to the deep-rooted connection between the local communities and the biodiversity of the region [25].

Moreover, the nutritional composition of these wild edible fruits makes them an essential part of the local diet, providing valuable vitamins, minerals, and antioxidants. Their consumption contributes to the overall well-being of the communities, especially in areas where access to other forms of nutrition may be limited. Many wild edible fruits are rich in vitamins (such as vitamin C, vitamin A, vitamin K, and various B vitamins) and minerals (including potassium, calcium, magnesium, and iron), which are essential for overall health and well-being [29]. Some wild fruits contain healthy fats, such as omega-3 and omega-6 fatty acids, which are important for brain health, heart health, and overall cellular function. They are often packed with antioxidants like polyphenols, flavonoids, and carotenoids, which help neutralize harmful free radicals in the body, reducing the risk of chronic diseases [30].

The deep ties between the Western Ghats communities and these wild edible fruits underscore the importance of preserving traditional knowledge and sustainable harvesting practices. Understanding the intricate relationship between these fruits, nutrition, and traditional medicine is crucial for both the conservation of biodiversity and the well-being of the local communities (Table 1).

Antioxidant Properties of Wild Edible Fruits in Western Ghats

Wild edible fruits play a crucial role in the diets of rural communities, particularly in hilly and mountainous regions, where they may be the only available source of food. Wild edible fruits not only serve as a vital source of nutrition for rural communities in the Western Ghats but also offer significant health benefits due to their

Table 1 Bioactive compounds in wild edible fruits in Western Ghats

S. no	Binomial name	Common name	Vernacular name	Family	Therapeutic use	Bioactive compounds	References
1.	*Averrhoa carambola* L.	Star fruit	Tamarattai	Oxalidaceae	Fever, cough, diarrhea, chronic headache, inflammatory skin disorders (eczema), and fungal skin infections	Zeta-carotene, phytofluene, mutatoxanthin	[31]
2.	*Annona muricata* L.	Soursop	Mullu-sitha-pazham	Annonaceae	Treat diarrhea and neuralgia, eliminate worms and parasites, increase milk production in lactating women, and reduce fever	Myrcene, undecane, decane, 2,3-dihydrobenzofuran	[32]
3.	*Annona reticulata* L.	Custard apple	Ramaseetha pazham	Annonaceae	Anthelmintic, analgesic, anti-inflammatory, antipyretic, wound healing, and cytotoxic effects	Myrcene, acetoin, 2-acetylfuran, carvacrol	[33]

4.	*Berberis tinctoria* Lesch.,	Nilgiri barberry	Oosikala	Berberidaceae	Skin diseases, wound healing, inflammation, menorrhagia, diarrhea, jaundice, eye infections, and snake bites	Berbamine, palmatine, berberine, jatrorrhizine	[34]
5.	*Cyphomandra betacea* (Cav.) Sendtn.	Tree tomato	Mara thakali	Solanaceae	Improve immune system and circulatory. Use for colds, sore throat, high blood pressure, liver disease	Geranylacetone, ethyl vinyl ketone, 3-methylbutanal, acetoin	[35]
6.	*Gardenia gummifera* L.f.	Gummy cape jasmine	Kambil	Rubiaceae	Stomach problems, indigestion, constipation, menstrual cycle problem, muscle pain, dental problems, parasitic worms	Gardenin E, dikamaliartane D, wogonin, gardenin A	[36]

(continued)

Table 1 (continued)

S. no	Binomial name	Common name	Vernacular name	Family	Therapeutic use	Bioactive compounds	References
7.	*Rhodomyrtus tomentosa* (Alton) Hassk.	Rose myrtle	Thavutu pazham	Myrtaceae	Treat colic diarrhea, dysentery, abscesses, hemorrhage, as a tonic after childbirth	Tomentosone A, tomentosone B	[37]
8.	*Ziziphus oenopolia* (L.) Mill.	Jackal jujube	Suraimullu	Rhamnaceae	Antimicrobial, antidiabetic, antihepatotoxic, antiulcer, antiplasmodial, anticancer, wound healing, anthelmintic	Zizyphine A, zizyphine C, betulinic acid	[38]
9.	*Rubus racemosus* Roxb.	Black raspberry	Mullu pazham	Rosaceae	Antidiabetic, abortifacient, muscle relaxant, and free radical scavenging agents	Cyanidin-3-glucoside, cyanidin-3-rhutinoside	[39]

10.	*Canthium coromandelicum* (Burm.f.) Alston	Wild jessamine	Karay chedi	Rubiaceae	Treat different ailments including wound, cholesterol, diabetes, treat intestinal worms in children	Mannitol	[40]
11.	*Elaeocarpus tectorius* (Lour.) Poir.	Stinkwood	Bikkipalzam	Elaeocarpaceae	Treatment of rheumatism, pneumonia, ulcers, piles, and leprosy	3-Heptanol, hexadecanoic acid, 5-hydroxymethylfurfural	[41]
12.	*Persicaria chinensis* L. gross	Chinese knotweed	Sivappu kumbakodaali	Polygonaceae	Treat ulcer, eczema, stomach ache, and various inflammatory skin diseases	3-O-methylellagic acid, kaempferol, beta-sitosterol, batatifolin	[42]
13.	*Elaeagnus kologa* Schltdl.	Bastard oleaster	Poolanpazham	Elaeagnaceae	Antitumor, antimicrobial, antiviral, antifungal, antioxidant	Beta-sitosterol	[43]

(continued)

Table 1 (continued)

S. no	Binomial name	Common name	Vernacular name	Family	Therapeutic use	Bioactive compounds	References
14.	*Morus alba* L.	White mulberry	Kamblichedi	Moraceae	Antiphlogistic, liver protective, kidney protective, hypotensive, diuretic, anti-cough and analgesic agent.	Campesterol, beta-sitosterol, linoleic acid	[44]
15.	*Glycosmis pentaphylla* (Retz.) DC.	Gin berry	Boomi pazham	Rutaceae	Treats cough, fever, chest pain, anemia, jaundice, liver disorders, inflammation, bronchitis, rheumatism, urinary tract infections	Glycophymine, arborinine, mupamine, dictamnine	[45]
16.	*Zanthoxylum asiaticum* (L.) Appelhans, Groppo & J.Wen	Orange climber	Kindu mullu	Rutaceae	Used against rheumatism, pain, wound bleeding, and bruises	Berberine, pimpinellin, toddasin	[46]

17.	*Schleichera oleosa* (Lour.)	Kusum tree	Kumbadiri	Sapindaceae	Analgesic, antibiotic, used against skin inflammations, ulcers, itching, acne, and other skin infections	Stearic acid, oleic acid, brassicasterol, betulin	[47]
18.	*Garcinia indica* (Thouars) Choisy	Kokum	Murgalli	Clusiaceae	Treat inflammation, dermatitis, and diarrhea and to promote digestion	Isogarcinol, hydroxycitric acid, garcinol, cyanidin 3-glucoside	[48]
19.	*Flacourtia indica* (Burm.f.) Merr.	Indian plum	Sottaikala	Salicaceae	Treat functional disorders like rheumatoid arthritis and gout, diarrhea	Ascorbic acid, 6-O-(e)-p-coumaroyl glucopyranose, cianidanol	[49]
20.	*Ficus exasperata* Vahl	Sandpaper fig	Maramthinniathi	Moraceae	Treat high blood pressure, rheumatism, arthritis, intestinal pains and colics, epilepsy, bleeding, and wounds	Beta-amyrin, bergaptol, beta-sitosterol	[50]

antioxidant properties. These fruits are rich in phytochemicals such as flavonoids, polyphenols, and vitamin C, which have been linked to a range of health-promoting effects, including reducing the risk of chronic diseases such as cancer, cardiovascular diseases, and diabetes.

The unique environmental conditions of the Western Ghats, including diverse ecosystems and rich biodiversity, contribute to the abundance of wild edible fruits with potent antioxidant properties. Additionally, the traditional knowledge and practices of local communities in harvesting and consuming these fruits have sustained their utilization as a valuable food resource for generations [51]. Research studies have shown that the antioxidant compounds in these wild edible fruits exhibit free radical scavenging activity and have the potential to mitigate oxidative stress and inflammation in the human body. Furthermore, efforts to identify and characterize the specific antioxidant compounds present in these fruits can provide valuable insights for their potential use in functional foods, nutraceuticals, and pharmaceutical applications [52].

Understanding the antioxidant properties of these wild edible fruits in the Western Ghats not only sheds light on their nutritional significance but also underscores the importance of preserving the ecosystem and traditional knowledge that enable their sustenance. By studying the antioxidant properties of wild edible fruits in the Western Ghats, we can gain valuable insights into their potential health benefits and explore their applications in developing natural antioxidant-based products that contribute to human health and well-being.

Relationship Between Antioxidant Intake and Disease Prevention

The relationship between fruit antioxidant intake and disease prevention has been extensively studied in numerous prospective cohort studies. These studies consistently show that a higher intake of fruits, which are rich in antioxidants, is associated with a reduced risk of various diseases, including cardiovascular disease, certain types of cancer, and age-related macular degeneration. Moreover, the protective effects of fruit antioxidants have been linked to their ability to neutralize harmful free radicals and reduce oxidative stress in the body. Furthermore, fruits contain various bioactive compounds such as flavonoids, carotenoids, and vitamin C, which have been shown to exhibit anti-inflammatory properties and contribute to the overall health benefits associated with fruit consumption [53]. Thus, incorporating fruits into one's diet is a crucial step in promoting optimal health and reducing the risk of chronic diseases.

In addition to their antioxidant properties, fruits also provide a wide array of essential nutrients such as fiber, vitamins, and minerals that play a crucial role in maintaining optimal health. The combination of these nutrients and antioxidants in fruits works synergistically to support the body's immune function, protect against chronic inflammation, and promote overall well-being [54]. Furthermore, the diverse range of antioxidants found in different fruits highlights the importance of consuming a variety of fruits to maximize the potential health benefits. For example, berries are known for their high levels of anthocyanins, while citrus fruits are abundant in vitamin C. By incorporating a colorful selection of fruits into one's diet, individuals can ensure that they are receiving a broad spectrum of antioxidants and nutrients that can contribute to disease prevention and long-term health [55].

It is important to recognize that while the evidence supporting the positive relationship between fruit antioxidant intake and disease prevention is compelling, individual factors such as overall diet, lifestyle, and genetic predisposition also play a significant role in determining health outcomes. Nevertheless, the consistent findings from research studies underscore the significance of incorporating a variety of antioxidant-rich fruits into a balanced diet to support health and potentially reduce the risk of chronic diseases. Therefore, further research is needed to fully understand the complexities of fruit antioxidants and their effects on human health. Research in this area also suggests that the form in which fruits are consumed may impact the body's ability to absorb and utilize their antioxidants. For example, whole fruits contain not only antioxidants but also dietary fiber, which can moderate the release and absorption of sugars and other nutrients in the body. On the other hand, fruit juices may lack certain beneficial compounds present in whole fruits, and their high sugar content could potentially offset some of the health benefits associated with fruit consumption [55, 56].

Incorporating a variety of fruits into one's diet also contributes to overall dietary diversity, which is linked to improved nutrient intake and health outcomes. Additionally, the inclusion of fruits in meals and snacks can help displace less nutrient-dense foods, contributing to better overall dietary patterns. Building on the existing body of research, ongoing studies are exploring the potential synergistic effects of consuming fruits alongside other foods, such as nuts, seeds, and whole grains, to enhance the body's ability to absorb and utilize antioxidants. Understanding these interactions could provide valuable insights into optimizing dietary patterns for disease prevention and overall health [57].

In conclusion, while the relationship between fruit antioxidant intake and disease prevention is a compelling area of study, it's important to consider the broader context of overall dietary patterns and lifestyle factors. Emphasizing the consumption of a variety of antioxidant-rich fruits as part of a balanced diet remains a valuable approach for promoting health and well-being.

Conservation Challenges Facing Wild Edible Fruit Species

Conservation efforts play a vital role in safeguarding the diversity and enduring well-being of wild edible fruit species. Without conservation initiatives, these species encounter various obstacles such as loss of habitat, excessive harvesting, climate variations, and invasive species. These hurdles can cause reductions in population size and genetic variation, rendering wild edible fruit species more susceptible to extinction. Habitat destruction resulting from deforestation, urbanization, and agricultural expansion poses a significant risk to their survival by limiting the available space for growth and reproduction. Excessive harvesting is also a critical concern due to increased demand for wild edible fruits attributed to their nutritional and medicinal properties. Unregulated harvesting has the potential to diminish the populations of these species and disrupt natural regeneration processes, thereby exacerbating their decline [58]. Wild edible fruit species are additionally threatened by climate change which directly impacts suitable habitats through temperature fluctuations and altered rainfall patterns leading to effects on growth cycles. Furthermore, invasive species have the capacity to compete with native varieties ultimately endangering their existence.

To overcome these challenges related to conservation efforts, it is imperative that sustainable harvest practices be put into action; furthermore, protection/conservation measures must be established alongside raising awareness about conserving wild edible fruit varieties. Collaborative endeavors involving local communities along with governmental agencies are essential for ensuring long-term preservation of these esteemed culturally valuable plant types [59].

Opportunities for Further Research and Development on Wild Edible Fruits

Wild edible fruits are a valuable and underutilized resource that has the potential to contribute to food security and improve nutrition. Exploring the potential of wild edible fruits for food security and nutrition requires a multidisciplinary approach that encompasses various aspects, such as ecological conservation, genetic diversity, nutritional value, and cultural significance. Understanding the ecological context in which these fruits grow is essential for sustainable harvesting and conservation efforts. Research on the genetic diversity of wild edible fruits can provide insights into their adaptability to changing environmental conditions and aid in the development of improved varieties. Furthermore, delving into the nutritional value of these fruits will shed light on their potential contribution to addressing malnutrition and diet-related health issues. This can involve studies on macronutrient and micronutrient content, as well as bioactive compounds with potential health benefits. Additionally, considering the cultural significance of wild edible fruits in different communities can provide valuable insights into traditional knowledge, harvesting practices, and culinary uses, offering opportunities for preserving and promoting cultural heritage while harnessing the benefits of these fruits.

Moreover, there is a need for research on post-harvest handling and processing techniques to extend the shelf life of these fruits and make them more accessible for consumption. Investigating innovative uses of wild edible fruits in food products and creating value-added products can also open up opportunities for income generation and economic development, especially in rural and marginalized communities. In conclusion, further research and development on wild edible fruits present exciting opportunities to address food security, nutrition, cultural preservation, and economic empowerment, making it a compelling area for future exploration and innovation. It is also important to consider the potential impact of climate change on the availability and distribution of wild edible fruits. Understanding how these fruits may be affected by changes in temperature, precipitation, and other environmental factors can help in developing strategies to ensure their continued availability for consumption and economic use. Additionally, exploring the potential for domestication and cultivation of certain wild edible fruit species can contribute to their sustainable production and commercialization. This can involve research into propagation techniques, agronomic practices, and breeding programs aimed at improving yield and quality. Furthermore, engaging with local communities and indigenous knowledge holders is crucial for the ethical and sustainable utilization of wild edible fruits. Collaborative research efforts that integrate traditional knowledge with scientific expertise can lead to the development of strategies that respect and support local food systems and cultural practices.

Overall, continued research and development in the field of wild edible fruits offer promising avenues for scientific discovery, conservation, and socioeconomic development. By integrating ecological, nutritional, cultural, and economic perspectives, we can unlock the full potential of these valuable resources for the benefit of both people and the environment.

Importance of Wild Edible Fruits from the Western Ghats in Promoting Health

Wild edible fruits from the Western Ghats play a crucial role in promoting health and well-being. These fruits are not only a source of nutrition but also hold cultural and ecological significance for the communities residing in this region. The diverse range of wild edible fruits found in the Western Ghats provides essential vitamins, minerals, and antioxidants that contribute to a healthy diet. Moreover, the consumption of these fruits has been linked to traditional medicinal practices, where they are believed to possess therapeutic properties. The unique flavors and textures of wild edible fruits also add variety to local diets and contribute to the culinary heritage of the region. Furthermore, the sustainable harvesting and consumption of wild edible fruits from the Western Ghats support biodiversity conservation and sustainable livelihoods for local communities. By recognizing the importance of these fruits, it becomes evident that they are not just a source of food but also a vital component of the region's heritage and ecosystem.

The Western Ghats' rich biodiversity of wild edible fruits offers a wide array of flavors, textures, and nutritional benefits, making them a valuable addition to the local cuisine. Additionally, the sustainable harvesting and consumption of these fruits contribute to the preservation of the region's unique ecosystem and support the livelihoods of indigenous communities. Incorporating wild edible fruits from the Western Ghats into mainstream agriculture and culinary practices can not only enhance the nutritional diversity of diets but also promote sustainable agricultural practices and conservation efforts. Understanding and valuing the significance of these fruits can lead to the development of policies and initiatives that prioritize their conservation and responsible utilization, safeguarding this essential component of the region's heritage and ecological balance. Wild edible fruits from the Western Ghats have also gained attention for their potential economic value. As interest in organic and locally sourced products grows, these fruits present an opportunity for sustainable entrepreneurship and economic development within the region. By integrating these fruits into value-added products such as jams, jellies, and juices, local communities can expand their market reach and generate additional income. Furthermore, the cultivation and commercialization of these wild edible fruits can contribute to the region's tourism industry, attracting visitors interested in exploring the unique flavors and cultural significance of these fruits. This presents an opportunity to not only showcase the rich biodiversity of the Western Ghats but also to educate the public about the importance of conservation and sustainable harvesting practices. In conclusion, the significance of wild edible fruits from the Western Ghats extends beyond their nutritional and cultural value to encompass economic opportunities and environmental conservation. Recognizing and promoting the potential of these fruits can contribute to the overall well-being of the region while safeguarding its ecological balance and heritage.

Call to Action for Conservation Efforts and Further Exploration of Their Potential Benefits

The necessity of urging action for the preservation of our planet's biodiversity and natural resources cannot be overstated. Through proactive measures to protect ecosystems, we can guarantee a sustainable future for generations to come. It is imperative to raise awareness about conservation's significance and encourage individuals, communities, and organizations to actively engage in environmental stewardship. Furthermore, conservation efforts offer potential benefits that extend beyond environmental protection. Research indicates positive economic, social, and health impacts from conservation. For instance, preserving natural habitats can create opportunities for ecotourism, leading to job creation and local economy growth. Additionally, preserving green spaces and forests can contribute to better air quality as well as improved water quality with consequent public health advantages.

It remains crucial to continue exploring these potential benefits while advocating policies supporting such endeavors since collaboration will make a meaningful difference in safeguarding the planet for future generations. Supporting ongoing research on the benefits of conservation efforts is indispensable in promoting them further. A collaboration involving scientists, environmentalists, policymakers, and local communities holds valuable insights into how conservation supports different sectors of society and the environment.

Through research, evidence supports the economic benefits including ecosystem services value and sustainable resource management levels. Furthermore, studies concerning social and health impacts provide data crucial for developing policies prioritizing people's well-being. Notably, the collaborative initiatives foster partnerships between governments, nongovernmental organizations, and businesses implementing practical conservation schemes. With input from various stakeholders' comprehensive approaches addressing linkages between environmental, social, &economic challenges are developed. Thus, it is clear that endorsing research and collaborative initiatives are essential for gaining a deep understanding of conservation benefits and promoting inclusive practices lead succeeding creating a sustainable, resilient future for all.

References

1. Shailasree S, Sampathkumara KK, Niranjana SR, Prakash HS. Bioactive potential of medicinal plants from western ghats region, India. J Herbs Spices Med Plants. 2014;20(3):221–34.
2. Sardeshpande M, Shackleton C. Wild edible fruits: a systematic review of an under-researched multifunctional NTFP (non-timber forest product). Forests. 2019;10(6):467.
3. Maikhuri RK, Rao KS, Saxena KG. Bioprospecting of wild edibles for rural development in the central Himalayan mountains of India. Mt Res Dev. 2004;24(2):110–3.
4. Tabassam Q, Ahmed S, Mehmood T, Iqbal M, Saeed S, Ahmad MM, Sultan A, Ur Rehman MM, Qadir R, Ullah A, Rehman NA. Phytochemical investigation and effective therapeutic potential of plants extracts against breast and ovarian cancer cell lines: compounds from zizyphus mauritiana and triticum aestivum. Cell Mol Biol. 2021;67(1):147–52.
5. Pai V, Satish BN, Hareesh TS, Kushalappa CG, Nayak BG. Ethnobotanical survey of wild edible plants in different landscapes of Kodagu. Int J Curr Microbiol App Sci. 2019;8(6):2103–9.
6. Pradhan R, Nayak AP, Reddy S. Potential wild edible plants and its significance in livelihood of indigenous people of male Mahadeshwara Hills, Karnataka. Econ Aff. 2020;65(4):589–602.
7. Khoo HE, Azlan A, Kong KW, Ismail A. Phytochemicals and medicinal properties of indigenous tropical fruits with potential for commercial development. Evid Based Complement Alternat Med. 2016;2016:7591951.
8. Jaime L, Santoyo S. The health benefits of the bioactive compounds in foods. Food Secur. 2021;10(2):325.
9. Samtiya M, Aluko RE, Dhewa T, Moreno-Rojas JM. Potential health benefits of plant food-derived bioactive components: an overview. Food Secur. 2021;10(4):839.
10. Patil BS, Jayaprakasha GK, Chidambara Murthy KN, Vikram A. Bioactive compounds: historical perspectives, opportunities, and challenges. J Agric Food Chem. 2009;57(18):8142–60.
11. Sathyavathi R, Janardhanan K. Wild edible fruits used by Badagas of Nilgiri District, Western Ghats, Tamilnadu, India. J Med Plants Res. 2014;8(2):128–32.
12. Kruijssen F, Sudha M. Enhancing biodiversity management and utilization to improve livelihoods–a case study of 'kokum' in India. In: International workshop on tropical and subtropical fruits. 787; 2007. p. 165–72.
13. Sivakumar A, Murugesan M. Ethnobotanical studies on the wild edible plants used by the tribals of Anaimalai Hills, the Western Ghats. Anc Sci Life. 2005;25(2):69–73.
14. Kasodekar AK, Jadhav AD, Bhagat RB, Pawar RM, Gupta VS, Kadoo NY. The importance of conserving fragmented forest patches with high diversity of flowering plants in the northern Western Ghats: an example from Maharashtra, India. J Threat Taxa. 2019;11(7):13833–49.
15. Kumar R, Devi KR. Conservation of freshwater habitats and fishes in the Western Ghats of India. Int Zoo Yearb. 2013;47(1):71–80.
16. Bhattarai U, Tetali P, Kelso S. Contributions of vulnerable hydrogeomorphic habitats to endemic plant diversity on the Kas Plateau, Western Ghats. SpringerPlus. 2012;1:1–9.
17. Balasubramanian M, Sangha KK. Valuing ecosystem services applying indigenous perspectives from a global biodiversity hotspot, the Western Ghats, India. Front Ecol Evol. 2023;11:1026793.
18. Rao PB. Diversity of some important wild edible plants of Kumaun Uttarakhand: a review. Agric Rev. 2021;42(4):371–80.

19. Rawat S, Das KK. Value addition of underutilized fruits of arid region: a review. IJCS. 2020;8(4):1655–9.
20. Navia ZI, Suwardi AB. The importance of tropical edible fruit plants for tribal communities in East Aceh region, Indonesia. In: IOP conference series: earth and environmental science. 637 IOP Publishing; 2021. p. 012003.
21. Shai KN, Ncama K, Ndhlovu PT, Struwig M, Aremu AO. An exploratory study on the diverse uses and benefits of locally-sourced fruit species in three villages of Mpumalanga Province, South Africa. Food Secur. 2020;9(11):1581.
22. Alam MK, Rana ZH, Islam SN, Akhtaruzzaman M. Comparative assessment of nutritional composition, polyphenol profile, antidiabetic and antioxidative properties of selected edible wild plant species of Bangladesh. Food Chem. 2020;320:126646.
23. Koubala BB, Bayang JP, Wangso H, Kolla MC, Laya A. Variation of phenolics (bound and free), minerals, and antioxidant activity of twenty-eight wild edible fruits of twenty-three species from far north region of Cameroon. Biomed Res Int. 2021;12:2021.
24. Shrinet K, Singh RK, Chaurasia AK, Tripathi A, Kumar A. Bioactive compounds and their future therapeutic applications. In: Natural bioactive compounds. Academic Press; 2021. p. 337–62.
25. Afsana C, Biswas D. Study of some ethno-medicinal wild edible plants used for several strong diseases in District Jashpur (Chhattisgarh). Int Res J Multidiscip Scope. 2020;1:52–60.
26. Joseph R. Ethnobotanical uses of wild edible fruits of Vitaceae in Kerala. European J Med Plants. 2020;31(11):72–6.
27. Sharma R, Goraya GS, Seth MK. Diversity and traditional knowledge of some wild edible fruits used by rural communities of district Mandi, Himachal Pradesh. Int J Adv Res. 2020;8(7):132–9.
28. Karuppusamy S, Muthuraja G, Rajasekaran KM. Antioxidant activity of selected lesser-known edible fruits from Western Ghats of India. Indian J Nat Prod Resour. 2011;2:174–8.
29. Muniyandi K, George E, Sathyanarayanan S, George BP, Abrahamse H, Thamburaj S, Thangaraj P. Phenolics, tannins, flavonoids and anthocyanins contents influenced antioxidant and anticancer activities of Rubus fruits from Western Ghats, India. Food Sci Hum Wellness. 2019;8(1):73–81.
30. Kurniawati E, Wibowo FS, Rusmeilina R. AKTIVITAS PENANGKAPAN RADIKAL BEBAS PADA KOMBINASI EKSTRAK ETANOL DAUN MANGGA (Mangifera indica L.) DAN DAUN SIRSAK (Annona muricata L.). Cendekia J Pharm. 2021;5(1):92–7.
31. Narain N, Bora PS, Holschuh HJ, Da SMV. Physical and chemical composition of carambola fruit (Averrhoa carambola L.) at three stages of maturity. Ciencia y Tecnología Alimentaria. 2001;3(3):144–8.
32. Márquez Cardozo CJ, Villacorta Lozano V, Yepes Betancur DP, Ciro Velásquez HJ, Cartagena Valenzuela JR. Physiological and physico-chemical characterization of the soursop fruit (Annona muricata L. cv. Elita). Revista Facultad Nacional de Agronomía Medellín. 2012;65(1):6477–86.
33. Jamkhande PG, Wattamwar AS. Annona reticulata Linn. (Bullock's heart): plant profile, phytochemistry and pharmacological properties. J Tradit Complement Med. 2015;5(3):144–52.
34. Smilin AG, Maheshu V, Gincy MM, Joji C, Sasikumar JM. Antioxidant and antihemolytic activities of common Nilgiri barberry (Berberis tinctoria Lesch.) from South India. Int Food Res J. 2012;19:1601–7.
35. Ghosal M, Chhetri PK, Ghosh MK, Mandal P. Changes in antioxidant activity of Cyphomandra betacea (Cav.) Sendtn. fruits during maturation and senescence. Int J Food Prop. 2013;16(7):1552–64.
36. Vinaykumar NM, Mahmood R, Krishna V, Ravishankara B, Shastri SL. Antioxidant and in vivo hepatoprotective effects of Gardenia gummifera Lf fruit methanol extract. Clin Phytoscience. 2020;6:1–4.

37. Lai TN, André C, Rogez H, Mignolet E, Nguyen TB, Larondelle Y. Nutritional composition and antioxidant properties of the sim fruit (Rhodomyrtus tomentosa). Food Chem. 2015;168:410–6.
38. Nahrin A, Junaid M, Afrose SS, Barua A, Akter Y, Alam MS, Sharmin T, Ferdousy S, Hosen SM. Ziziphus oenoplia Mill.: a systematic review on ethnopharmacology, phytochemistry and pharmacology of an important traditional medicinal plant. Mini Rev Med Chem. 2022;22(4):640–60.
39. Kumar PR, Vaidhyalingam V. Pharmacognostical and preliminary phytochemical studies on the aerial parts of Rubus racemosus (Roxb). Res J Pharmacogn Phytochem. 2010;2(5):381–5.
40. Prabodhani M, Wansapala J. Nutritional and in vitro bioactive compounds analysis of Canthium coromandelicum (Burm. F.) Alston leaves. Int J Food Sci Nutr. 2018;3(6):268–71.
41. Amit Dadhich AD, Anirudha Rishi AR, Gargi Sharma GS, Subhash Chandra SC. Phytochemicals of Elaeocarpus with their therapeutic value: a review. Int J Pharma Bio Sci. 2013;4(3):591–8.
42. Lai SM, Sudhahar D, Anandarajagopal K. Evaluation of antibacterial and antifungal activities of Persicaria chinensis leaves. Int J Pharm Sci Res. 2012;3:2825–30.
43. Durairaj B, Dhanabal M. In vitro evaluation of Elaeagnus conferta Roxb.-As a potent antioxidant and anti-inflammatory agent. Indian J Nat Prod Resour. 2023;14. https://doi.org/10.56042/ijnpr.v14i4.5414.
44. Yuan Q, Zhao L. The Mulberry (Morus alba L.) Fruit–a review of characteristic components and health benefits. J Agric Food Chem. 2017;65(48):10383–94.
45. Kuladip G, Varsha JR. Proximate and mineral analysis of fruit of Zanthoxylum rhetsa DC. and Glycosmis pentaphylla (Retz.) DC.: most useful ethnomedicinal plants in Kolhapur district. Int J Life Sci. 2018;9:48–52.
46. Mutinda ES, Kimutai F, Mkala EM, Waswa EN, Odago WO, Nanjala C, Ndungu CN, Gichua MK, Njire MM, Gituru RW, Hu GW. Ethnobotanical uses, phytochemistry and pharmacology of pantropical genus Zanthoxylum L.(Rutaceae): an update. J Ethnopharmacol. 2023;303:115895.
47. Khamphukdee C, Chulikhit Y, Monthakantirat O, Maneenet J, Boonyarat C, Daodee S. Phytochemical constituents and Antioxidative activity of Schleichera oleosa fruit. Trop J Nat Prod Res. 2021;5(8):1445–9.
48. Tharachand C, Selvaraj CI, Abraham Z. Comparative evaluation of anthelmintic and antibacterial activities in leaves and fruits of Garcinia cambogia (Gaertn.) desr. and Garcinia indica (Dupetit-Thouars) choisy. Braz Arch Biol Technol. 2015;58:379–86.
49. Makuttan S, Fernandes J. Anti-diabetic potential of fruit extracts of Flacourtia indica (Burm. F.) Merr-An in-vitro study. Indian J Nat Prod Resour. 2022;13(4):491–6.
50. Abotsi WM, Woode E, Ainooson GK, Amo-Barimah AK, Boakye-Gyasi E. Antiarthritic and antioxidant effects of the leaf extract of Ficus exasperata P. Beauv(Moraceae). Pharmacogn Res. 2010;2(2):89.
51. Dhalaria R, Verma R, Kumar D, Puri S, Tapwal A, Kumar V, Nepovimova E, Kuca K. Bioactive compounds of edible fruits with their anti-aging properties: a comprehensive review to prolong human life. Antioxidants. 2020;9(11):1123.
52. Aune D. Plant foods, antioxidant biomarkers, and the risk of cardiovascular disease, cancer, and mortality: a review of the evidence. Adv Nutr. 2019;10:S404–21.
53. Lapuente M, Estruch R, Shahbaz M, Casas R. Relation of fruits and vegetables with major cardiometabolic risk factors, markers of oxidation, and inflammation. Nutrients. 2019;11(10):2381.
54. Wallace TC, Bailey RL, Blumberg JB, Burton-Freeman B, Chen CO, Crowe-White KM, Drewnowski A, Hooshmand S, Johnson E, Lewis R, Murray R. Fruits, vegetables, and health: A comprehensive narrative, umbrella review of the science and recommendations for enhanced public policy to improve intake. Crit Rev Food Sci Nutr. 2020;60(13):2174–211.
55. Zhang F, Batley J. Exploring the application of wild species for crop improvement in a changing climate. Curr Opin Plant Biol. 2020;56:218–22.

56. Kour S, Bakshi P, Sharma A, Wali VK, Jasrotia A, Kumari S. Strategies on conservation, improvement and utilization of underutilized fruit crops. Int J Curr Microbiol App Sci. 2018;7(03):638–50.
57. Omotayo AO, Aremu AO. Underutilized African indigenous fruit trees and food–nutrition security: opportunities, challenges, and prospects. Food Energy Secur. 2020;9(3):e220.
58. Gautam RS, Shrestha SJ, Shrestha I. Wild edible fruits of Nepal. Int J Res Appl Sci Biotechnol. 2020;8(3):289–304.
59. Garg M, Kunwar MS, Kashyap H. Enhancing the livelihood of Himalayan communities through action research and transforming wild produce into high-value products. Curr Sci. 2021;10:818–24.

In Silico Analysis of *Syzygium calophyllifolium* Phytocompounds with Type 2 Diabetes Mellitus Targets

Pavithran Balasubramanian, Saikumar Sathyanarayanan, Rahul Chandran, Heidi Abrahamse, and Parimelazhagan Thangaraj

Abstract

This study explores the potential of phytochemical compounds derived from *Syzygium calophyllifolium* for the treatment of diabetes mellitus through a comprehensive computational approach. Using in silico techniques, including molecular docking and ADMET analysis, we investigated the binding patterns and pharmacokinetic properties of selected compounds with the insulin receptor tyrosine kinase. Our results indicate that 1,3-propanediamine, n′-(3-(dimethylamino)-n-n-dimethyl exhibits promising binding affinity and efficacy comparable to the standard drug acarbose, suggesting its potential as an effective treatment option. Additionally, we elucidated the structural characteristics and functional roles of the insulin receptor, shedding light on the mechanism of insulin signal transduction and action. Furthermore, we highlight the significance of phytochemicals as a valuable source for drug discovery, emphasizing their potential in providing novel therapeutic solutions with enhanced bioavailability and reduced toxicity. Overall, our findings underscore the utility of computational techniques in identifying promising drug candidates, paving the way for the development of effective treatments for diabetes mellitus and other complex diseases.

P. Balasubramanian · S. Sathyanarayanan (✉)
Department of Botany, PSG College of Arts & Science, Coimbatore, Tamil Nadu, India

R. Chandran · H. Abrahamse
Laser Research Centre, Faculty of Health Sciences, University of Johannesburg, Johannesburg, South Africa

P. Thangaraj
Department of Botany, Bharathiar University, Coimbatore, Tamil Nadu, India

P. Thangaraj (ed.), *Pomotherapeutic Insights on Wild Edible Fruits*,
https://doi.org/10.1007/978-981-96-0543-9_5

Keywords

Syzygium calophyllifolium · Phytocompounds · Diabetes mellitus · ADMET · Molecular docking

Introduction

Plants have long served as a rich source of medicinal compounds, utilized both in traditional healing practices and as active ingredients in modern medicine [1]. Since ancient times, the utilization of medicinal plants has laid the foundation for contemporary medical advancements [2]. Approximately ten percent of vascular plants worldwide possess medicinal properties, with estimates ranging from 350,000 to nearly half a million species [2]. Initially, trial-and-error methods were employed to identify plants with therapeutic benefits, a process integral to discerning their efficacy in treating ailments and promoting well-being. Traditional medicine, encompassing a diverse array of knowledge, skills, and practices rooted in cultural beliefs and experiences, plays a crucial role in maintaining health and addressing both physical and mental illnesses [2].

Phytochemicals encompass a diverse array of plant-derived compounds with unique structures capable of fostering health benefits. Synthesized exclusively by plants, these natural substances are not involved in energy metabolism, anabolic, or catabolic processes but are produced within specific cell types. A wealth of epidemiological evidence underscores the extensive health-promoting potential inherent in phytochemicals. Among these compounds, flavonoids, tannins, anthocyanins, and other phenolic constituents found in plant-based foods demonstrate significant antioxidant properties [3]. The exploration of natural antioxidant reservoirs within plant-based foods is pivotal for promoting health. Moreover, compounds exhibiting antioxidant attributes often exhibit notable biological activities, including antimicrobial effects. Consequently, the antimicrobial properties of these compounds hold significant promise for extending the shelf life of food products [4].

Efforts to isolate and screen novel compounds for their bioactivity in traditional medicine have been pursued, with the extraction procedure standing out as a critical step in the research and development of plant secondary metabolites [5]. The identification and exploration of phenolic compounds or extracts from various plants have emerged as a significant focus in health and medicinal research [6]. Phytocompounds, found abundantly in plants, exhibit a range of bioactive properties that contribute to reducing the risk of diseases like diabetes. Specifically, these compounds possess antioxidant and anti-inflammatory properties, which help mitigate oxidative stress and inflammation, both of which are implicated in the development and progression of diabetes. Additionally, certain phytocompounds have been shown to improve insulin sensitivity and regulate blood glucose levels, thereby reducing the risk of diabetes and its complications. Moreover, phytocompounds may also influence various cellular pathways involved in glucose metabolism, lipid metabolism, and insulin signaling, further contributing to their potential in reducing the risk of diabetes. Overall, the multifaceted effects of phytocompounds make them promising candidates for dietary interventions aimed at diabetes prevention and management.

Diabetes mellitus, a metabolic disorder characterized by hyperglycemia and disruptions in carbohydrate, protein, and fat metabolisms, stems from inadequate insulin production or utilization [7]. Recent data from the World Health Organization (WHO) indicates that the global prevalence of diabetes exceeds 400 million individuals, with projections suggesting a rise to over 700 million by 2045 [8]. Hyperglycemia instigates the production of reactive oxygen species (ROS), pivotal in diabetic complications [9]. In addition to oxidative stress, diabetes disrupts insulin function, leading to increased hepatic glucose production. Given concerns regarding adverse effects associated with synthetic medications, there is increasing interest in plant-derived therapies in developed nations as alternative diabetes treatments. Many traditional systems of medicines recommend exploring plants with hypoglycemic properties as potential sources for new oral antidiabetic drugs or dietary adjuncts to current therapies [10]. In light of the escalating global burden of diabetes, there is a growing interest in exploring phytocompounds as potential therapeutic agents for the treatment of Diabetes mellitus.

In silico analysis, a computational approach to drug discovery is increasingly recognized for its pivotal role in identifying novel therapeutic targets and compounds for managing diabetes. By leveraging sophisticated algorithms and molecular modeling techniques, researchers can predict the interactions between potential drug candidates and biological targets implicated in diabetes pathogenesis. This enables the rapid screening of vast chemical libraries to identify molecules with the greatest likelihood of efficacy and safety profiles. Moreover, in silico analysis facilitates the optimization of lead compounds, allowing for the design of derivatives with enhanced pharmacological properties and reduced side effects. Additionally, this approach enables the exploration of complex biological pathways involved in diabetes, shedding light on novel drug targets that may have been overlooked using traditional methods. Overall, in silico analysis serves as a valuable tool in the arsenal of drug discovery efforts for diabetes, offering accelerated and cost-effective means to identify promising therapeutic interventions and advancing our understanding of this complex metabolic disorder [11, 12]. Employing in silico methodologies enables the concurrent screening of vast libraries comprising millions of phytocompounds and drugs against various diseases. Additionally, molecular dynamics-based screening aids in predicting the efficacy, stability, and toxicity profiles of significant drug candidates or phytocompounds, guiding the design of in vivo experiments. Thus, this study aimed to identify potential phytoconstituents of *Syzygium calophyllifolium* through in silico approaches, with the goal of developing innovative therapeutics for managing diabetes mellitus.

Materials and Methods

Plant Selection and Component Identification

The selection of *Syzygium calophyllifolium* was based on an extensive literature survey highlighting its significant medicinal value. Five compounds were chosen for in silico analysis, guided by a literature review and the known biological activity of *Syzygium calophyllifolium* [13]. Acarbose served as the standard control in the docking experiments.

In Silico Analysis of *Syzygium calophyllifolium*

Active constituents from *Syzygium calophyllifolium* were subjected to computational analysis using a Lenovo system equipped with an Intel Core i3 processor and 4GB RAM. Software tools, including PyMOL, PubChem, PyRX, and AutoDock Vina, were utilized.

Ligand Selection and Preparation

The phytochemical structure of *Syzygium calophyllifolium* was obtained in SDF file format from the PubChem database. Conversion to PDB format was accomplished using Open Babel software [14].

Identification of Active Sites in Target Proteins

Target proteins for the inhibition of inflammation were identified, providing binding sites for ligands. The 3D structure of the aldose reductase in complex target (PDB ID 1US0) was obtained from the Protein Data Bank (PDB) [15, 16].

Protein Data Bank (PDB)

The Protein Data Bank (PDB) serves as a vital repository for three-dimensional structural data of biological molecules. Structural information obtained from PDB, including the 3D structure of the target protein, was utilized for docking analysis [17].

PubChem

PubChem, maintained by the National Center for Biotechnology Information (NCBI), is a database containing chemical molecules. Natural compound structures selected from PubChem were saved as 3D SDF files [18].

ADME Toxicity

ADME/Tox properties of compounds were predicted using the SwissADME portal [19, 20]. The Lipinski rule of five was employed to assess drug-likeness [21].

Molecular Docking Study

Molecular docking calculations were performed using MGL tools with AutoGrid and AutoDock Vina. Crystalized 3D structures obtained from PDB were utilized to prepare receptor and ligand files. Lamarckian genetic algorithms were employed for docking, with the lowest energy conformations selected as binding modes [22, 23].

PyMOL

PyMOL, an open-source molecular visualization system, was utilized for generating high-quality 3D images of ligands and proteins. It extends the Python programming language for advanced customization [24].

Results and Discussions

In Silico Docking Studies on Identified Compounds from *Syzygium calophyllifolium*

Selection and Preparation of Ligands

Five compounds were chosen for in silico analysis based on a review of the literature and the observed biological activity of *Syzygium calophyllifolium* [13]. Acarbose served as the standard control in the docking experiment. Acarbose, a complex oligosaccharide, functions as an inhibitor of various enzymes responsible for breaking down complex carbohydrates in the intestines. It targets pancreatic alpha-amylase and membrane-bound alpha-glucosidases, including intestinal glucoamylase, sucrase, maltase, and isomaltase. These enzymes play crucial roles in metabolizing complex starches and oligo-, tri-, and disaccharides into simple sugars that can be absorbed. By inhibiting their activity, acarbose restricts the absorption of dietary carbohydrates, thereby curbing the subsequent postprandial rise in blood

Table 1 Bioactive compounds of *Syzygium calophyllifolium* [13]

Name of the compound	PubChem CID	Molecular formula	Molecular weight
N-(3-(-aziridyl) propyl)-3-dimethylaminopropylamine	547145	C10H23N3	185.31 g/mol
3-Piperidinamine	148119	C5H12N2	100.16 g/mol
L-valine, 3-mercapto	12178514	C5H12ClNO2S	185.67 g/mol
1,3-Propanediamine, n′-(3-(dimethylamino)-n-n-dimethyl	110461	C14H35N5	273.46 g/mol
N-(3-(n-aziridyl) propylidene)-3-methylaminopropylamine	547145	C10H23N3	185.31 G/MOL
Acarbose	41774	C25H43NO18	645.6 g/mol

glucose and insulin levels. Consequently, acarbose is used alongside diet, exercise, and other pharmacotherapies to manage blood sugar levels in patients with type 2 diabetes. The 2D structures of the selected phytochemicals, along with their molecular weight, molecular formula, and PubChem CID, were obtained from the PubChem database and are presented in Table 1.

Protein Data Bank (PDB)

The target protein, aldose reductase (AR) in its complex three-dimensional (3D) structure, was retrieved from the Protein Data Bank (PDB) database (www.rcsb.pdb), with the structure and molecular properties illustrated in Fig. 1. The PDB serves as a repository for 3D structural data of significant biological macromolecules like proteins and nucleic acids, with the PDB ID being 1US0. Aldose reductase, a pivotal enzyme in the polyol pathway, catalyzes the nicotinamide adenine dinucleotide phosphate (NADPH)-dependent reduction of glucose to sorbitol, resulting in the excessive accumulation of intracellular reactive oxygen species (ROS) in various tissues affected by diabetes mellitus, including the heart, vasculature, neurons, eyes, and kidneys. For instance, hyperglycemia-induced oxidative stress via the polyol pathway is a key mechanism. Recent research has highlighted the potential of AR inhibitors in preventing or delaying the onset of cardiovascular complications, such as ischemia/reperfusion injury, atherosclerosis, and atherothrombosis. This review will delve into the crucial roles of AR in the pathogenesis of cardiovascular diseases and other diabetic complications, as well as explore the emerging therapeutic potential of AR inhibitors in mitigating diabetes mellitus complications [25–27].

ADME Properties of Selective Compounds

The drug absorption, distribution, metabolism, excretion, and toxicity (ADME) properties of the selected compounds were evaluated using Swiss ADME. This

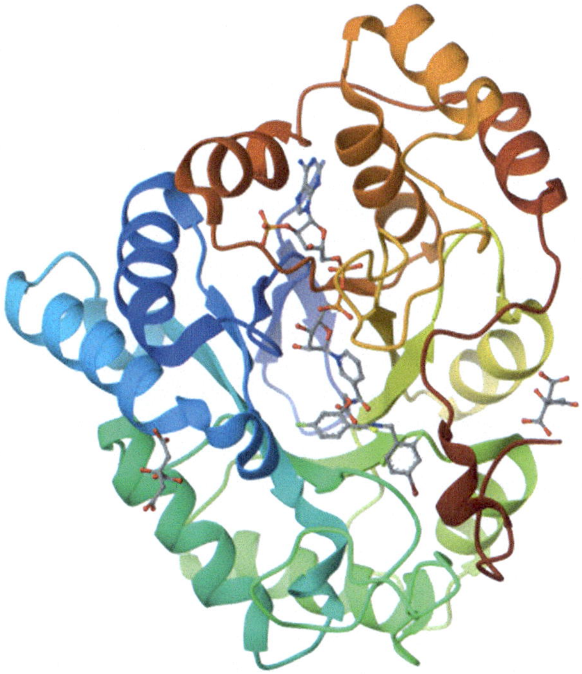

Fig. 1 Aldose reductase in complex (PDB ID: 1US0) [16]

comprehensive analysis encompasses various aspects crucial for drug development. One such parameter assessed is the skin permeability value (Kp) in cm/s, which indicates the ability of molecules to be absorbed through the skin. The Kp values obtained for all compounds fell within the normal range, suggesting low skin permeability. Furthermore, the evaluation of gastrointestinal (GI) and blood-brain barrier (BBB) permeation provides insights into the absorption and distribution of drug molecules within the body. The in silico prediction results of absorption, distribution, metabolism, and excretion (ADME) parameters for the studied compounds are summarized in Table 2, offering valuable information for understanding their pharmacokinetic profiles and potential therapeutic applications.

The Swiss ADME prediction parameters indicate that the compounds exhibit high gastrointestinal (GI) absorption and blood-brain barrier (BBB) permeation. Additionally, drug metabolism is governed by a range of cytochrome enzymes (CYPs), with CYP1A2, CYP2C19, CYP2C9, CYP2D6, and CYP3A4, playing crucial roles in the biotransformation of drug molecules. Notably, the Swiss ADME prediction confirms that all studied alkaloids adhere to Lipinski's rule of five, suggesting their potential oral activity. Furthermore, the total polar surface area (TPSA) values of the compounds fall within the range of 18–102 Å, well below the limit of 160 Å, indicating favorable hydrogen bonding potential and bioavailability. Moreover, the calculated numbers of rotatable bonds (NRB) for the compounds are less than 10, suggesting their conformational stability. These findings collectively

Table 2 ADME properties of the ligands

Name of the ligand	Molecular mass less than 500 Dalton	High lipophilicity (expressed as LogP less than 5)	Less than 5 hydrogen bond donors	Less than 10 hydrogen bond acceptors	Molar refractivity should be between 40 and 130	Lipinski's rule of five violation
N-(3-(-aziridyl) propyl)-3-dimethylaminopropylamine	185	−0.15	1	3	60.58	0
3-Piperidinamine	100	−0.59	2	2	33.46	0
L-valine, 3-mercapto	185	−0.98	2	3	45.56	0
1,3-Propanediamine, n′-(3-(dimethylamino)-n-n-dimethyl	273	−0.58	3	5	83.42	0
N-(3-(n-aziridyl) propylidene)-3-methylaminopropylamine	309.00	5.10	0	2	146.78	1

contribute to a comprehensive understanding of the pharmacokinetic properties and potential therapeutic efficacy of the studied compounds.

Molecular Docking Analysis

The analysis of binding patterns between the protein and ligands revealed variability based on the nature of the ligand. Docking results for both standard and bioactive compounds are presented in Table 3, along with Fig. 2. The binding energy of selected phytochemical compounds with the enzyme, as determined by AutoDock, is detailed in the table. Notably, docking studies indicate that the ligands bind to the active site region of the AR target with favorable binding energy, occupying the same hydrophobic pocket as the standard.

Computer docking techniques serve as invaluable tools in drug design and mechanism elucidation. Flexible docking programs such as AutoDock and molecular operating environment (MOE) enable the prediction of favorable protein–ligand complex structures with reasonable accuracy and speed. When employed prior to experimental screening, these docking programs act as potent computational filters, effectively reducing the labor and cost required for the development of efficacious medicinal compounds.

Following experimental screening, these computational techniques can provide invaluable insights into bioactivity mechanisms. Molecular models were constructed to elucidate binding modes through docking using the AutoDock program. Interactions between N-(3-(-aziridyl)propyl)-3-dimethylaminopropylamine, 3-piperidinamine, L-valine, 3-mercapto, 1,3-Propanediamine, n′-(3-(dimethylamino)-n-n-dimethyl, N-(3-(n-aziridyl)propylidene)-3-methylamino propylamine, and the control molecule acarbose with the insulin receptor AR protein

Table 3 Interaction of compounds with aldose reductase in complex

Name of the ligand	Binding affinity	rmsd/ub[a]	rmsd/lb[a]
N-(3-(-aziridyl)propyl)-3-dimethylaminopropylamine	−5.5	6.146	1.752
3-Piperidinamine	−4.9	3.153	1.832
L-valine, 3-mercapto	−4.6	2.613	2.178
1,3-Propanediamine, n′-(3-(dimethylamino)-n-n-dimethyl	−6.7	7.98	1.513
N-(3-(n-aziridyl) propylidene)-3-methylaminopropylamine	−5.5	1.976	1.315
Acarbose	−6.9	2.062	0.803

[a]rmsd—root mean square deviation

were investigated. The structure of each drug was drawn and subsequently subjected to energy optimization for further analysis.

The utilization of computational techniques following experimental screening presents a significant advancement in understanding bioactivity mechanisms [28]. By employing molecular modeling, insights into binding modes were elucidated through docking simulations using the AutoDock program [23]. This approach enabled the investigation of interactions between various compounds, including N-(3-(-aziridyl)propyl)-3-dimethylaminopropylamine, 3-piperidinamine, L-valine, 3-mercapto, 1,3-propanediamine, n′-(3-(dimethylamino)-n-n-dimethyl, N-(3-(n-aziridyl)propylidene)-3-methylamino propylamine, and the control molecule acarbose, with the insulin receptor AR protein. Each drug's structure was meticulously drawn and subsequently subjected to energy optimization, facilitating a deeper analysis of their binding affinities and potential therapeutic efficacy [29]. This integrated approach not only enhances our understanding of the molecular mechanisms underlying drug-protein interactions but also provides valuable insights into the design and development of novel therapeutic agents targeting diabetes mellitus and related complications.

Conclusion

In conclusion, our study provides valuable insights into the potential of phytochemical compounds derived from *Syzygium calophyllifolium* as novel therapeutic agents for the management of diabetes mellitus. Through a comprehensive computational analysis, we demonstrated the efficacy of 1,3-propanediamine, n′-(3-(dimethylamino)-n-n-dimethyl in binding to the AR, indicating its potential as an effective treatment option. Moreover, our investigation into the structural characteristics and functional roles of the insulin receptor enhances our understanding of insulin signal transduction and action, offering avenues for further research in this field. Additionally, we emphasize the significance of phytochemicals in drug discovery, highlighting their advantages such as enhanced bioavailability and reduced toxicity compared to synthetic drugs. Overall, our findings underscore the

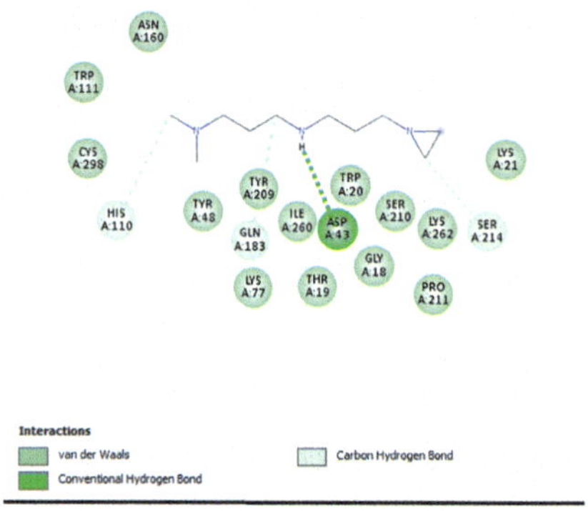

a) N-[3-[-Aziridyl]propyl]-3-dimethylaminopropylamine

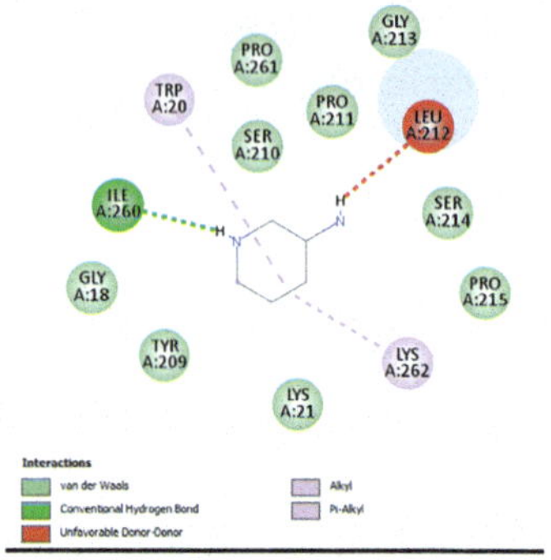

b) 3-Piperidinamine

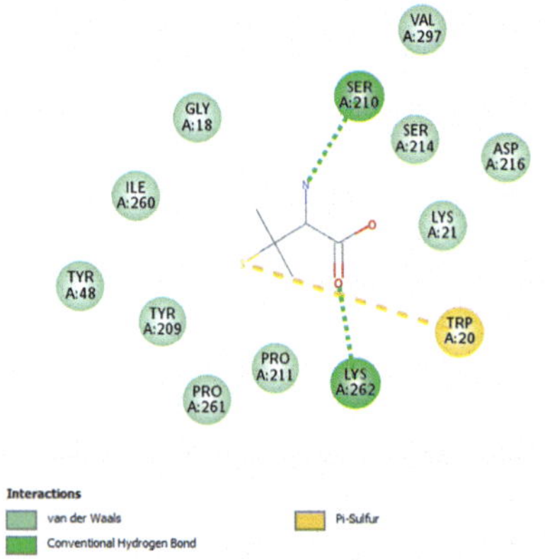

c) L-valine, 3-mercapto

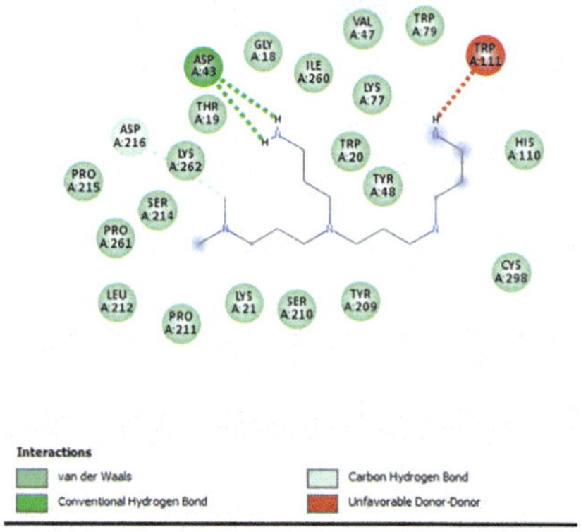

d) 1,3-Propanediamine, n'-[3-(dimethylamino)-n-n-dimethyl

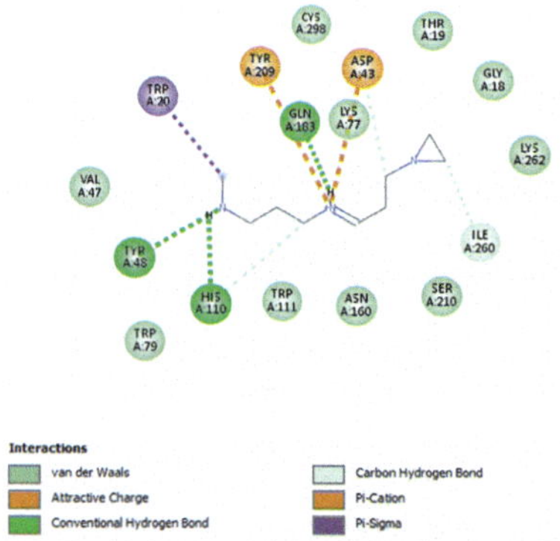

e) N-[3-[n-aziridyl]propylidene]-3-methylaminopropylamine

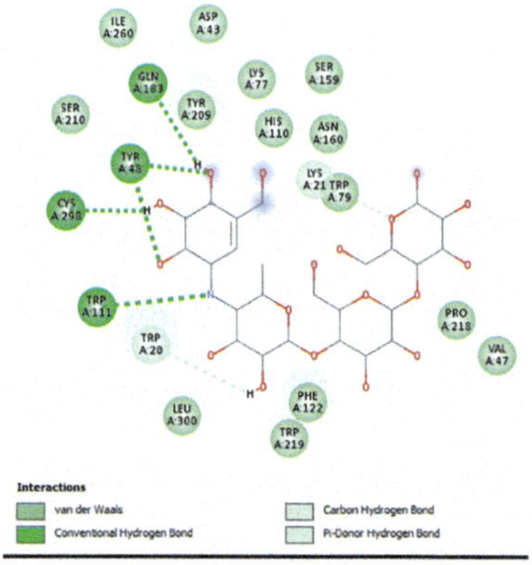

f) Acarbose

Fig. 2 Interaction of compounds with aldose reductase in complex protein

importance of computational techniques in identifying promising drug candidates and provide a foundation for future experimental studies to validate the efficacy and safety of these compounds. Through collaborative efforts between computational and experimental research, we can accelerate the development of effective treatments for diabetes mellitus and improve patient outcomes.

Acknowledgments The author Saikumar Sathyanarayanan is thankful to the management authorities of PSG College of Arts and Science, Coimbatore, who provided his support throughout the study.

Conflict of Interest The authors have no conflict of interest.

Funding This research did not receive any specific grant from funding agencies in the public, commercial, or not-for-profit sectors.

References

1. Farnsworth NR, Akerele O, Bingel AS, Soejarto DD, Guo Z. Medicinal plants in therapy. Bull World Health Organ. 1985;63(6):965.
2. Salmerón-Manzano E, Garrido-Cardenas JA, Manzano-Agugliaro F. Worldwide research trends on medicinal plants. Int J Environ Res Public Health. 2020;17(10):3376. https://doi.org/10.3390/ijerph17103376. PMID: 32408690. PMCID: PMC7277765.
3. Salah N, Miller NJ, Paganga G, Tijburg L, Bolwell GP, Rice-Evans C. Polyphenolic flavanols as scavengers of aqueous phase radicals and as chain-breaking antioxidants. Arch Biochem Biophys. 1995;322(2):339–46.
4. Bakkali F, Averbeck S, Averbeck D, Idaomar M. Biological effects of essential oils—a review. Food Chem Toxicol. 2008;46(2):446–75.
5. Savithramma N, Rao ML, Suhrulatha D. Screening of medicinal plants for secondary metabolites. Middle-East J Sci Res. 2011;8(3):579–84.
6. Dai J, Mumper RJ. Plant phenolics: extraction, analysis and their antioxidant and anticancer properties. Molecules. 2010;15(10):7313–52.
7. Ahmad N, et al. Efficient regeneration and antioxidant potential in regenerated tissues of Piper nigrum L. Plant Cell Tissue Organ Cult. 2010;102:129–34.
8. Harding JL, Weber MB, Shaw JE. The global burden of diabetes. Textbook of Diabetes; 2024. p. 28–40.
9. Dewanjee S, et al. Antidiabetic activity of Diospyros peregrina fruit: effect on hyperglycemia, hyperlipidemia and augmented oxidative stress in experimental type 2 diabetes. Food Chem Toxicol. 2009;47(10):2679–85.
10. Alam S, et al. Antidiabetic phytochemicals from medicinal plants: prospective candidates for new drug discovery and development. Front Endocrinol (Lausanne). 2022;13:800714.
11. Smith AB, Jones CD. Computational approaches in drug discovery: an overview. J Med Chem. 2021;64(14):1000–17.
12. Johnson EF, Gentry PR. Translational toxicology: defining a new therapeutic discipline. Springer; 2018.
13. Sathyanarayanan S, Chandran R, Thankarajan S, Abrahamse H, Thangaraj P. Phytochemical composition, antioxidant and anti-bacterial activity of Syzygium calophyllifolium Walp fruit. J Food Sci Tech 2018;55:341–50.
14. Li Q, Shah S. Structure-based virtual screening. In: Protein Bioinformatics: From Protein Modifications and Networks to Proteomics; 2017. p. 111–24.

15. Packyanathan SJ, Prakasam G. Antibacterial effect of clove oil against clinical strains of Escherichia coli. J Pharm Sci Res. 2017;9(7):1203.
16. Howard EI, Sanishvili R, Cachau RE, Mitschler A, Chevrier B, Barth P, Lamour V, Van Zandt M, Sibley E, Bon C, Moras D, Schneider TR, Joachimiak A, Podjarny A. Ultrahigh resolution drug design I: Details of interactions in human aldose reductase–inhibitor complex at 0.66 Å. Proteins, 2004;55:792–804. https://doi.org/10.1002/prot.20015
17. Dikshit R. Cancer mortality in India: a nationally representative survey. Lancet. 2012;379(9828):1807–16.
18. Maffei M. Dietary supplements of plant origin: nutrition and health approach. CRC Press; 2003. p. 256.
19. López-Vallejo F, Caulfield T, Martínez-Mayorga K, Giulianotti MA, Nefzi A, Houghten RA, Medina-Franco JL. Integrating virtual screening and combinatorial chemistry for accelerated drug discovery. Comb Chem High Throughput Screen. 2011;14(6):475–87.
20. SwissADME: a free web tool to evaluate pharmacokinetics, drug-likeness and medicinal chemistry friendliness of small molecules. Sci Rep 2017;7:42717
21. Lipinski CA. Avoiding investment in doomed drugs. Curr Drug Discov. 2001;1:17–9.
22. Meenambiga SS, Venkataraghavan R, Biswal RA. In silico analysis of plant phytochemicals against secreted aspartic proteinase enzyme of Candida albicans. J Appl Pharm Sci. 2018;8(11):140–50.
23. Trott O, Olson A. J. AutoDock Vina: improving the speed and accuracy of docking with a new scoring function, efficient optimization, and multithreading. J Comput Chem 2010;31(2):455–61.
24. DeLano WL. The PyMOL molecular graphics system. http://www.pymol.org; 2002.
25. Zhang H, et al. Structure of the human aldose reductase AKR1B1 bound to the medical plant extract tolrestat. Acta Crystallogr F Struct Biol Commun. 2017;73(9):514–20.
26. Petrash JM. All in the family: aldose reductase and closely related aldo-keto reductases. Cell Mol Life Sci. 2004;61(7–8):737–49.
27. Bhatia A, et al. Aldose reductase modulates acute and chronic adipose inflammation. Sci Rep. 2019;9(1):1–15.
28. Grinter SZ, Zou X, Campbell AJ. Computational approaches to target protein-small molecule interaction. Curr Opin Chem Biol. 2019;52:56–61.
29. Morris GM, Huey R, Lindstrom W, Sanner MF, Belew RK, Goodsell DS, Olson AJ. AutoDock4 and AutoDockTools4: automated docking with selective receptor flexibility. J Comput Chem. 2009;30(16):2785–91.

An Insight to Intensify the Potential Prospects of Underutilized Wild Species of *Piper*

M. Pugalenthi, M. A. Deepa, M. Pradheeba, S. Narthanaa, and T. Parimelazhagan

Abstract

Piper species, belonging to the Piperaceae family, are renowned for their diverse medicinal properties and have been integral to traditional medicine systems across various cultures. These plants, which include the well-known black pepper (*Piper nigrum*), exhibit a wide range of pharmacological activities such as anti-inflammatory, antimicrobial, antioxidant, anticancer, and analgesic properties. The bioactive compounds in *Piper* species, notably alkaloids, flavonoids, lignans, and amides, contribute significantly to these medicinal effects. Piperine, the most studied alkaloid, enhances the bioavailability of various drugs and nutrients, making it a valuable component in therapeutic formulations. The utilization of *Piper* species extends beyond medicinal applications. They are used in culinary practices, particularly as spices, and in the cosmetics industry for their aromatic properties. The potential of *Piper* plants in developing novel drugs and therapeutic agents is immense, given their extensive range of bioactivities. Recent scientific research has focused on isolating and characterizing these bioactive compounds, understanding their mechanisms of action, and evaluating their efficacy in clinical settings. The sustainable cultivation and utilization of Piper species can lead to significant advancements in pharmacology and therapeutics, highlighting their importance in both traditional and modern medicine.

Keywords

Piper sp. · Wild *Piper* sp. · Medicinal Piper · Underutilized Piper

M. Pugalenthi (✉) · M. A. Deepa · M. Pradheeba · S. Narthanaa
Department of Botany, Government Arts College(A), Coimbatore, India

T. Parimelazhagan
Department of Botany, Bharathiar University, Coimbatore, India

P. Thangaraj (ed.), *Pomotherapeutic Insights on Wild Edible Fruits*,
https://doi.org/10.1007/978-981-96-0543-9_6

Introduction

Medicinal plants have gained considerable popularity in various traditional systems of medicine due to their extensive pharmacological potential and fewer side effects on biological systems [1]. These plants serve as crucial sources of natural bioactive compounds [2]. The investigation of medicinal plants and natural products has been a long-standing focus in the field of drug discovery and development [3]. Bioactive compounds play a pivotal role in the discovery of new drugs, and it is estimated that approximately 60% of approved drugs are of natural origin [4]. Plant-based drugs have been successfully employed in the treatment of a wide range of ailments, including skin diseases, tuberculosis, diabetes, jaundice, hypertension, mental disorders, cancer, AIDS and other infectious diseases [5].

Currently, there is a growing trend toward returning to the roots of medicine [6]. This is reflected in the development of new approaches for analyzing folk herbs that are used in various cultures [7]. These scientific advancements have confirmed the pharmacological properties of plants that have been used for centuries for food, medicine, or spiritual purposes [8]. The heritage of using certain plants in traditional medicine is being continuously corroborated through scientific inquiry, as evidenced by studies [9].

The current world is immensely dependent on drug resistance, which poses a significant challenge to the healthcare sector worldwide, affecting both developed and developing nations where traditional medications are commonly relied upon by local communities [10]. In response, biologists, chemists, and pharmacologists are directing their efforts toward discovering new drug compounds derived from natural sources [11]. Medicinal plants, widely available across the globe, have garnered considerable interest from pharmacologists seeking to identify potential drug molecules [12]. Various parts of plants, including roots, stems, flowers, and leaves have been utilized for centuries as repellents and agents with anticoagulant, antifungal, and antibacterial properties [13]. Many of these medicinal plants have also been recognized for their ability to promote wound healing [14]. The collaborative efforts of traditional healers and traditional medicine practitioners play a crucial role in identifying plants specific to certain geographic regions [15].

General Outline on *Piper* Species

The genus *Piper*, consisting of nearly 1000 species, is mainly distributed in the tropical and subtropical regions of the world, with the greatest diversity occurring in the American tropics (700 spp.), followed by Southern Asia (300 spp.) [16]. Plant taxonomists have generally avoided classifying species within the *Piper* genus due to the large number of species, which exhibit both unisexual and bisexual flowers, as well as widespread distributions [17]. The taxa's extremely variable vegetative characters, closely packed spike inflorescences, and minute flowers make it one of the most challenging to classify systematically [18]. Over 1000 species of *Piper* have been described in recent decades without any reference to subgeneric categories, making it the largest genus of angiosperms lacking a functional classification [19]. Consequently, many new taxa have been described, but none of them truly merit formal recognition [20]. Therefore, it is essential to conduct a thorough morphological assessment of *Piper* species and select the attributes that will provide more systematic groupings of taxa with succinct characteristics [21].

One of the widely distributed plant genera in pantropical regions is the genus *Piper*, which is commonly known as pepper [17]. The presence of oil cells in the structures of almost all *Piper* species classifies them as aromatic plants [20]. In addition to their well-known uses as culinary spices, the secondary metabolites isolated from *Piper* plants have been found to have a wide range of beneficial effects on human health [22].

Numerous studies have documented the considerable antioxidant activity exhibited by the *Piperaceae* family. These plants have gained widespread utilization in traditional medicinal practices in China, Indonesia, India, and Korea, exhibiting a multitude of biological properties, including antifungal, antioxidant, anti-inflammatory, antimicrobial, and antidiabetic activities [23].

The bioactive constituents present in *Piper* species can serve as a reservoir of novel pharmaceutical targets. These bioactive components, found in the seeds, leaves and stem bark, encompass phenolic, tannins, saponins, alkaloids, flavonoids, glycosides, and terpenoids [22]. Amide alkaloids, a distinct compound, is prevalent in *Piper* species, with a majority of these compounds featuring a *Piper* ring (pyridine) or pyrrole ring (pyrrolidine). Moreover, certain *Piper* species exhibit characteristic compounds such as open-chain alkamides, cyclohexylamine, aristolactams, and ceramides [24].

Piper plants are renowned for their significance in both medicinal and economic contexts. They are widely utilized as a source of indigenous medicine, vegetables, spices, and ornamentals and hold great importance in traditional religious rituals [23]. The *Piper* species thrive best in tropical conditions with cool shade, considerable humidity and an ample supply of soil moisture and rainfall. They are often cultivated as intercrops with trees for support [25]. *Piper* species can be propagated through vegetative means, with four primary methods such as branch repositioning, prostration of branches in the soil, fragmentation of branches, and fragmentation of rhizomes [26]. While some well-known *Piper* species, such as *P. nigrum*, *P. longum*, and *P. betle* have established cultivation practices, most *Piper* species are from the wild and are not cultivated. As a result, they are often exploited from their natural habitats, which poses a threat of endangerment. Therefore, there is a need to bring these neglected wild *Piper* species under cultivation. Many of them have the potential to be used commercially and cultivating them would be beneficial for their conservation [17]. To cultivate economically important wild *Piper* species, techniques from well-known cultivated species can be adopted, taking into account the phenology for harvesting. Despite their widespread occurrence in almost all tropical habitats and their numerous uses, the *Piper* genus remains poorly investigated [18].

Piperlongumine, also known as piplartine, is an amide alkaloid that has been extensively studied and is found in several species of the *Piper* plant family (*Piper*aceae) [27]. It has demonstrated the potential anticancer properties, as well as beneficial effects in the treatment of the parasitic infection schistosomiasis, which is caused by helminth flatworms of the genus *Schistosoma* [28, 29].

Compounds derived from *Piper tuberculatum* fruits have also shown antiplasmodial and antileishmanial activities, making them important for addressing neglected tropical diseases in pantropical regions where these plants are native [30, 31]. Piperlongumine has been shown to exhibit anti-inflammatory effects in the central nervous system (CNS), which can be beneficial in preventing and treating inflammation-related brain diseases such as Alzheimer's disease (AD), Parkinson's disease (PD) and Huntington's disease [32]. The potential approach in prevention and treatment of these diseases is the normalization of microglia activity, and Piperlongumine has been shown to inhibit the production of nitric oxide (NO) and prostaglandin E2 (PGE2) induced by lipopolysaccharide (LPS), reduce the expression of inducible nitric oxide synthase and cyclooxygenase-2, as well as proinflammatory cytokines such as tumor necrosis factor (TNF)-α and interleukin-6 (IL-6) and suppress the nuclear factor kappa B (NF-κB) signaling pathway [33, 34].

The antimicrobial activity of *Piper* plants has been demonstrated in the treatment of chronic periodontitis, as well as in the treatment of gastric pathogen *Helicobacter pylori* [35]. Additionally, *Piper* plants have shown activity as antidiabetic, antiulcer, diuretic and local anesthetic agents [23, 36]. The majority of information regarding the various biological activities of *Piper* plants has been derived from in vitro studies, while in vivo and toxicology studies are somewhat limited. However, these plants possess multi-targeting potential against various diseases.

Distribution of *Piper* Species in India

In India, there are two main centers of diversity of *Piper* species, namely, the Eastern Himalayas, including the North Eastern States and the Southern Deccan [37]. North East India, which falls within the Indo-Burma Biodiversity hotspot, one of the global biodiversity hotspots recognized, is blessed with a wide range of physiographic and climatic conditions. It also falls in the biogeographic junction of the Indian, Indo-Malayan, and Indo-Chinese biogeographic regions [38, 39]. The region supports a rich flora of *Piper* species that are mostly different from their counterparts in the Western Ghats [40]. The Peninsular Indian segment of the *Piper* genus encompasses approximately 21 species, which are primarily found in the lower levels of tropical evergreen forests. These species have a fragmented distribution in the Northeastern Himalayan region, the Western Ghats, and the Eastern Ghats [37, 41]. According to Hooker (1886) in his Flora of British India, a total of 33 species were recorded, of which 20 species were from North East (NE) India and its adjacent regions. Candolle (1912) reported 60 species from NE India [42, 43].

Till date, a total of 84 species have been documented in the entirety of the Indian region [37]. Phytogeographical investigations conducted previously proposed that the *Piper* species found in Peninsular India can be traced back to ancestral forms originating from the Gondwanan region (Indo-Asian), which subsequently migrated to the Indian subcontinent during the mid-Cretaceous period [41, 44, 45].

The primary rationale behind the utilization of *Piper* species as a therapeutic alternative based on natural products is the abundance of bioactive compounds and diverse reported biological activities in these plants. Hence, the copious presence of compounds in *Piper* species can serve as a foundation for drug discovery [22, 27].

In this chapter, a comprehensive analysis was conducted on the selected *Piper* genus from the Floral databases of Tamil Nadu, encompassing their botanical descriptions, geographical distribution, ethnobotanical applications, scientific validation and economic significance.

Piper nigrum L.

P. nigrum, known by various names such as Kali Mirch in Urdu and Hindi, Pippali in Sanskrit, Milagu in Tamil, as well as peppercorn, white pepper, green pepper, black pepper, and Madagascar pepper, holds a significant position. In India, it is known as "the Home of the Black Pepper." Black pepper is also known as "the Black Gold" due to its immense value in terms of commerce, economy and trade. The common pepper found in the market is the fruit of *P. nigrum*, which comprises small, round green berries that turn bright red when fully ripe [46].

Description

Stout arborescent climbers reach heights of up to 8 m. Rooting is found at the nodes and it lacks hair. The leaves are simple and opposite, characterized by their succulent or fleshy and soft nature. The petioles measure between 0.5 and 4 cm in length and are hairless. The lamina is approximately 6–15 cm long and 2.5–12 cm wide, displaying an ovate-lanceolate or ovate-elliptic shape. The base of the lamina is oblique or cuneate, while the apex is acuminate. The leaves are entire and membranous, with 5–9 pairs of secondary nerves. The flowers are polygamous and usually monoecious, growing in slender spikes that measure around 10–25 cm. These spikes hang down, with the bracts of the female spikes being cup-shaped and without raised margins. The bracteoles form a cup around the ovary and the fruiting spikes are fleshy and interrupted, measuring up to 18 cm in length. The peduncles are hairless and measure between 1 and 2.5 cm in length. There are two stamens, one on each side of the ovary, with thick, short filaments and reniform anthers. The ovary is globose and there are typically 3 or 4, occasionally 5, persistent stigmas. The drupes are sessile and measure approximately 3–4 mm in diameter. They are red when ripe and turn black when unripe, assuming a pyriform or globose shape. The flowering and fruiting period occurs from March to August [47].

Distribution

Black pepper holds a significant place in India's agricultural l and scape as one of the most widely cultivated crops. Believed to have originated in the hills of Southern-Western India, its cultivation primarily thrives in Southern India, specifically in the states of Tamil Nadu and Kerala. Beyond India, black pepper has made its presence known in various regions, including Singapore, Acheen, and Lampong districts of Sumatra, as well as Sri Lanka, France, Indonesia, Thail, South America, and the West Indies. Indigenous to the Malabar Travancore coasts, along the western edge of India, black pepper has established itself in evergreen forests, reaching altitudes of up to 1500 m. This versatile spice has become a global commodity with a rich history deeply rooted in the diverse landscapes it inhabits [48].

Ethnobotanical Uses

In the realm of classical literature, numerous Unani scholars have documented diverse therapeutic applications of black pepper. Its effectiveness in treating cholera, dyspepsia, flatulence, diarrhea, and various gastrointestinal disorders has been extensively elucidated. Furthermore, it serves as a valuable constituent in tooth powders. The text "Ilaj-ul-Ghurba" recommends a pill for the treatment of syphilis, comprised of black pepper (*P. nigrum*), *Calotropis gigantea*, and jaggery. Topically, black pepper is utilized for boils, as well as for alleviating conditions, such as sore throat, alopecia, skin ailments, and piles. The combination of finely powdered black pepper and sesame oil, when appropriately blended and heated, has demonstrated efficacy when applied to areas affected by paralysis. A concoction consisting of black pepper leaves and leaves of *Cassa occidentalis* is particularly beneficial for addressing night blindness. Additionally, black pepper is employed for the treatment of gonorrhea, and it also serves as an antidote for scorpion stings. According to Dioscorides, Al-Jama-al-Mufradatal-Advia-wo-al-Aghzia mentions the advantageous use of sharbat (syrup) or lauq (paste) in treating cough and chest pain. The combination of onion, salt, and black pepper, when topically applied to the bald area, exhibits effectiveness in the treatment of alopecia. Additionally, when combined with vinegar, it proves to be beneficial for oral health [48, 49].

Scientific Validation and Economic Values

Piper species, including *Piper nigrum* (black pepper), are known for producing secondary metabolites that play crucial roles in both body metabolism and defense mechanisms against various agents such as insect-feeding plants and animals. One notable group of active compounds in black pepper is the *Piper* amides, extracted from different *Piper* species, including *P. nigrum*. Piperine, a pungent alkaloid, is a key component of black pepper. Additionally, important sesquiterpenes like β-caryophyllene and nerolidol are present, with the former exhibiting anesthetic activity and the latter being used as a flavoring agent. India stands out as the primary source of traditional reports on the versatile uses of *Piper nigrum* (black pepper) in both human and veterinary medicine. In human health, it is frequently employed for menstrual and ear-nose-throat disorders, while in livestock, it proves effective in addressing gastrointestinal issues. The seeds and fruits of *P. nigrum* are commonly used, typically in powdered form, pills or tablets, and paste. *Piper nigrum* and its bioactive compounds exhibit significant pharmacological properties, including antimicrobial activity against a wide range of pathogens by inhibiting biofilm formation, bacterial efflux pumps, and motilities. The spice also showcases antioxidant effects, demonstrating its ability to scavenge various reactive oxygen and nitrogen species and enhance antioxidant enzymes in vivo. Additionally, *Piper nigrum* displays promising anticancer effects against cell lines from different organs, employing cytotoxicity, apoptosis, autophagy, and interference with signaling pathways. The spice is recognized for its antidiabetic and hypolipidemic activities, evidenced by reductions in cholesterol, triglycerides, and low-density lipoprotein and an increase in high-density lipoprotein. Its medicinal significance extends to anti-inflammatory, analgesic, anticonvulsant, and neuroprotective effects. While *Piperin*e emerges as a major bioactive compound, others like *piperic* acid, Piperlonguminine, pellitorine, *Piperolein* B, *Pipe*ramide, *Piperettine*, and (−)-kusunokinin also exhibit biological potency. Although numerous pharmacological studies have been conducted in vitro, there is a pressing need for more in vivo experiments utilizing pharmacokinetic and pharmacodynamic approaches. In conclusion, *Piper nigrum* surpasses its reputation as the "King of spices" and should be acknowledged as a source of medicinal agents, offering a diverse array of bioactive compounds with potential applications in nutraceuticals and pharmaceuticals [50–58].

Piper longum L.

Piper longum L., also known as long pepper, has a rich history dating back to ancient times. Initially documented by Hippocrates, who recognized its medicinal properties rather than categorizing it as a spice, long pepper made its way to Greece in the sixth or fifth century BC. It held significant importance and was well-known as a spice well before the European discovery of the New World [52].

Description

Piper longum, commonly known as long pepper, is characterized by its small shrub structure featuring a substantial woody root and numerous creeping, jointed stems that thicken at the nodes. The leaves are alternately arranged, spreading, lack stipules and vary significantly in size. The lower leaves measure 5–7 cm in length, while the uppermost ones are shorter at 2–3 cm. The plant produces solitary spikes of flowers and its fruits form in fleshy spikes measuring 2.5–3.5 cm in length and 5 mm in thickness. These oblong, blunt, and blackish-green fruits are collected and dried to create the commercial form of pippali, with the root referred to as pippalimula. Three grades of pippalimula exist, with Grade I, featuring thick roots and underground stems, fetching a higher price compared to Grade II or III, which consist of thin roots, stems, or broken fragments. The commercial drug primarily comprises transversely cut pieces, ranging from 5 to 25 mm in length and 2 to 7 mm in diameter. These cylindrical pieces may be straight or slightly curved, with distinct, swollen internodes displaying leaf, and rootlet scars. The surface of the drug is a dirty light brown, and it possesses a distinctive odor and a pungent, bitter taste that induces numbness on the tongue. This detailed description provides insight into the botanical characteristics and commercial aspects of *Piper longum* [47].

Distribution

Piper longum, or long pepper, thrives in the evergreen forests of India and is cultivated in specific regions such as Assam, Tamil Nadu, and Andhra Pradesh. This plant is notably grown on a large scale, particularly in areas with limestone soil and heavy rainfall, where the relative humidity is consistently high. Cultivating long pepper in these conditions allows it to flourish and contribute to producing this valuable spice with its unique characteristics and uses [53, 54, 59].

Ethnobotanical Uses

The fruits of *Piper longum* contain the alkaloid *Piperine*, which imparts the enhanced pungency. It is applied topically, to soothe and alleviate muscular pains and inflammation. The infusion of *P. longum* root is traditionally used during postpartum to induce the expulsion of the placenta. The entire plant and its various parts, especially the fruit, are used in traditional practices, comprehensive information on its usage is not extensively compiled. Traditional applications include its use in managing diseases such as cancer, inflammation, depression, diabetes, obesity, and hepatotoxicity, making it a versatile and accessible remedy in traditional medicine [60].

Scientific Validation and Economic Uses

The *P. longum* plant is utilized as a spice and medicine in India and other countries for centuries. In Ayurveda medicinal system, there are about 135 Ayurvedic formulations, where it is used as a major ingredient and rejuvenator with various therapeutic properties. It is used for stimulating the appetite and relieving gas from the intestines. In the traditional medicinal system, the plant is used to treat numerous diseases, such as epilepsy, pleural effusion, spleen disorders, dementia, diarrhea, dysentery, insomnia, antivenom, etc. There were also available reports in which it has pharmacological activities such as antiulcer, anti-inflammatory, anticancer, neuroprotective, and others [61].

Piper attenuatum Buch.-Ham. ex Miq

Piper attenuatum (Fig. 1a) is an Indian medicinal plant belonging to the family *Piperaceae*. It is also known by various synonyms such as *Chavica diffusa* (Vahl) Miq., *Piper malamiris* Roxb., *Piper karok* Blume, and *Piper diffusum*. In English, it is referred to as flat-branched pepper or oval-leaved pepper plant. In Malayalam, it is known as Kattumulaka. In Tamil, it is called arenukam and kattu-milaku. These names reflect the plant's recognition and use in different regions and cultures [62].

Description

The described plant is a shrubby root-climber, reaching up to 2.5 m in length. Its stems and branches are robust but become compacted and notched when dry, displaying a greenish-yellow color and a glabrous surface. The lower internodes measure 8.0–8.5 cm, while the upper ones are 2.0–3.5 cm long. Stipules are subulate and measure 0.5–0.65 cm. The leaves are broadly cordate, with the lower ones measuring 8.0–9.5 × 7.0–10 cm and slightly wider than long, orbicular-ovate. The upper leaves are 4.5–7.0 × 2.5–5.0 cm, entire, pellucid, tapering with a mucro, with a cordate-truncate base. The leaves are membranous, relatively glabrous above and hairy on veins. There are generally 7 main nerves, sometimes 7–9 from the base, with laterals disparate except for the central ones. Petioles are 2.5–6.5 cm long, angled, greenish-yellow, and hairy upwards. The plant is dioecious, with male spikes about 0.03 cm long, clustered and stamens numbering 3, sessile, adnate to elongated basifixed bracts (1.0–0.2 cm), winged and decurrent (4.0–5.0 cm long) on the rachis. Female spikes (up to 10 cm long) have adnate bracts bearing their ends on ovoid ovaries (0.05 cm long) with an obscurely 4-stigma structure. The drupes are 0.2 × 0.15 cm, globose, loosely aggregated, glabrous, and sessile and measure 0.35–0.4 cm across, appearing black. Flowering and fruiting occur from August to March [63].

Distribution

It is native to the states of Andhra Pradesh, Tamil Nadu, and Telangana. It is used as a substitute for black pepper.

Ethnobotanical Uses

Distinct parts of *P. attenuatum* have been traditionally employed in herbal medicine for various therapeutic purposes. The plant has found use in treating conditions such as migraines and muscular pain and has been applied as a rubefacient, providing relief for localized discomfort. The wood of the plant has been utilized to address throat pain, showcasing its diverse medicinal applications. Additionally, the root

Fig. 1 Underutilized wild species of *Piper*. (**a**) *Piper nigrum*; (**b**) *P. longum*; (**c**) *P. attenuatum*; (**d**) *P. arnottianum*; (**e**) *P. mullesua*; (**f**) *P. hamiltonii*; (**g**) *P. galeatum*; (**h**) *Piper betle*; (**i**) *P. rushagandham*; (**j**) *P. hymenophyllum*; (**k**) *P. schmidtii*; (**l**) *P. wightii*

part of *P. attenuatum* has demonstrated diuretic activity. The leaves of *P. attenuatum* have been harnessed for their wound-healing properties, indicating their potential to support the body's natural recovery processes. Overall, the multifaceted uses of different plant parts underscore *P. attenuatum* significance in traditional medicine and its potential to address various health conditions [63].

Scientific Validation and Economic Uses

Phytochemical analyses of *P. attenuatum* have revealed the presence of various bioactive compounds. Pipoxide chlorohydrins are identified as a major component, contributing significantly to the plant's chemical composition. Additionally, the leaves of *P. attenuatum* contain (−)-galbelgin, aliphatic liquor, and 8-hentriacontanol. Aristolactams, including cepharanone B, Piperolactam A, Piperolactam D, and cephradine A, have been reported in the aerial parts of the plant. The roots of *P. attenuatum* have been found to contain alkamides, with identified compounds such as *Piper*ine, Piperlonguminine and guineensine. In the petroleum ether extract of *P. attenuatum*, a novel chain alcohol named 14-benzo [1, 3] dioxol-5-yl-tetradecan-2-ol has been identified, along with beta-sitosterol, kadsurin A, kadsurin B and (+)-crotepoxide. Additionally, the leaf oil of *P. attenuatum* comprises β-caryophyllene and β-cubebene. These diverse phytoconstituents contribute to the plant's pharmacological properties and may explain its traditional uses in herbal medicine for various conditions. The entire plant is reported to contain a unique phytoconstituent called crotepoxide. This compound has shown significant antitumor activity against Lewis lung carcinoma cell lines, suggesting a potential role in cancer treatment. The presence of compounds with reported antitumor activity further highlights the potential therapeutic benefits of *P. attenuatum*. In conclusion, *P. attenuatum* stands out as an important medicinal plant in India, offering potential therapeutic benefits due to its diverse phytoconstituents [64].

Piper arnottianum (Miq.) C. DC

Piper arnottianum (Fig. 1b) has other synonyms, which were widely known as *Chavica arnottiana* Miq.; *Piper hapnium* Buch.-Ham. ex Hook. f.; *Piper siriboa* Herb. Madr. ex Itiner. This population was found to be trailing on *Ailantus* sp., *Aporosa lindleyana,* and various other tree species. Notably, male plants of *P. arnottianum* were observed newly, although they were relatively scarce in comparison to the female plants.

Description

The plant is a climbing shrub characterized by warty stems. Its leaves are alternate, measuring 6–12 × 3–6 cm and have an ovate shape. The base of the leaves is cordate and the apex is either acute or acuminate. The upper side of the leaves is glabrous, while the underside is puberulous along the nerves. The leaves have a basal count of 5–7 nerves, with petioles extending up to 3 cm in length. Leaves on flowering shoots are small and oblique at the base. The flowers are unisexual, arranged in erect cylindrical spikes with a yellow color. The peduncle supporting the spikes can reach up to 1.5 cm in length. Male spikes are longer than female ones. The male flowers have two stamens, while the female flowers have a globose ovary and 3–5 stigmas. The fruiting spikes measure 4–6 cm in length, contributing to the overall morphology of this climbing shrub [57].

Distribution

Piper arnottianum, a species of *Piper*, is classified as an endangered species. Its occurrence is primarily documented in the low-lying hills of the Western Ghats. Found in moist deciduous forests and is Endemic to Southern Western Ghats.

Ethnobotanical Uses

Not much evaluated and the fruiting spike of this plant is not acknowledged in the market in its current state due to the inadequate availability of this plant.

Scientific Validation and Economic Uses

The phytochemical and pharmacognostic characteristics of the fruiting spike of *Piper arnottianum*, which bears a resemblance to the long pepper and is used as an adulterant, have been examined for the first time. Through analysis, it was found that the spikes contain 6.716% of a brown aromatic volatile oil with apiole as the main constituent along with other sesquiterpenes such as β-bisabolene, (E) β-farnesene, β-caryophyllene oxide, nerolidol, and (Z) β-farnesene, among others. Furthermore, the fruits also contain Piperine, acacetin, and phenolic acids. The key pharmacognostic characteristics include large polygonal starch grains, rectangular oil cells, uniseriate filaments, and sclereids. The presence of apiole as the primary component of the oil, along with the aforementioned pharmacognostic characteristics and the absence of Piperlongumine, serves to differentiate the fruiting spike of *P. arnottianum* from that of *P. longum* [65, 66].

Piper mullesua Buch.-Ham. ex D. Don

Piper mullesa (Fig. 1c) is known as Kattuthipili in common in Tamil Nadu and Namar in Arunachal Pradesh. The synonym of this plant is *Piper mullesuanum* (Buch.-Ham.) St. Lag. They are commonly called Hill Pepper. They are mostly distributed in the Evergreen Shola forests of the Western Ghats of south India and endemic to Peninsular and North East India [67].

Description

The habit of this plant is slender climbers. Leaves are elliptic, acuminate at apex, acute at base, 9 × 4 cm in length, 3–5 ribbed at base, secondary nerves are prominent, glabrous. The flowers were called spikes. Female spike 7 × 4 mm, erect, oblong; peduncle 3 mm long; male spike 4 cm long, erect, slender; bracts peltate, orbicular; stamens 3. Fruits are Berries yellow-red in color. Flowering and fruiting is mostly seen in the month of January which continues till April.

Distribution

Piper mullesua is found in the shola forest of the Nilgiris, Coimbatore, Dindigul, Theni, and Tirunelveli. Endemic to Southern Western Ghats.

Ethnobotanical Uses

In Chinese folk medicine, the plant, known as Duan-Jv or *Piper mullesua*, is utilized to address a variety of health issues. Its applications include treating bleeding, bone fractures, injuries from falls, rheumatoid arthritis, rheumatic arthralgia, acroanesthesia, asthma, colds, stomach aches, abdominal pain, toothaches, swelling, and pain of furuncles, dysmenorrhea, menoxenia, pyrosis, as well as snake and insect bites. The whole plant paste is applied topically to cure various diseases. It is notable for curing cold.

Scientific Validation and Economic Uses

Various compounds have been discovered in *Piper mullesua*, including amides like retrofractamide A, chingchengenamide A, N-isobutyl-16-phenylhexadeca-2E,4E-dienamide, and N-isobutyldeca-2E,4E-dienamide. Lignans, such as (−)-nectandrin A, nectandrin B, galgravin, asarinin, fargesin, and sesamin, have also been found, showing antifeedant properties. Moreover, myristicin, a phenylpropanoid, has been isolated for its insecticidal effects. Several arylalkenyl carboxylic acid esters have been identified. Cubebene, (Z)-asarone, bicyclogermacrene, dillapiole, apiole, beta-elemene, sarisan, (Z) bisabolene, (Z,Z) farnesene, and epi-bisabolol were the compounds isolated from the selected plant with their potent pharmaceutical evaluation against various diseases like anticancer, antifeedant, etc [67, 68]

Piper hamiltonii C. DC

Piper hamiltonii is commonly found thriving within shola forests, typically climbing upon trees such as *Cinnamomum wightii* Meissner, *Litsea floribunda* (Blume) Gamble, *Neolitsea foliosa* (Nees) Gamble, *Meliosma simplicifolia* (Roxb.) Walp., and *Phoebe wightii* Meissner. These forests are typically situated at altitudes ranging from 1800 to 2200 m above sea level. It is often confused with *P. nigrum* [69].

Description

P. hamiltonii is a perennial, dioecious climbing shrub with subterete branches bearing swollen nodes with climbing roots. Its leaves are simple, alternate, broadly elliptic to ovate, with dimensions ranging from 8 to 14.5 cm in length and 5.3 to 10.4 cm in width. The leaf surface is bullate, pale when dry and coriaceous, with a glabrous texture on both sides. The leaf base is unequal or nearly equal, with recurved margins and an attenuate or shortly caudate-acuminate apex. The petioles are 1.2 to 2.8 cm long and grooved, while the lanceolate stipules measure 0.8 to 1 cm in length and are cuspidate at the apex, eventually falling off. Its inflorescences consist of leaf-opposing spikes that droop, with slender peduncles measuring 1.3 to 3.5 cm long. The bracts are peltate and orbicular, ranging from 4 to 5 cm in diameter and 2 to 2.2 cm in width, with a hyaline margin. The male spike is slender with a glabrous rachis, ranging from 5 to 12.5 cm long, bearing 3 to 4 stamens with 2-locular anthers that dehisce longitudinally. The female spike measures 4 to 9.5 cm long, with an ovoid ovary containing a single locule and solitary ovule and three cylindrical, reflexed stigmas. The fruiting spike is loose and interrupted, with a curved tip and the persistent bracts almost form a cup when mature, appearing papery. The drupe is ovoid, measuring 0.5 to 0.7 cm in length and 0.3 to 0.4 cm in width, loosely arranged with the persistent stigma, and turns yellowish or red when mature [69].

Distribution

P. hamiltonii is distributed in Sikkim, Terai, West Bengal, Khasi hills of North Eastern India, South Andaman Islands and Western Ghats—Kerala and Tamil Nadu [69].

Ethnobotanical Uses

Used as diuretic and carminative properties [69].

Scientific Validation and Economic Uses

As per the survey of review of literature, no reports on the scientific validation or economic uses were found.

Piper galeatum (Miq.) C. DC

Piper galeatum (Miq.) C. DC. is commonly called helmet pepper. The synonyms of this plant are *Mulder agaleata* Miq. and *Mulder awightiana* Miq.

Description

Piper galeatum are stout climbing shrubs with glabrous characteristics. Their leaves, measuring between 10 and 16 cm in length and 3 and 6 cm in width, are elliptic in shape and feature acuminate tips and acute bases. They are typically five-ribbed with a petiole ranging from 1 to 1.5 cm in length. The spikes can extend up to 9 cm in length and are leaf-opposed, drooping, and slender, with the male spike typically longer and slimmer. The peduncle measures approximately 1 cm in length, appearing reddish and glabrous while the bracts form a fleshy globose cup. Flowers are distant, with stamens numbering 2. The berries are globose, measuring around 8 × 8 mm, smooth in texture, and yellow in color [70].

Distribution

Global distribution and they are endemic to Southern Western Ghats. They are distributed in Kerala, Idukki, Kollam, Palakkad, and Thiruvananthapuram and Tamil Nadu: Coimbatore, Tirunelveli, and Theni [70].

Ethnobotanical Uses

As per the survey of review of literature, no reports on ethnobotanical uses were found.

Scientific Validation and Economic Uses

Phytochemical analysis of its fruits extract has revealed the presence of ten sesquiterpenes and monoterpenes, including β-elemene, δ-elemene, α-humulene, β-caryophyllene, α-copaene, α-ionone, 10-(acetylmethyl)-3-carene, dihydrocarvyl acetate, 1-p-menthen-8-yl acetate, and linalyl acetate. A novel amide, 1-(3′-hydroxy-5′-methoxycinnamoyl)-piperidine, is also identified. There are no report available regarding the economic uses of this plant [70].

Piper betle L.

Piper betle L. (Fig. 1h), also known as *Piper betel* Blanco, is a perennial creeping plant. Native to central and eastern Peninsular Malaysia, it has spread to East Africa and various tropical regions of Asia. This plant holds significant commercial value and is cultivated extensively in countries such as India, Bangladesh, Sri Lanka, Thailand, Taiwan, Malaysia, and several other Southeast Asian Nations. It is referred to as the "green gold of India," and it sustains the livelihoods of nearly 20 million individuals engaged in its cultivation, transportation, handling, processing, and preparation of betel leaves [71].

Description

P. betle is a dioecious root climber capable of reaching heights between 3 and 10 m, depending on available climbing support. It features lateral branches distributed along its entire length, typically extending a few feet from the ground. Its stems are swollen and articulate, exhibiting dichotomous branching and rooting at the nodes. The stout stems are nearly terete, slightly flattened and initially light green with short raised whitish streaks and pinkish stripes along the nodes. Internodes typically measure around 12 cm in length and 1.2 cm in diameter. The leaves are characterized by a simple blade, arranged alternately and spirally, lacking stipules. Petioles are pubescent and channeled and measure 2–5 mm in length. Leaf blades are glabrous, coriaceous, and fleshy and range from greenish to yellowish, exhibiting a shiny appearance. They are broadly ovate, with a width of 7–8.5 cm and a length of 9–11 cm, possessing a cordate base and an acuminate apex. Leaf margins are entire and narrowly recurved, with reticulate venation and 7–9 veins arranged in two or three pairs originating from the midrib, with one pair elevated from the base. The inflorescence consists of axillary spikes reaching up to 5.5 cm in length. Male inflorescences form cylindrical, pendulous catkins measuring 10 cm in length and 2 cm in diameter. Female spikes are also cylindrical and pendulous, ranging from 2.5 to 4 cm in length and 0.5 cm in diameter. Individual flowers are very minute and unisexual, reduced, with each consisting of a couple of stamens and stigmas inserted into the axil of each bract. The bracts are orbicular and peltate, arranged densely in a spiral series. The mature inflorescence emits a strong aromatic scent. Fruiting spikes measure 3–5 cm in length and are orange in color and droop, embedded on the rachis of the mature inflorescence [57].

Distribution

It is worldwide in distribution.

Ethnobotanical Uses

Betel leaf, historically referenced in ancient Vedic literature, is a cornerstone of traditional medicine across various cultures. Its applications span a wide array of health concerns, including digestive issues, oral hygiene, respiratory ailments, and even spiritual ceremonies. In Ayurvedic and Unani practices, betel leaf is esteemed for its anthelmintic, aphrodisiac, and cardiac tonic properties, among others. The practice of chewing betel quid, a preparation of betel leaf, areca nut, lime and flavorings, is prevalent among millions in South Asia and the Pacific region. Additionally, betel leaf holds cultural significance in ceremonies and festivals such as Thai rituals and Laleng community gatherings. Communities in Nepal and Southeast Asia also incorporate betel leaf into their traditional remedies for various ailments and as a means of social and cultural expression. Overall, the ethnomedicinal uses of betel leaf and its enduring importance in traditional healthcare practices are well documented [71, 72].

Scientific Validation and Economic Uses

Piper betle, extensively studied for its phytochemical composition, harbors a diverse array of biologically active compounds. Phenols within the leaves contribute to their characteristic pungent aroma. Phytochemical analyses have identified alkaloids, flavonoids, tannins, sterols, phenols, glycosides, saponins, and terpenoids in aqueous and methanol extracts. Recent investigations also revealed starch, diastases, sugars, and essential oils, with safrole, chavibetol acetate, and allyl pyrocatechol diacetate being prominent constituents. Betel leaf oil is rich in safrole and eugenol. Hydroxychavicol is the most important bioactive compound found in this plant. They are antiproliferative, anticancerous, neuropharmacological, analgesic, antioxidant, antiulcerogenic, hepatoprotective, antifertility, antibacterial, antifungal, and many more. Nowadays, immense attention has been paid to nano-formulations and their applications [71–73].

Piper rukshagandhum J. Mathew

Piper rukshagandhum J. Mathew is a new species discovered in Achankovil forest, Kerala section of the Southern Western Ghats, India. The name "rukshagandhum" is derived from the Sanskrit word "rukshagandha," which means pungency. The fruits of this plant were very pungent in nature. The conservation status of this plant according to IUCN comes under the "critically endangered" category [74].

Description

Branched evergreen woody climber, dioecious, cylindrical stem, smooth, rooting at nodes; nodes swollen, young nodes pinkish in hue; internodes 1–7 cm long, slender, smooth, green, and slightly angled in youth, turning brown and woody with lenticels upon maturity. Leaves are arranged alternately; petioles 2–3 cm long; leaf blades 8–16 cm long, 6–9 cm wide, broadly ovate, leathery and smooth on both surfaces, tapering to a long-pointed tip, base asymmetrical, leathery, margin smooth, veins arranged in a pinnate pattern with 2–3 prominent veins, all originating from the base or near it. The inflorescence is spikelike, positioned opposite to leaves, solitary, and hanging. Male inflorescence is 15 cm long, consisting of 100–120 flowers, tinted blush red/pink; peduncle 1.5–2.5 cm long, smooth; receptacle attached directly, round, smooth, aligned with the axis, curved back; bracts fused to the axis, bent back, forming a fleshy cup with a narrow slit-like opening, smooth. Stamens 2, short, measuring 0.3–0.5 mm; filaments thick, smooth; anthers bilocular, opening through a longitudinal slit at the tip. Female inflorescence is 3–7 cm long, with 16–22 flowers, blush red/pink in color; peduncle 1.5–2.5 cm long, smooth; receptacle sessile, globose, smooth and parallel to the axis, recurved; bracts attached to the axis, curved back, transformed into a fleshy cup with a narrow slit-like opening, smooth. Carpel solitary; ovary round, attached directly; stigmas 4, white, sessile. Fruit is drupe, measuring 6–8 mm in diameter, round, attached directly, initially green with a purple tip, ripening to red. Contains a single seed [74].

Distribution

Piper rukshagandhum is a climbing plant found along the edges of river valleys within the Agasthyamala Biosphere Reserve. It typically utilizes trees such as *Elaeocarpus variabilis* Zmarzty and *Chionanthus mala-elengi* (Dennst.) P.S.Green subsp. *mala-elengi* as support structures, both of which are native to peninsular India. It shares its habitat with species like *Oxyceros rugulosus* (Thwaites) Tirveng., *Eugenia mooniana* Wight, and *Calamus gamblei* Becc. While *Calamus gamblei* is endemic to the Western Ghats, the latter two are species with distributions extending into the Indo-Sri Lankan region [74].

Piper rukshagandhum has not been thoroughly investigated and documented for its potential medicinal applications, and conservation is necessary as this plant is newly discovered.

Piper hymenophyllum Miq

Piper hymenophyllum (Fig. 1e) has a synonym of *Piper nilghirianum* C. DC., and it is commonly known as "umbrella pepper." It is found in semievergreen and evergreen forests.

Description

Slender climber with ovate leaves measuring up to 13 × 6 cm, rounded to subcordate at the base, often with uneven sides and acuminate at the apex. The leaves are typically seven-ribbed and membranous, with a petiole about 1 cm long. Its inflorescence is in the form of a slender, drooping spike, approximately 20 cm long, with the male spike being shorter. The bracts are oblong and hairy, adnate to the rachis with free margins. Male flowers usually have two or three stamens. The berry is about 5 mm long, ovoid, and beaked, with a three-lobed stigma [75].

Distribution

It is global in distribution. This plant species is found in various parts of Karnataka, including Chikmagalur, Hassan, Mysore, and S. Kanara. Similarly, it is also present in several districts of Kerala, such as Idukki, Kannur, Kollam, Malappuram, Palakkad, Pathanamthitta, and Thiruvananthapuram. In Tamil Nadu, *Piper hymenophyllum* can be found in Coimbatore, Dindigul, Madurai, Namakkal, Nilgiri, Theni, Tirunelveli, Tiruchirappalli, and Virudhunagar districts. This distribution indicates the wide range of habitats, where this species can thrive within the southern regions of India [75].

Ethnobotanical Uses

It is used by Kani tribals for gastric problems [75].

Scientific Validation and Economic Uses

The fruits of *Piper hymenophyllum* have been analyzed, using GCMS analysis revealing the presence of 15 compounds in the oil, constituting 98.65% of its composition. The compounds such as phytol are followed by dihydro terpineol and α-terpineol. Other significant compounds including trans-piperitol, endo-fenchol at, camphene at, and γ-terpinen were found in this plant. It possesses strong antioxidant, antibacterial, antifungal, antipyretic, antiproliferative, anti-inflammatory, and neuropharmacological activities [75].

Piper schmidtii Hook. Fill

Piper. schmidtii Hook. Fill is found in the shola forest, and they are endemic to the Southern Western Ghats.

Description

It is a climbing shrub where the stem is up to 1.5 cm thick and alternate leaves, often asymmetric, with dark green glossy upper sides and very pale to white lower sides. The leaves are ovate-elliptic and coriaceous, typically with five main veins and prominent reticulate venation. Male spikes, ranging from 1 to 3, measure 2.5 to 7 cm with pendent stalks and circular bracts. Stamens usually occur in pairs, characterized by thick filaments and anthers opening via vertical cross slits. Female spikes, 5–7 cm in length, elongate and may twist spirally when in fruit. Bracts of female spikes are square or roundish with raised margins and are free all around. The berries of this plant are typically red or yellow.

Distribution

This plant species is found in various regions of South India. In Karnataka, it can be spotted in Mysore. In Kerala, it is present in districts like Idukki, Kollam, Kottayam, and Palakkad. In Tamil Nadu, it is observed in the Nilgiri region [76].

Ethnobotanical Uses

P. schmidtii Hook. Fill is ethnobotanically reported to cure various diseases such as toothaches, headaches, stomach aches, cough, rheumatism, colds, cough, and bronchitis-related problems [76].

Scientific Validation and Economic Uses

This plant possesses antioxidant, antimicrobial and anti-inflammatory properties, which could serve as an excellent anti-inflammatory agent analogously reducing oxidative stress [76].

Piper wightii Miq

Piper wightii Miq (Fig. 1f) has a synonym *Piper mullesuanum* (Buch.-Ham.) St. Lag. The vernacular name is Kattukurumulagu. It is commonly found in shola forest.

Description

It is a robust, smooth climber that features ovate leaves measuring 14–17 cm in length and 6–8 cm in width. The leaves are acuminate at the apex and round to subcordate at the base, with 3–5 prominent ribs. They are leathery in texture, with the midrib bearing 2–3 lateral nerves similar to the ribs. The petiole is approximately 2 cm long. The stout spike, about 8 cm long, hangs droopingly and is supported by a 2 cm long peduncle. The bracts are curved and hooded on one side. Flowers are distantly arranged and stamens are typically in pairs. The berry is reddish, approximately 7 mm across, globose, and smooth, with three or four stigmatic lobes.

Distribution

This plant species is endemic to the Southern Western Ghats and can be found in various regions of Tamil Nadu, including the Nilgiris, Coimbatore, Dindigul, Theni, and Tirunelveli districts [67].

Ethnobotanical Uses

This plant is used to cure various diseases, such as toothaches, headaches, stomach aches, cough, rheumatism, colds, cough, and bronchitis problems [67].

Scientific Validation and Economic Uses

The *P. wightii* fruit ethyl acetate extracts were found to possess significant anticancer activity against DLA-induced solid tumors. In vivo, antioxidant study showed that ethyl acetate extract of fruits of *P. wightii* inhibits the oxidation and maintains the activity of antioxidant enzymes such as superoxide dismutase (SOD), catalase (CAT), glutathione peroxidase (GPx), lipid peroxidation (LPO), and reduced (GSH) glutathione to the normal level in the liver [67].

Conclusion

This chapter comprises a comprehensive exploration into various aspects of commercially exploited and unexplored wild pepper species that are endemic to the Western Ghats providing valuable insights for their intensified utilization. Through detailed discussions of their botanical description, geographical distribution, and ethnobotanical and scientific validation, this chapter sheds light on the potential for enhancing the cultivation and conservation efforts of these valuable species. By understanding their unique traits, habitat requirements, and economic importance, stakeholders can develop strategies to promote sustainable cultivation, preserve biodiversity, and harness the untapped potential of wild pepper species for various purposes, including culinary, medicinal, and ecological. With a focus on its potentialities and availabilities, this chapter serves as a valuable resource for researchers, practitioners, and policymakers seeking to maximize the benefits derived from wild pepper species while ensuring their long-term viability and ecological resilience.

References

1. Patwardhan B, Vaidya ADB, Chorghade M. Ayurveda and natural products drug discovery. Curr Sci. 2004;86(6):789–99.
2. Fabricant DS, Farnsworth NR. The value of plants used in traditional medicine for drug discovery. Environ Health Perspect. 2001;109(Suppl 1):69–75. https://doi.org/10.1289/ehp.01109s169.
3. Newman DJ, Cragg GM, Snader KM. Natural products as sources of new drugs over the period 1981–2002. J Nat Prod. 2003;66(7):1022–37. https://doi.org/10.1021/np030096l.
4. Newman DJ, Cragg GM. Natural Products as Sources of New Drugs from 1981 to 2014. J Nat Prod. 2016;79(3):629–61. https://doi.org/10.1021/acs.jnatprod.5b01055.
5. Rates SMK. Plants as source of drugs. Toxicon. 2001;39(5):603–13. https://doi.org/10.1016/S0041-0101(00)00154-9.
6. Tilburt JC, Kaptchuk TJ. Herbal medicine research and global health: an ethical analysis. Bull World Health Organ. 2008;86(8):594–9. https://doi.org/10.2471/BLT.07.042820.
7. Heinrich M, Gibbons S. Ethnopharmacology in drug discovery: an analysis of its role and potential contribution. J Pharm Pharmacol. 2001;53(4):425–32. https://doi.org/10.1211/0022357011775712.
8. Cragg GM, Newman DJ. Plants as a source of anti-cancer agents. J Ethnopharmacol. 2005;100(1–2):72–9. https://doi.org/10.1016/j.jep.2005.05.011.
9. Khan MS, Ahmad I, Aqil F. Ongoing and emerging avenues for naturally based antimicrobials and resistance-modifying agents. J Clin Microbiol. 2009;47(3):751–2. https://doi.org/10.1128/JCM.01622-08.
10. Ventola CL. The antibiotic resistance crisis: part 1: causes and threats. P T. 2015;40(4):277–83.
11. Harvey AL, Edrada-Ebel R, Quinn RJ. The re-emergence of natural products for drug discovery in the genomics era. Nat Rev Drug Discov. 2015;14(2):111–29. https://doi.org/10.1038/nrd4510.
12. Newman DJ, Cragg GM. Natural products as sources of new drugs over the nearly four decades from 01/1981 to 09/2019. J Nat Prod. 2020;83(3):770–803. https://doi.org/10.1021/acs.jnatprod.9b01285.
13. Cowan MM. Plant products as antimicrobial agents. Clin Microbiol Rev. 1999;12(4):564–82. https://doi.org/10.1128/CMR.12.4.564.
14. Pieroni A, Quave CL. Wild food plants: folk traditions of use and collection in the Mediterranean region. In: Ethnobotany and biocultural diversities in the balkans. Springer, New York 2014. p. 41–56. doi:https://doi.org/10.1007/978-1-4939-1492-0_4.
15. Balick MJ, Cox PA. Plants, people, and culture: the science of ethnobotany. New York: Scientific American Library; 1996.
16. Jaramillo MA, Manos PS. Phylogeny and patterns of floral diversity in the genus Piper (Piperaceae). Am J Bot. 2001;88(4):706–16. https://doi.org/10.2307/2657078.
17. Quijano-Abril MA, Callejas-Posada R, Miranda-Esquivel DR. Areas of endemism and distribution patterns for Neotropical Piper species (Piperaceae). J Biogeogr. 2006;33(7):1266–78. https://doi.org/10.1111/j.1365-2699.2006.01489.x.
18. Smith JF, Stevens AC, Tepe EJ, Davidson C. Placing the origin of two species-rich genera in the late cretaceous with later species divergence in the tertiary: a phylogenetic, biogeographic and molecular dating analysis of Piper and Peperomia (Piperaceae). Syst Biol. 2008;57(4):635–56. https://doi.org/10.1080/10635150802302401.
19. Bornstein AJ. A monograph of the Cuban species of Piper (Piperaceae). Ann Mo Bot Gard. 1989;76(3):1046–64. https://doi.org/10.2307/2399713.
20. Dyer LA, Palmer ADN. Piper: a model genus for studies of phytochemistry, ecology, and evolution. New York: Springer Science & Business Media; 2004. doi:https://doi.org/10.1007/978-0-306-48587-0.
21. Callejas R. A new species of Piper from Colombia and a key to the species of the Choco biogeographic region. Brittonia. 2001;53(4):553–60. https://doi.org/10.1007/BF02812712.

22. Parmar VS, Jain SC, Bisht KS, Jain R, Taneja P, Jha A, et al. Phytochemistry of the genus Piper. Phytochemistry. 1997;46(4):597–673. https://doi.org/10.1016/S0031-9422(97)00328-2.
23. Duraisamy S, Krishnasamy K, Abdul RK. Piper species: A comprehensive review on chemical composition, traditional uses, and pharmacological importance. J Ethnopharmacol. 2021;281:114495. https://doi.org/10.1016/j.jep.2021.114495.
24. Koul S, Koul JL, Taneja SC, Dhar KL, Atal CK. Alkaloids of Piper longum from the stem bark. Phytochemistry. 1990;29(12):3915–7. https://doi.org/10.1016/0031-9422(90)85352-B.
25. Ravindran PN, Babu KN. Black Pepper (*Piper nigrum*). CRC Press; 2004.
26. Prabhakaran P. Vegetative propagation of pepper (Piper nigrum L.). Indian J Arecanut Spices Med Plants. 2005;7(2):65–9.
27. Bezerra DP, Pessoa C, de Moraes MO, Saker-Neto N, Silveira ER, Costa-Lotufo LV. Overview of the therapeutic potential of piplartine (piperlongumine). Eur J Pharm Sci. 2013;48(4–5):453–63. https://doi.org/10.1016/j.ejps.2012.12.004.
28. Raj L, Ide T, Gurkar AU, Foley M, Schenone M, Li X, et al. Selective killing of cancer cells by a small molecule targeting the stress response to ROS. Nature. 2011;475(7355):231–4. https://doi.org/10.1038/nature10167.
29. Okonko IO, Soleye FA, Amusan TA, Ogun AA, Ogunnusi TA, Ejembi J, et al. Prevalence of schistosomiasis among children in some rural communities of abeokuta, nigeria: a parasitology and health education prospective study. J Rural Trop Public Health. 2009;8:79–84.
30. Melo MGD, Araújo ECC, Carvalho KMB, et al. Antileishmanial activity of the essential oil from *Piper tuberculatum* Jacq. (Piperaceae). J Ethnopharmacol. 2014;155(2):1105–12. https://doi.org/10.1016/j.jep.2014.06.036.
31. de Lima Júnior RCP, de Lavor ÉM, Fernandes AWC, et al. Antiplasmodial activity of piplartine: a promising alkaloid against malaria. Malar J. 2015;14:219. https://doi.org/10.1186/s12936-015-0737-7.
32. Han X, Zhang W, Yang X, Wheeler CG. Piperlongumine inhibits inflammatory cytokine production in LPS-activated microglia. J Neuroinflammation. 2017;14(1):7. https://doi.org/10.1186/s12974-016-0770-1.
33. Liu Z, Li M, Xiong Q, Zhao D. Piperlongumine attenuates LPS-induced neuroinflammation via the NF-κB pathway in a mouse model of Parkinson's disease. Neurochem Res. 2019;44(5):1123–32. https://doi.org/10.1007/s11064-019-02766-w.
34. Lee JY, Kim YK, Seo MK, et al. Piperlongumine inhibits LPS-induced neuroinflammation and improves neuronal survival in primary microglial cells and organotypic hippocampal slice culture. Biomol Ther (Seoul). 2018;26(2):210–7. https://doi.org/10.4062/biomolther.2017.196.
35. Hu Z, Zhang Y, Li Z, Yu Y, Kang W, Han Y, et al. Effect of Helicobacter pylori infection on chronic periodontitis by the change of microecology and inflammation. Oncotarget. 2016;7(41):66700–12. https://doi.org/10.18632/oncotarget.11449. PMID: 27602578; PMCID: PMC5341831.
36. Reddy GD, Jamil K. Diuretic activity of leaves of *Chenopodium album* Linn. in rats. Indian J Nat Prod Resour. 2006;2(3):94–6.
37. Ravindran PN, Nirmal Babu K, Shiva KN. Botany and crop improvement of black pepper. In: Ravindran PN, Nirmal Babu K, editors. Black pepper (Piper nigrum). CRC Press; 2003. p. 23–142.
38. Myers N, Mittermeier RA, Mittermeier CG, da Fonseca GA, Kent J. Biodiversity hotspots for conservation priorities. Nature. 2000;403(6772):853–8. https://doi.org/10.1038/35002501.
39. Champion HG, Seth SK. A revised survey of the forest types of India. Government of India Press; 1968.
40. Bijoy G, Vinod KV. Diversity of Piper species in India and its conservation strategies. Indian J Plant Genet Resour. 2009;22(1):1–14.
41. Renner SS, Ricklefs RE. Dioecy and its correlates in the flowering plants. Am J Bot. 1995;82(5):596–606. https://doi.org/10.2307/2445416.
42. Hooker JD. The Flora of British India. 5London: L. Reeve & Co.; 1886.
43. Candolle ACP de. Piperaceae in prodromus systematis naturalis regni vegetabilis. Paris: Treuttel and Wurtz; 1912.

44. Raven PH, Axelrod DI. Angiosperm biogeography and past continental movements. Ann Mo Bot Gard. 1974;61(3):539–673. https://doi.org/10.2307/2395021.
45. Kumar S, Malhotra S, Prasad AK, Van der Eycken EV, Bracke ME, Stetler-Stevenson WG, Parmar VS, Ghosh B. Anti-inflammatory and antioxidant properties of Piper species: a perspective from screening to molecular mechanisms. Curr Top Med Chem. 2015;15(9):886–93. https://doi.org/10.2174/1568026615666150319112438.
46. Sen S, Chakraborty R. The use of spices in ancient India. J Ethnobiol Ethnomed. 2017;13(1):1–12.
47. Gamble JS. The flora of the presidency of Madras. London: Adlard & Son; 1921. p. 136.
48. Ma Z, Zhang Y. Antioxidant properties and pharmacological potential of P. nigrum. Mol Nutr Food Res. 2011.
49. Takooree H, Aumeeruddy MZ, Rengasamy KRR, Venugopala KN, Jeewon R, Zengin G, et al. A systematic review on black pepper (Piper nigrum L.): from folk uses to pharmacological applications. Crit Rev Food Sci Nutr. 2019;59(1):210–43.
50. Beek TA. Chemical studies of *Piper nigrum* L. fruits. Nat Prod Rep; 2002.
51. Srinivasan K. Black pepper and its pungent principle-piperine: a review of diverse physiological effects. Crit Rev Food Sci Nutr. 2007.
52. Dalby A. Dangerous tastes: The story of spices. University of California Press. 2000.
53. Majumdar B. Medicinal plants in India: Importance and cultivation. New India Publishing Agency. 2013.
54. Khare CP. Indian medicinal plants: an illustrated dictionary. Springer; 2007.
55. Nadkarni KM. Indian Materia Medica. Popular Prakashan; 1976.
56. Chopra RN, Nayar SL, Chopra IC. Glossary of Indian medicinal plants. CSIR; 1956.
57. Sastri BN. The wealth of India: Raw materials. CSIR: Publications and Information Directorate; 1962.
58. Panda H. Handbook on ayurvedic medicines with formulae, processes and their uses. Asia Pacific Business Press Inc; 2004.
59. Biswas P, Ghorai M, Mishra T, Gopalakrishnan AV, Roy D, Mane AB, et al. Piper longum L.: a comprehensive review on traditional uses, phytochemistry, pharmacology, and health-promoting activities. Phytother Res. 2022;36(12):4425–76.
60. Dwivedi A, Dwivedi S. Ethnobotanical uses of *Piper longum*. J Nat Rem. 2007;7(2):152–6. https://doi.org/10.18311/jnr/2007/128.
61. Yadav BS, Jain RK. Piper longum: Ethnobotany and its medicinal uses. Asian J Chem. 1995;7(1):55–60. https://doi.org/10.14233/ajchem.1995.123.
62. Manohar SH, Viswanathan KR. Ayurvedic pharmacopoeia of India. Government of India, Ministry of Health and Family Welfare; 2005.
63. Pandey A, Tripathi P. Traditional and medicinal uses of *Piper attenuatum*. Int J Herb Med. 2014;2(3):12–5.
64. Kim YJ, Deok J, Kim S, Yoon DH, Sung GH, Aravinthan A, et al. Anti-inflammatory effect of *Piper attenuatum* methanol extract in LPS-stimulated inflammatory responses. Evid Based Complement Alternat Med. 2017:4606459. https://doi.org/10.1155/2017/4606459. Epub 2017 Jul 25. PMID: 28811826; PMCID: PMC5547706.
65. Anilkumar KR, Sudharshan S. *Piper arnottianum*: ethnobotany and pharmacognosy. J Med Plants. 2012;56(3):123–35.
66. Pandey A, Tripathi P. Phytochemistry and pharmacological studies of *Piper arnottianum*. Int J Herb Med. 2014;2(3):12–5.
67. Murugesan K. Scientific validation of wild endemic pepper species piper mullesua Buch. -Ham. Ex D. Don and Piper wightii Miq for their antioxidant and anticancer properties. [Doctoral dissertation]. Bharathiyar University; 2022.
68. Chen J, et al. Chemical constituents and bioactivities of piper mullesua Buch.-Ham. ex D. Don. J Ethnopharmacol. 2011.
69. Geethakumary, Sivadas D, Pandurangan AG. Piper hamiltonii C. DC. (Piperaceae)—a new distributional record for Peninsular India. Int J Plant Anim Environ Sci. 2014;4:18–20.

70. Kumar R, Rajani SR, George V. Chemical composition of the leaf oil of Piper galeatum. J Med Arom Plants Sci. 2006;28:48–9.
71. Biswas P, Anand U, Saha SC, Kant N, Mishra T, Masih H, et al. Betelvine (Piper betle L.): a comprehensive insight into its ethnopharmacology, phytochemistry, and pharmacological, biomedical and therapeutic attributes. J Cell Mol Med. 2022;26(11):3083–119.
72. Nayaka NMDMW, Sasadara MMV, Sanjaya DA, Yuda PESK, Dewi NLKAA, Cahyaningsih E, Hartati R. Piper betle (L): recent review of antibacterial and antifungal properties, safety profiles, and commercial applications. Molecules. 2021;26(8):2321. https://doi.org/10.3390/molecules26082321. PMID: 33923576; PMCID: PMC8073370.
73. Rekha VP, Kollipara M, Gupta BR, Bharath Y, Pulicherla KK. A review on Piper betle L.: nature's promising medicinal reservoir. Am J Ethnomed. 2014;1(5):276–89.
74. Mathew J, George KV, Yohannan R. Piper rukshagandhum (Piperaceae): a new species from southern Western Ghats, India. Telopea. 2016;19:23–9.
75. Hoang VD, Nguyen PH, Tran MH, Huynh NT, Nguyen HT, Min BS, To DC. Identification of anti-inflammatory constituents from Vietnamese Piper hymenophyllum. Rev Bras Farmacogn. 2020;30:312–6.
76. Pradheeba M, Pugalenthi M, Deepa MA, Kumar SV, Vasukipridharshini G. Evaluation of phytochemical profile and in vitro antioxidant, anti-bacterial and anti-inflammatory activity of Piper schmidtii Hook. fil. A wild edible fruit. Asian Pac J Health Sci. 2022;9(3):191–7.

Unveiling the Anti-inflammatory Properties of *Ficus* Fruits: A Promising Frontier for Potential Therapeutic Benefits

Madhu Bala Durairajan, Francis Jegan Raj, S. Anish Ruban, and Parimelazhagan Thangaraj

Abstract

This book chapter delves into the emerging potential of *Ficus* fruits as a source of anti-inflammatory agents with significant therapeutic implications. *Ficus* species have been traditionally utilized in various cultures for their medicinal properties, yet their anti-inflammatory potential remains underexplored. Through a comprehensive review of existing literature and experimental evidence, this review elucidates the diverse bioactive compounds present in *Ficus* fruits, such as flavonoids, phenolic compounds, and triterpenoids, which exhibit potent anti-inflammatory effects. It not only examines the ethnopharmacological importance of *Ficus* fruits in various cultural backgrounds, but it also examines the mechanistic pathways underlying the anti-inflammatory action of these compounds, including modulation of pro-inflammatory cytokines, inhibition of inflammatory enzymes, and regulation of oxidative stress. The synergistic interactions among these bioactive constituents contribute to the overall efficacy of *Ficus* fruits in attenuating inflammatory responses. Importantly, the findings highlight the therapeutic promise of *Ficus* fruits in mitigating inflammatory disorders, including but not limited to arthritis, cardiovascular diseases, and neurodegenerative conditions. This research underscores the necessity for further investigations to unlock the full therapeutic potential of *Ficus* fruits and pave the way for the development of novel anti-inflammatory therapies.

Keywords

Ficus fruits · Ethnopharmacology · Anti-inflammatory · Inflammation Bioactive compounds

M. B. Durairajan · F. J. Raj · S. Anish Ruban · P. Thangaraj (✉)
Bioprospecting Laboratory, Department of Botany, Bharathiar University, Coimbatore, Tamil Nadu, India

P. Thangaraj (ed.), *Pomotherapeutic Insights on Wild Edible Fruits*,
https://doi.org/10.1007/978-981-96-0543-9_7

Introduction

Ficus fruits, commonly known as figs, have a rich history of consumption and medicinal use around the world. These fruits originate from various species of the *Ficus* genus, which includes over 800 species distributed across tropical and subtropical regions. *Ficus* species are grouped under the division Magnoliophyta and belong to the family Moraceae [1]. Figs have been integral to human diets and cultures for millennia, with evidence of their consumption dating back to ancient civilizations such as those of Egypt, Greece, and Rome. Figs are widely appreciated for their nutritional value. They are a good source of dietary fiber, vitamins (including A, K, and various B vitamins), and minerals (such as potassium, calcium, magnesium, and iron). Additionally, figs contain antioxidants like polyphenols and flavonoids, which contribute to their potential health benefits. Culturally, figs hold significance beyond their culinary use [2]. They are often associated with fertility, abundance, and longevity in many societies. In religious contexts, figs are mentioned in various scriptures and have symbolic meanings in different faiths. In traditional medicine systems, figs are recognized for their medicinal properties. In Ayurveda, the traditional medicine of India, figs are used to treat digestive disorders, respiratory ailments, and skin conditions. Similarly, in traditional Chinese medicine (TCM), figs are prescribed to address coughs, constipation, and other health issues. Modern scientific research has corroborated some of the traditional uses of figs. Studies have shown that fig extracts possess antimicrobial, anti-inflammatory, antidiabetic, and anticancer properties. They may also have cardiovascular benefits and contribute to overall health and well-being. Overall, figs continue to be valued for their culinary versatility, nutritional content, and potential therapeutic effects in various cultures around the world, maintaining their status as a beloved fruit with diverse applications [3].

Distribution and Cultivation of *Ficus* Fruits

Ficus species are native to tropical and subtropical regions of the world, including Asia, Africa, Australia, and the Americas. They thrive in warm, humid climates with abundant sunlight and well-drained soil. They are cultivated in many countries across different continents due to their popularity as both food and ornamental plants. Major fig-producing countries include Turkey, Greece, Italy, Spain, Morocco, India, Egypt, Iran, the United States, Brazil, and Australia. There are over 800 species of *Ficus*, ranging from small shrubs to large trees. Some species are cultivated for their edible fruits, while others are prized for their ornamental value, shade, or medicinal properties. Figs have cultural significance in many societies and are often associated with traditions, folklore, and religious practices [4]. In some cultures, fig trees are considered sacred and symbolize fertility, abundance, and longevity. Commercial fig cultivation involves both traditional methods and modern agricultural practices. Figs are grown in orchards and plantations, with careful attention to soil quality, irrigation, pruning, and pest management. Cultivation techniques vary depending on the species and environmental conditions. Figs are typically harvested when fully ripe, as they do not ripen further once picked. Fresh figs are consumed raw or processed into various products such as dried figs, fig paste, jams, preserves, and baked goods. Dried figs are especially popular for their extended shelf life and concentrated sweetness. While figs prefer warm climates, some species and cultivars are adaptable to a range of environmental conditions, including semiarid regions and temperate zones. Cold-hardy varieties can withstand frost and are grown in cooler climates with proper protection during winter months [5].

Nutritional Composition of *Ficus* Fruits

Figs are a nutritious fruit that provides a range of essential nutrients and can be a valuable addition to a balanced diet. They are an excellent option for enhancing general health and well-being because they are especially high in fiber, vitamins, minerals, and antioxidants. *Ficus* fruit nutritional composition might vary slightly based on variety, ripeness, and production techniques. Fig fruits can be consumed raw, dried, canned, or in various preserved forms such as jam and sweets, which can be consumed on their own or used to make cakes and other treats. Of all the dried fruits, figs have the highest nutritional grade according to USDA data since they are a good source of vitamins and minerals. Figs have the highest fiber content, are heavy in calories, proteins, and iron. It has more calcium than milk, making it the richest source. Raw figs are 79% water, 19% carbohydrates, and 1% protein and contain negligible fat [6]. For every 100 grams (74 calories), they provide 14% of the daily value, or DV, of dietary fiber, making them a moderate source. Figs contain 64% carbohydrates, 3% protein, and 1% fat when they are dehydrated to 30% water content. Dried figs are a good source (more than 20% DV) of dietary fiber and the key mineral manganese (26% DV) in a 100-gram portion that provides 249 calories. They are also a moderate source of calcium, iron, magnesium, potassium, and vitamin K [7].

Ethnopharmacological Exploration of *Ficus* Fruits

Indigenous people around the world have historically used *Ficus* fruit species to heal ailments, including diabetes. Numerous research conducted over the past few decades have revealed that both in vitro and in vivo models have demonstrated the powerful antidiabetic properties of crude extracts and isolated compounds from a variety of *Ficus* species, including *F. benghalensis, F. religiosa, F. glumosa, F. deltoidea, F. racemosa, and F. carica* [1]. Indigenous people have been using *Ficus* fruit species for thousands of years to cure diabetes and other ailments. Various *Ficus* species, particularly *F. benghalensis, F. religiosa, F. glumosa, F. deltoidea, F. racemosa*, and *F. carica*, have been shown in several studies over the past few decades to exhibit strong antidiabetic properties in both in vitro and in vivo models when their crude extracts and isolated compounds were used [8]. Traditionally, the leaves are utilized to stabilize palm oil and maybe improve its keeping characteristics [9]. *Ficus* plants were used in indigenous system of medicine like Ayurveda and homeopathy [9–11]. Currently, about 20 species of *Ficus* genus are traditionally used as natural remedies [12] (Table 1).

Significance of Exploring *Ficus* Fruits as Anti-inflammatory Agents

Figs are rich in various bioactive compounds with known anti-inflammatory properties, including phenolic compounds, flavonoids, and carotenoids. These compounds exert their effects by modulating inflammatory pathways and reducing the production of pro-inflammatory mediators. Studies have shown that bioactive compounds found in figs can inhibit key inflammatory pathways in the body, such as the nuclear factor-kappa B (NF-κB) pathway. NF-κB is a central regulator of inflammation, and its inhibition can help alleviate inflammatory responses associated with various chronic diseases. Nuclear factor-kappa B (NF-κB) is a central regulator of inflammation, controlling the expression of genes involved in the inflammatory response. Studies have shown that bioactive compounds present in figs can inhibit the activation of NF-κB, thereby reducing the production of pro-inflammatory cytokines and chemokines. *Ficus* extracts have been found to suppress the production of various pro-inflammatory mediators, including interleukins (IL-1β, IL-6, IL-8), tumor necrosis factor-alpha (TNF-α), prostaglandins (PGE2), and nitric oxide (NO). By inhibiting the release of these molecules, fig extracts help attenuate the inflammatory response and alleviate associated symptoms. *Ficus* extracts may modulate the cyclooxygenase (COX) and lipoxygenase (LOX) pathways, which are involved in the synthesis of pro-inflammatory prostaglandins and leukotrienes, respectively. By inhibiting the activity of COX enzymes, figs can reduce the production of inflammatory prostaglandins, contributing to their anti-inflammatory effects. In addition to their anti-inflammatory effects, figs also possess strong antioxidant properties. Antioxidants help neutralize harmful free radicals in the body, which are known to contribute to inflammation and oxidative stress. By scavenging free radicals, figs can help mitigate inflammatory damage and promote overall health. Figs contain compounds that can suppress the production of inflammatory mediators like cytokines, prostaglandins, and leukotrienes. By reducing the levels of these molecules, figs can help dampen inflammation and alleviate associated symptoms. Chronic inflammation is implicated in the pathogenesis of various diseases, including cardiovascular disease, diabetes, cancer, and neurodegenerative disorders. By serving as natural anti-inflammatory agents, figs have the potential to complement conventional therapies in the management of these conditions, potentially reducing the risk of disease progression and improving overall health outcomes. Preclinical studies involving animal models have provided evidence supporting the anti-inflammatory properties of fig extracts. These studies have demonstrated reductions in inflammatory markers and improvements in inflammatory conditions following fig supplementation. While clinical research in humans is limited, preliminary evidence

Table 1 Ethnopharmacological uses of *Ficus* fruits

Species	Plant part used	Usage	References
F. Abutilifolia	Leaves	Edema	[13]
F. indica	Mashed for juice	Liver pain, vesicle pain	[14]
F. guianensis	Fruit	Broken limb	[14]
F. killpii	Fruit	Bruises	[14]
F. insipida	Latex mixed in water	Intestinal parasites	[14]
F. asperifolia	Dried fruits	Anthelmintic, laxative	[15]
F. benghalensis	Fruit	Diarrhea, dysentery, leprosy, gastric ulcers, fever, erysipelas inflammations, diabetes, leucorrhea, lumbago, ulcers pains, piles, rheumatism, nose diseases, lumbago, gonorrhea, spermatorrhea, aphrodisiac, tonic	[16, 17]
F. benjamina	Leaves	Skin disorders, stomachic, hypotensive, and antidysentery	[18]
F. carica	Bark, leaves, latex, fruits, & leaves	Nutritional, skin disorders, warts, insect bites, calluses, thrones, sunburn, eye, throat, skin diseases, hemorrhoids, constipation, cough, kidney stones, respiratory diseases, asthma, high cholesterol, hypoglycemic, laxative, stimulant, antitussive, emmenagogue, resolvent, antiseptic, weakness, constipation, vision problems, sore throat, diabetes, tuberculosis	[19–21]
F. cordata	Whole plant	Leukoderma, eye wash	[22]
F. deltoidea	Leaves & root	Aphrodisiac, tonic, rheumatism, diabetes, wounds, sores	[23]
F. exasperata	Flower, root, leaves, & bark	Analgesic, antiarthritic, diuretic, anthelmintic, antipyretic, abortive, general weakness, malnutrition, parasitic infection, ophthalmic, and oral infections, nasopharyngeal diseases, rheumatism, gout, enema, kidney disease, diarrhea, dysentery, hemorrhoids, venereal diseases, Therapeutical meal (second trimester)	[9]
F. glumosa	Leaves & bark	Chest pain, bronchitis, pneumonia, cough, gingivitis, toothaches	[24]
F. hispida	Leaves & fruit	Antihemorrhagic, aphrodisiac, tonic, lactagogue, emetic and diarrhea	[25]
F. ingens	Stem	Cough, fever, & diarrhea	[13]
F. insipida	Latex	Anthelmintic, intestinal parasites	[14]
F. natalensis	Root	Promote labor, cataract, retained placenta, hiccup	[26]

(continued)

Table 1 (continued)

Species	Plant part used	Usage	References
F. palmata	Fruit & latex	Bleeding wound, lung diseases, constipation, piles	[27]
F. platyphylla	Bark	Malaria, tuberculosis, stomach troubles, pile	[28]
F. polita	Fruit, leaves, bark, root	Dyspepsia, infections, abdominal pains, diarrhea	[29]
F. racemosa	Fruit, latex, bark, leaves, root	Sore throat, leprosy, Rickets, diarrhea, hemorrhoids, stomachache, cholera, mumps, aphrodisiac, Bilious affection, dysmenorrhea, dysentery, gonorrhea, diabetes, hemoptysis, constipation, menorrhagia, visceral obstruction	[30–32]
F. religiosa	Leaves, bark, fruit, seeds, latex	Body pain, toothache, constipation, migraines, gastric disorders, skin diseases, gonorrhea, diarrhea, dysentery, hemorrhoids, inflammations, hiccup, vomiting, aphrodisiac, diarrhea, asthma, cough, earache, toothache, migraine, gastric problems, hematuria, asthma	[33, 34]
F. retusa	Fruit	Improve the body stamina, cure sterility in men.	[35]
F. sycomorus	Stem	Cough, skin diseases, tuberculosis, cold, other chest problems	[36]
F. tinctoria subsp. gibbosa	Fruit	Fever	[37]
F. Thonningii	Leaves, bark, roots	Rheumatism, stomach offset, bronchitis, urinary tract infections, urinary schistosomiasis, wounds, bleeding gums, gingivitis, diarrhea, cysts, ulcers, preventing miscarriages, nose bleeding	[38, 39]
F. virgata	Leaves	Constipations, bladder diseases	[12]

suggests that figs may have potential as anti-inflammatory agents in various inflammatory diseases [40].

Mechanism of Inflammation

Inflammation is a complex biological response that occurs in the body's tissues in response to harmful stimuli, such as pathogens, damaged cells, or irritants. This intricate process involves a variety of molecular and cellular mediators that work together to protect the body and promote healing. One of the key players in inflammation is the immune system, which employs a network of cells, including white blood cells and immune signaling molecules, to identify and eliminate the source of injury or infection. The immune system's ability to coordinate these responses is crucial for effectively combating threats to our health. During the inflammatory response, immune cells release chemical signals known as cytokines, which attract other immune cells to the site of inflammation. These cytokines activate various inflammatory signaling pathways, including the NF-κB pathway and the JAK/STAT pathway, which further amplify the immune response. Additionally, inflammation is marked by increased blood flow to the affected area, resulting in redness and warmth, as well as increased permeability of blood vessels. This allows immune cells to easily move through the blood vessels and migrate to the site of inflammation. Inflammatory cells, such as neutrophils and macrophages, play a crucial role in eliminating pathogens and removing cellular debris. They engulf and destroy invading microorganisms, release antimicrobial substances, and help facilitate tissue repair. Furthermore, reactive oxygen species and reactive nitrogen species are produced during inflammation [40]. These reactive molecules have antimicrobial properties and can directly kill pathogens. In addition to immune cells and signaling molecules, inflammatory responses also involve the production of pro-inflammatory mediators called eicosanoids. Eicosanoids are lipid-based molecules that can have both pro-inflammatory and anti-inflammatory effects in the body. They are synthesized from arachidonic acid and include prostaglandins, leukotrienes, and thromboxanes. These molecules can regulate various aspects of the inflammatory response, including vasodilation, chemotaxis, and the recruitment of immune cells. Inflammatory responses are tightly regulated to prevent excessive or prolonged inflammation, which can lead to tissue damage and chronic diseases. The resolution of inflammation involves several mechanisms that act to dampen the immune response and promote tissue healing. These mechanisms include the production of anti-inflammatory cytokines, such as IL-10 and TGF-β, which help suppress immune cell activity and promote tissue repair. Resolution of inflammation also involves the clearance of inflammatory cells and debris by phagocytic cells, such as macrophages. In vitro stimulation with specific cytokines can induce the differentiation and expansion of Th17 cells, a subset of CD4+ T cells involved in inflammation [41].

The sequence of events in inflammation is orchestrated by a series of interrelated pathways, beginning with the recognition of the harmful stimuli by the immune system. This triggers the release of inflammatory mediators, such as cytokines, chemokines, and eicosanoids, which promote the recruitment of immune cells to the site of injury. These immune cells, including neutrophils, macrophages, and lymphocytes, then work together to remove the noxious agents and initiate tissue repair. Some of the mechanisms of inflammation include the release of inflammatory mediators, such as cytokines and eicosanoids, the recruitment of immune cells to the site of inflammation, the production of reactive species that kill pathogens, and the resolution of inflammation to promote tissue healing. Overall, the mechanisms of inflammation involve a complex interplay between immune cells, signaling molecules, and mediators [42].

Cellular Interactions in Inflammation

During inflammation, various immune cells interact to coordinate the immune response. These immune cells include macrophages, neutrophils, dendritic cells, and lymphocytes. Macrophages play a central role in initiating and orchestrating the inflammatory response. They phagocytose pathogens and release pro-inflammatory cytokines, such as tumor necrosis factor-alpha and interleukin-1, which recruit and activate other immune cells. These interactions between different immune cells amplify the inflammatory response and facilitate the clearance of pathogens or damaged tissue. Furthermore, dendritic cells are also important in inflammation as they capture and present antigens to T cells, initiating adaptive immune responses. The activation and interaction of these immune cells are tightly regulated by various signaling molecules, such as chemokines and adhesion molecules, which guide the migration and communication between immune cells. These cellular interactions and signaling molecules are crucial for the amplification and resolution of inflammation. These mechanisms of inflammation involve complex cellular interactions and signaling pathways that ultimately lead to the recruitment and activation of immune cells, the release of pro-inflammatory cytokines, and the resolution of inflammation through the clearance of pathogens or damaged tissue. Regulatory T cells are another critical component of the immune regulatory network that contributes to the control of excessive inflammation. T cells suppress the activation and function of effector T cells and other immune cells, thus preventing immune-mediated tissue damage and maintaining immune homeostasis Overall, the mechanisms of inflammation involve the activation of pattern recognition receptors, such as Toll-like receptors and inflammasomes, which trigger the release of pro-inflammatory cytokines and chemokines [43].

Understanding Pro-inflammatory Mediators and Resolution Mechanisms in Inflammation

The resolution of inflammation is an active and tightly regulated process that is essential for restoring tissue homeostasis and promoting healing. Specialized pro-resolving lipid mediators, such as lipoxins, resolvins, and protectins, play a key role in orchestrating the resolution of inflammation. These mediators exert anti-inflammatory and pro-resolving effects by dampening pro-inflammatory signaling, enhancing the clearance of apoptotic cells and debris, and promoting the non-phlogistic recruitment of immune cells to facilitate tissue repair. Understanding the intricate mechanisms of inflammation is essential for developing targeted therapeutic interventions to alleviate inflammation-related diseases. In the realm of autoimmune disorders, dysregulation of the immune response leads to the body's immune system attacking its own cells and tissues, resulting in conditions such as rheumatoid arthritis, systemic lupus erythematosus, and inflammatory bowel disease. The development of immunomodulatory therapies that target specific components of the immune response holds great promise in managing these conditions and improving the quality of life for affected individuals. Furthermore, cardiovascular diseases, including atherosclerosis and myocardial infarction, are closely linked to chronic inflammation. The role of inflammatory processes in the initiation and progression of atherosclerosis has been well-established, and the identification of therapeutic targets to modulate these inflammatory pathways has significant implications for the prevention and treatment of cardiovascular diseases. In the context of cancer, chronic inflammation has been associated with the promotion of tumor growth, angiogenesis, and metastasis. Understanding the interplay between the immune system and cancer cells is crucial for the development of novel immunotherapies that harness the body's immune response to target and eliminate cancer cells [44].

Several molecules and pathways have been identified that contribute to the resolution of inflammation. For example, the Annexin A1 protein and its bioactive peptides possess potent anti-inflammatory and pro-resolving properties. Annexin A1 modulates leukocyte trafficking, phagocytosis, and cytokine production to promote the resolution of inflammation and tissue regeneration. By gaining a deeper understanding of the molecular and cellular mechanisms that regulate inflammation, new therapeutic targets and strategies can be identified to modulate immune responses and promote the resolution of inflammation. Once inflammatory mediators are released, immune cells, including neutrophils, macrophages, and lymphocytes, work together to remove the noxious agents and initiate tissue repair. Neutrophils are among the first immune cells to migrate to the site of inflammation, where they engulf and destroy invading microorganisms. Macrophages play a crucial role in the phagocytosis of cellular debris and the production of anti-inflammatory cytokines during the resolution phase of inflammation. Furthermore, the interactions between various immune cells are crucial in regulating the inflammatory response. T cells, including T helper 1 cells and regulatory T cells, play a pivotal role in orchestrating a balanced immune reaction. While T helper 1 cells produce pro-inflammatory cytokines to combat pathogens, regulatory T cells function to maintain immune homeostasis and prevent excessive inflammation [45].

The production of pro-inflammatory mediators, such as eicosanoids, including prostaglandins and leukotrienes, regulates various aspects of the inflammatory response, including vasodilation and the recruitment of immune cells. Concurrently, the resolution of inflammation involves the production of anti-inflammatory cytokines, such as IL-10 and transforming growth factor-β, which help suppress immune cell activity and promote tissue repair. The dysregulation of the immune response in autoimmune disorders leads to the body's immune system attacking its own cells and tissues, resulting in conditions such as rheumatoid arthritis, systemic lupus erythematosus, and inflammatory bowel disease. The development of immunomodulatory therapies that target specific components of the immune response holds great promise in managing these conditions and improving the quality of life for affected individuals. These therapies aim to restore immune tolerance and prevent the immune system from mistakenly attacking healthy tissues [46]. Chronic inflammation plays a significant role in cardiovascular diseases, including atherosclerosis and myocardial infarction. The role of inflammatory processes in the initiation and progression of atherosclerosis has been well-established, and the identification of therapeutic targets to modulate these inflammatory pathways has significant implications for the prevention and treatment of cardiovascular diseases. Targeted therapies that aim to reduce inflammation within the cardiovascular system have the potential to decrease the risk of cardiovascular events and improve patient outcomes [47]. Chronic inflammation has been associated with the promotion of tumor growth, angiogenesis, and metastasis in cancer. Understanding the interplay between the immune system and cancer cells is crucial for the development of novel immunotherapies that harness the body's immune response to target and eliminate cancer cells. Immunotherapies that bolster the body's immune response against cancer cells, such as immune checkpoint inhibitors and CAR T-cell therapy, hold promise for revolutionizing cancer treatment and improving patient survival rates [48]. The development of therapeutic strategies to modulate inflammatory responses encompasses a wide range of approaches, including the targeting of specific molecular pathways, immune cell function, and the modulation of cytokine signaling. Biologic therapies, such as monoclonal antibodies that neutralize pro-inflammatory cytokines or target specific immune cell populations, have revolutionized the treatment of inflammatory diseases. Additionally, small molecule inhibitors that target key components of inflammatory pathways, including kinases and transcription factors, hold promise for the development of novel anti-inflammatory agents. These approaches offer new avenues for the treatment of inflammatory conditions by selectively modulating the dysregulated inflammatory responses [49].

Role of *Ficus* Fruits in Disease Pathogenesis through Inflammation

Ficus fruits have been traditionally used in various cultures for their potential anti-inflammatory properties. Several studies have shown that compounds found in *Ficus* fruits, such as flavonoids and polyphenols, exhibit anti-inflammatory effects by modulating various pathways involved in the inflammatory response. These compounds have been found to inhibit the production of pro-inflammatory cytokines, such as tumor necrosis factor-alpha and interleukin-6, and suppress the activation of inflammatory cells like macrophages and neutrophils. Furthermore, research has indicated that the bioactive compounds in *Ficus* fruits can also reduce the expression of inflammatory enzymes such as cyclooxygenase-2 and inducible nitric oxide synthase (iNOS), which are involved in the synthesis of inflammatory mediators. In addition to their direct anti-inflammatory effects, *Ficus* fruits are also known to possess antioxidant properties, which can help mitigate oxidative stress and its associated inflammatory effects in the body. These antioxidant properties of *Ficus* fruits can neutralize free radicals and reduce oxidative damage to cells and tissues, thereby preventing the initiation and progression of inflammatory processes. Given these findings, incorporating *Ficus* fruits into the diet may offer potential benefits in managing inflammatory conditions and could be a promising avenue for further exploration in the field of natural anti-inflammatory therapies. The role of *Ficus* fruits in inflammation and disease pathogenesis has been the subject of several studies. One study conducted on animal models demonstrated that the extract of *Ficus* fruits reduced the levels of inflammatory markers such as interleukin-6 and tumor necrosis factor-alpha. These findings suggest that *Ficus* fruits may play a role in regulating the inflammatory processes involved in disease pathogenesis. Furthermore, the bioactive compounds present in *Ficus* fruits have been shown to inhibit the activity of pro-inflammatory enzymes, such as cyclooxygenase and lipoxygenase, which are involved in the production of inflammatory mediators [50]. Overall, the role of *Ficus* fruits in inflammation and disease pathogenesis is a promising area of research that warrants further investigation to fully understand the mechanisms and potential therapeutic applications of these fruits.

Anti-inflammatory Properties of *Ficus* Fruits

Ficus fruits have long been recognized for their incredible health benefits and anti-inflammatory properties. These fruits are packed with essential vitamins, minerals, and antioxidants that play a significant role in reducing inflammation in the body. The high levels of polyphenols and flavonoids found in *Ficus* contribute to their potent anti-inflammatory effects. The anti-inflammatory properties of *Ficus* have been studied extensively, with research showing promising results in reducing inflammation in various conditions such as arthritis, diabetes, and cardiovascular diseases. Furthermore, the unique combination of fiber and natural sugars in *Ficus* helps regulate blood sugar levels, thereby reducing the risk of inflammation associated with high blood sugar. In addition to their internal benefits, *Ficus* are also used externally for their anti-inflammatory properties. *Ficus* have been traditionally used in natural remedies for skin conditions such as eczema and psoriasis due to their soothing. Furthermore, the high fiber content in *Ficus* may also contribute to their anti-inflammatory effects by promoting a healthy gut microbiome, which is increasingly recognized as being linked to lower levels of inflammation in the body. Additionally, the presence of antioxidants, such as flavonoids and phenolic compounds in *Ficus*, further supports their potential anti-inflammatory properties [51].

The phytocompounds present in *Ficus religiosa* have the potential to act as effective anti-inflammatory agents. Further research has revealed that *F. religiosa* contains a variety of compounds, such as flavonoids, polyphenols, and tannins, which have been shown to exhibit anti-inflammatory properties. These compounds have been found to inhibit the production of pro-inflammatory cytokines and enzymes, thereby reducing inflammation in the body. Additionally, studies have demonstrated that the extract of *F. religiosa* fruits is able to downregulate the expression of inflammatory genes in immune cells, further supporting its anti-inflammatory effects. This evidence highlights the potential of *F. religiosa* as a natural and potent source of anti-inflammatory compounds, with implications for the development of novel therapeutic agents for inflammatory conditions. Recent in vivo studies have also delved into the mechanism of action of these compounds found in *F. religiosa*. It has been discovered that the flavonoids and polyphenols in *F. religiosa* fruits exert their anti-inflammatory effects by modulating key signaling pathways involved in the inflammatory response. This includes the inhibition of nuclear factor kappa B signaling, which is a crucial regulator of immune and inflammatory responses [52].

Furthermore, the antioxidant activities of crude water and ethanolic extracts from *Ficus pandurata* have been investigated. These extracts have been shown to possess strong antioxidant properties, as demonstrated by their ability to scavenge free radicals and reduce oxidative stress. These antioxidant activities are attributed to the presence of phytochemicals such as phenols, flavonoids, and tannins. The presence of these phytochemicals in *F. pandurata* suggests its potential for mitigating oxidative damage in the body and protecting against various diseases associated with oxidative stress. Considering the promising findings from these in vivo studies, it is evident that both *F. religiosa* and *F. pandurata* possess significant therapeutic potential. The anti-inflammatory and antioxidant properties of compounds found in these *Ficus* fruits highlight their possible role in the development of natural remedies for inflammatory conditions and oxidative stress-related disorders. The next steps in research could focus on conducting clinical trials to further investigate the efficacy and safety of utilizing *F. religiosa* and *F. pandurata* as therapeutic agents. Additionally, identifying the specific bioactive compounds responsible for these effects and elucidating their mechanisms of action could pave the way for the development of targeted treatments derived from *Ficus* fruits. This avenue of exploration holds promise for the advancement of natural medicine and the potential discovery of novel pharmaceutical agents [53]. The comprehensive understanding of the molecular mechanisms and bioactive compounds in *Ficus* religiosa presents an exciting opportunity for developing targeted and efficient anti-inflammatory therapies. These findings underscore the potential of *Ficus* religiosa as a valuable natural resource for combating inflammatory conditions and offer promising prospects for the development of novel therapeutic interventions.

Human Clinical Trials on the Anti-inflammatory Properties of *Ficus* Fruits

In recent years, there has been a growing interest in the potential anti-inflammatory properties of *Ficus* fruits, also known as *Ficus*. Several preclinical studies have suggested that certain compounds found in *Ficus* may have the ability to reduce inflammation in the body. However, to truly understand the impact of *Ficus* on inflammation in humans, clinical trials are essential. One notable human clinical trial conducted by researchers at a renowned medical institution aimed to investigate the effects of consuming *Ficus carica* fruits on inflammatory markers in individuals with mild knee osteoarthritis. The randomized, double-blind, placebo-controlled trial involved a 12-week intervention period, during which participants consumed either a specified quantity of *Ficus* or a placebo. The results of this trial indicated that participants who consumed the *Ficus* intervention experienced a significant reduction in inflammatory markers compared to those in the placebo group. The results of the trial demonstrated a statistically significant reduction in inflammatory markers, such as C-reactive protein and interleukin-6, in the group that consumed *Ficus* compared to the placebo group. Furthermore, participants reported a subjective improvement in their knee pain and mobility, indicating a potential benefit of *Ficus* consumption in managing inflammatory conditions [54]. These findings provide compelling evidence of the anti-inflammatory properties of *Ficus* fruits in a clinical setting, shedding light on their potential therapeutic role in conditions characterized by inflammation. Further research in this field could uncover the specific bioactive compounds responsible for these effects and explore the broader applicability of *Ficus* in mitigating inflammation-related health issues.

Comparison with Conventional Anti-inflammatory Agents

When comparing natural anti-inflammatory agents with conventional pharmaceutical options, it is essential to consider their efficacy. Studies have demonstrated that natural remedies such as turmeric and ginger can be as effective as conventional nonsteroidal anti-inflammatory drugs in reducing inflammation and pain. Research has also shown that these natural options have fewer side effects compared to conventional medications, making them a safer choice for long-term use. Safety is another crucial factor to consider when evaluating anti-inflammatory agents. Natural remedies are often favored for their perceived safety profile, with fewer reported adverse effects compared to conventional NSAIDs. While some pharmaceutical medications may pose risks such as gastrointestinal irritation, cardiovascular complications, and kidney damage, natural anti-inflammatory agents are generally well-tolerated, with minimal adverse effects. The cost of treatment is a significant consideration for many individuals. Natural anti-inflammatory agents can often be more cost-effective than conventional pharmaceutical options. This can make them a more accessible choice for individuals who may face financial constraints or lack adequate insurance coverage for prescription medications. Additionally, the potential cost savings associated with using natural remedies may contribute to improved treatment adherence and overall patient satisfaction. In conclusion, when comparing natural anti-inflammatory agents with conventional options, factors such as efficacy, safety, and cost must be carefully weighed. It is evident that natural remedies offer promising benefits and may present as a viable alternative or complementary approach to conventional anti-inflammatory agents [55].

When compared to conventional anti-inflammatory agents, natural remedies often have a favorable safety profile. They are generally well-tolerated and have fewer reported adverse effects compared to conventional NSAIDs. This is an important consideration for individuals who prioritize safety and wish to minimize the risk of potential side effects. Additionally, the perceived safety of natural remedies may contribute to greater peace of mind for patients, leading to improved treatment adherence. It is important to acknowledge that while natural anti-inflammatory agents offer promising benefits, further research is necessary to fully understand their efficacy and safety profiles. Advancements in scientific studies and clinical trials can provide valuable insights into the potential of these natural remedies as a complementary or alternative approach to conventional anti-inflammatory treatments. Additionally, ongoing research can help identify any specific patient populations that may particularly benefit from the use of natural anti-inflammatory agents. As the field of natural remedies continues to evolve, collaboration between healthcare professionals, researchers, and individuals seeking alternative treatments is essential to ensure that comprehensive and accurate information is available. By fostering an open dialogue and sharing knowledge about the efficacy and safety of these natural options, individuals can make informed decisions about their anti-inflammatory treatment plans [56].

Pharmacokinetics and Bioavailability of Anti-inflammatory *Ficus* Fruit Compounds

Ficus fruit compounds have shown promising anti-inflammatory properties in preliminary studies. To fully understand their pharmacokinetics and bioavailability, further research is needed to investigate how these compounds are absorbed, distributed, metabolized, and excreted in the body. Understanding the pharmacokinetics of these compounds is crucial for determining the appropriate dosage and frequency of administration for maximum therapeutic benefit. Bioavailability studies will also be essential to assess the extent and rate at which these compounds reach the systemic circulation and target tissues after administration. Factors such as solubility, stability, and formulation will influence the bioavailability of these compounds and should be carefully considered in future research. Additionally, investigating potential interactions with other medications and the impact of food intake on the absorption of *Ficus* fruit compounds will be important for understanding their clinical significance. This comprehensive approach to studying the pharmacokinetics and bioavailability of *Ficus* fruit compounds will provide valuable insights for their potential use as anti-inflammatory agents. Pharmacokinetics of *Ficus* fruit compounds. Pharmacokinetics is a crucial area of study for understanding the behavior of *Ficus* fruit compounds in the human body. Absorption, distribution, metabolism, and excretion are key processes that determine the pharmacokinetic profile of these compounds. Investigating how these compounds are absorbed across different biological barriers, such as the gastrointestinal tract, and how they are distributed to various tissues and organs will provide important insights into their potential therapeutic effects. Metabolism studies will help in identifying the metabolic pathways of *Ficus* fruit compounds, including the formation of active or inactive metabolites. Understanding the routes of excretion and the elimination half-life of these compounds will also be essential for determining their overall presence and duration of action in the body [57].

In addition to pharmacokinetics, bioavailability studies will play a crucial role in evaluating the extent and rate of absorption of *Ficus* fruit compounds. Factors such as solubility, stability, and formulation will significantly impact the bioavailability of these compounds. Therefore, detailed investigations into these factors are necessary to optimize their delivery and absorption in the body. Assessing the impact of food intake on the bioavailability of *Ficus* fruit compounds is particularly important, as it will provide valuable insights into their clinical relevance and potential interactions with dietary components. Moreover, understanding the influence of coadministered medications on the bioavailability of *Ficus* fruit compounds will be essential for assessing their safety and efficacy in clinical settings. Overall, a comprehensive exploration of the pharmacokinetics and bioavailability of *Ficus* fruit compounds will not only enhance our understanding of their therapeutic potential as anti-inflammatory agents but also pave the way for their effective utilization in clinical practice [58].

Future directions in anti-inflammatory treatment research include investigating the mechanisms of action and potential synergistic effects of combining natural remedies with conventional drugs. This could potentially lead to the development of more effective and targeted treatment approaches for managing inflammation. Comparisons with conventional anti-inflammatory agents are necessary to evaluate the potential of natural remedies as a viable treatment option for chronic inflammatory diseases.

Implications for Healthcare and Future Research Directions

Ficus plants have long been used in traditional medicine for their anti-inflammatory properties. These properties have significant implications for healthcare, as they can potentially be used in the development of new treatments for various inflammatory conditions. In addition to their anti-inflammatory effects, *Ficus* plants also exhibit antioxidant and antimicrobial properties, further expanding their potential applications in healthcare. Future research directions in this field could include exploring the specific compounds within *Ficus* plants that are responsible for their anti-inflammatory effects. Understanding the underlying mechanisms of action could lead to the development of more targeted and effective treatments. Furthermore, investigating the potential synergistic effects of combining *Ficus* extracts with existing anti-inflammatory drugs could open up new possibilities for improving treatment outcomes. In addition, there is a need for clinical trials to evaluate the safety and efficacy of *Ficus*-based treatments in humans. Gathering robust scientific evidence on the benefits and potential side effects of these treatments is essential for their integration into mainstream healthcare. Furthermore, exploring the potential role of *Ficus* plants in preventing chronic inflammatory conditions, such as arthritis and inflammatory bowel diseases, could offer new avenues for disease prevention and management. Overall, the anti-inflammatory properties of *Ficus* plants have the potential to significantly impact healthcare and merit further exploration through rigorous scientific research and clinical studies.

Acknowledgments The authors thank the National Medicinal Plants Board, New Delhi, under the Ministry of AYUSH, Government of India (F.No.Z.18017/187/CSS/R&D/TN-02/2022-23-NMPB-IV A); Dt. 04.10.2022) for supporting the work.

References

1. Deepa P, Sowndhararajan K, Kim S, Park SJ. A role of Ficus species in the management of diabetes mellitus: a review. J Ethnopharmacol. 2018;215:210–32.
2. Sirisha N, Sreenivasulu M, Sangeeta K, Chetty CM. Antioxidant properties of Ficus species-a review. Int J PharmTech Res. 2010;4:2174–82.

3. Sasidharan SP, Yang X, Arunachalam K. An overview of ethnobotany, phytochemicals, and pharmacological properties of Ficus species. Bioprospect Trop Med Plants. 2023;481–509.
4. Isa MM, Jaafar MN, Kasim KF, Mutalib MF. Cultivation of fig (Ficus carica L.) as an alternative high value crop in Malaysia: a brief review. In: IOP conference series: materials science and engineering 2020. 864, No. 1 IOP Publishing; 2020. p. 012134.
5. Stover E, Aradhya M, Ferguson L, Crisosto CH. The fig: overview of an ancient fruit. HortScience. 2007;42(5):1083–7.
6. Del Caro A, Piga A. Polyphenol composition of peel and pulp of two Italian fresh fig fruits cultivars (Ficus carica L.). Eur Food Res Technol. 2008;226:715–9.
7. Yang XM, Yu W, Ou ZP, Ma HL, Liu WM, Ji XL. Antioxidant and immunity activity of water extract and crude polysaccharide from Ficus carica L. fruit. Plant Foods Hum Nutr. 2009;64:167–73.
8. Scherrer AM, Motti R, Weckerle CS. Traditional plant use in the areas of monte vesole and ascea, cilento national park (Campania, Southern Italy). J Ethnopharmacol. 2005;97(1):129–43.
9. Ahmed F, Ahmed KM, Abedin M, Karim A. Traditional uses and pharmacological potential of Ficus exasperata Vahl. Syst Rev Pharm. 2012;3(1):15.
10. Barolo MI, Mostacero NR, López SN. Ficus carica L.(Moraceae): an ancient source of food and health. Food Chem. 2014;164:119–27.
11. Singh D, Singh B, Goel RK. Traditional uses, phytochemistry and pharmacology of Ficus religiosa: a review. J Ethnopharmacol. 2011;134(3):565–83.
12. Salehi B, Prakash Mishra A, Nigam M, Karazhan N, Shukla I, Kiełtyka-Dadasiewicz A, Sawicka B, Głowacka A, Abu-Darwish MS, Hussein Tarawneh A, Gadetskaya AV. Ficus plants: state of the art from a phytochemical, pharmacological, and toxicological perspective. Phytother Res. 2021;35(3):1187–217.
13. Dambatta SH, Aliyu BS. A survey of major ethno medicinal plants of Kano North, Nigeria, their knowledge and uses by traditional healers. Bayero J Pure Appl Sci. 2011;4(2):28–34.
14. Bourdy G, DeWalt SJ, De Michel LC, Roca A, Deharo E, Muñoz V, Balderrama L, Quenevo C, Gimenez A. Medicinal plants uses of the Tacana, an Amazonian Bolivian ethnic group. J Ethnopharmacol. 2000;70(2):87–109.
15. Watcho P, Ngadjui E, Alango Nkeng-Efouet P, Benoît Nguelefack T, Kamanyi A. Evaluation of in vitro uterotonic activities of fruit extracts of Ficus asperifolia in rats. Evid Based Complement Alternat Med. 2011;2011:783413.
16. Bapuji JL, Ratnam SV. Traditional uses of some medicinal plants by tribals of Gangaraju Madugula Mandal of Visakhapatnam district, Andhra Pradesh. Ethnobotanical Leaflets. 2009;2009(3):2.
17. Joseph B, Raj SJ. Phytopharmacological and phytochemical properties of three Ficus species-an overview. Int J Pharma Bio Sci. 2010;1(4):246–53.
18. Kanaujia VK, Rirchhaiya HK, Kailasiya SD, Verma M, Yadav RD, Shivhare D. Evaluation of hepatoprotective activity on the leaves of Ficus benjamina Linn. J Nat Prod Plant. 2011;1:59–69.
19. Badgujar SB, Patel VV, Bandivdekar AH, Mahajan RT. Traditional uses, phytochemistry and pharmacology of Ficus carica: a review. Pharm Biol. 2014;52(11):1487–503.
20. Shahrajabian MH, Sun W, Cheng Q. A review of chemical constituents, traditional and modern pharmacology of fig (Ficus carica L.), a super fruit with medical astonishing characteristics. Polish J Agron. 2021;44:22–9.
21. Nuri ZN, Uddin MS. A review on nutritional values and pharmacological importance of Ficus carica. J Curr Res Food Sci. 2021;2(1):07–11.
22. Youssef RS. Medicinal and non-medicinal uses of some plants found in the middle region of Saudi Arabia. J Med Plants Res. 2013;7(34):2501–13.
23. Bunawan H, Amin NM, Bunawan SN, Baharum SN, Mohd NN. Ficus deltoidea Jack: a review on its phytochemical and pharmacological importance. Evid Based Complement Alternat Med. 2014;2014:902734.

24. Nana F, Sandjo LP, Keumedjio F, Ambassa P, Malik R, Kuete V, Rincheval V, Choudhary MI, Ngadjui BT. Ceramides and cytotoxic constituents from Ficus glumosa Del. (Moraceae). J Braz Chem Soc. 2012;23:482–7.
25. Ali M, Chaudhary N. Ficus hispida Linn.: a review of its pharmacognostic and ethnomedicinal properties. Pharmacogn Rev. 2011;5(9):96.
26. Mbougnia JF, Happi GM, Bitchagno GT, Awouafack MD, Lenta BN, Kouam SF, Tane P, Sewald N, Tene M. Chemical constituents from Ficus natalensis hochst (Moraceae) and their chemophenetic significance. Biochem Syst Ecol. 2021;95:104227.
27. Joshi Y, Joshi AK, Prasad N, Juyal D. A review on Ficus palmata (wild himalayan fig). J Phytopharmacol. 2014;3(5):374–7.
28. Chindo BA, Anuka JA, McNeil L, Yaro AH, Adamu SS, Amos S, Connelly WK, Lees G, Gamaniel KS. Anticonvulsant properties of saponins from Ficus platyphylla stem bark. Brain Res Bull. 2009;78(6):276–82.
29. Kuete V, Kamga J, Sandjo LP, Ngameni B, Poumale HM, Ambassa P, Ngadjui BT. Antimicrobial activities of the methanol extract, fractions and compounds from Ficus polita Vahl. (Moraceae). BMC Complement Altern Med. 2011;11:1–6.
30. Ahmed F, Urooj A. Traditional uses, medicinal properties, and phytopharmacology of Ficus racemosa: a review. Pharm Biol. 2010;48(6):672–81.
31. Bhalerao SA, Verma DR, Teli NC, Didwana VS, Thakur SS. Ficus racemosa Linn.: a comprehensive review. J Appl Chem. 2014;3(4):1423–31.
32. Manian R, Anusuya N, Siddhuraju P, Manian S. The antioxidant activity and free radical scavenging potential of two different solvent extracts of Camellia sinensis (L.) O. Kuntz, Ficus bengalensis L. and Ficus racemosa L. Food Chem. 2008;107(3):1000–7.
33. Kayani S, Ahmad M, Zafar M, Sultana S, Khan MP, Ashraf MA, Hussain J, Yaseen G. Ethnobotanical uses of medicinal plants for respiratory disorders among the inhabitants of Gallies–Abbottabad, Northern Pakistan. J Ethnopharmacol. 2014;156:47–60.
34. Chandrasekar SB, Bhanumathy M, Pawar AT, Somasundaram T. Phytopharmacology of Ficus religiosa. Pharmacogn Rev. 2010;4(8):195.
35. Ayyanar M, Ignacimuthu S. Ethnobotanical survey of medicinal plants commonly used by Kani tribals in Tirunelveli hills of Western Ghats, India. J Ethnopharmacol. 2011;134(3):851–64.
36. El Said SA, EL-Kamali HH, El-Desogi YO. Phytochemical analysis and traditional uses of two locally available Ficus species (beneghalensis and sycomorus). J Fac Sci Technol. 2020;7:155–62.
37. Uniyal B, Shiva V. Traditional knowledge on medicinal plants among rural women of the Garhwal Himalaya, Uttaranchal. Indian J Tradit Knowl. 2005;4(3).
38. Dangarembizi R, Erlwanger KH, Chivandi E. Effects of Ficus thonningii extracts on the gastrointestinal tract and clinical biochemistry of suckling rats. Afr J Tradit Complement Altern Med. 2014;11(2):285–91.
39. Fongang YS, Bankeu JJ, Ali MS, Awantu AF, Zeeshan A, Assob CN, Mehreen L, Lenta BN, Ngouela SA, Tsamo E. Flavonoids and other bioactive constituents from Ficus thonningii Blume (Moraceae). Phytochem Lett. 2015;11:139–45.
40. Megha KB, Joseph X, Akhil V, Mohanan PV. Cascade of immune mechanism and consequences of inflammatory disorders. Phytomedicine. 2021;91:153712.
41. Schett G, Neurath MF. Resolution of chronic inflammatory disease: universal and tissue-specific concepts. Nat Commun. 2018;9(1):3261.
42. Chen L, Deng H, Cui H, Fang J, Zuo Z, Deng J, Li Y, Wang X, Zhao L. Inflammatory responses and inflammation-associated diseases in organs. Oncotarget. 2018;9(6):7204.
43. Sreedhar R, Watanabe K, Arumugam S. General mechanisms of immunity and inflammation. In: Japanese Kampo Medicines For The Treatment Of Common diseases: focus on inflammation; 2017. p. 23.
44. Welsh P, Grassia G, Botha S, Sattar N, Maffia P. Targeting inflammation to reduce cardiovascular disease risk: a realistic clinical prospect? Br J Pharmacol. 2017;174(22):3898–913.
45. Arulselvan P, Fard MT, Tan WS, Gothai S, Fakurazi S, Norhaizan ME, Kumar SS. Role of antioxidants and natural products in inflammation. Oxidative Med Cell Longev. 2016;2016:5276130.

46. Netea MG, Balkwill F, Chonchol M, Cominelli F, Donath MY, Giamarellos-Bourboulis EJ, Golenbock D, Gresnigt MS, Heneka MT, Hoffman HM, Hotchkiss R. A guiding map for inflammation. Nat Immunol. 2017;18(8):826–31.
47. Sorriento D, Iaccarino G. Inflammation and cardiovascular diseases: the most recent findings. Int J Mol Sci. 2019;20(16):3879.
48. Pahwa R, Goyal A, Jialal I. Chronic inflammation. StatPearls Publishing; 2023.
49. Barua A, Kesavan K, Jayanthi S. Molecular docking studies of plant compounds to identify efficient inhibitors for ovarian cancer. Res J Pharm Technol. 2018;11(9):3811–5.
50. Cicchese JM, Evans S, Hult C, Joslyn LR, Wessler T, Millar JA, Marino S, Cilfone NA, Mattila JT, Linderman JJ, Kirschner DE. Dynamic balance of pro-and anti-inflammatory signals controls disease and limits pathology. Immunol Rev. 2018;285(1):147–67.
51. Mir PA, Mahajan S, Verma A, Kumar N, Arora M, Nagpal N. In-vitro antioxidant and anti-inflammatory potential of ficus infectoria fruits. Pharmacognosy. Research. 2022;14(2):153.
52. Utami W, Aziz HA, Fitriani IN, Zikri AT, Mayasri A, Nasrudin D. In silico anti-inflammatory activity evaluation of some bioactive compound from ficus religiosa through molecular docking approach. In: Journal of physics: conference series 2020. 1563, No. 1 IOP Publishing; 2020. p. 012024.
53. Lv H, Zhang X, Chen X, Xie Z, Hu C, Wen C, Jiang K. Phytochemical compositions and anti-oxidant and anti-inflammatory activities of crude extracts from Ficus pandurata H.(Moraceae). Evid Based Complement Alternat Med. 2013;2013:1.
54. Abe T, Koyama Y, Nishimura K, Okiura A, Takahashi T. Efficacy and safety of fig (Ficus carica L.) leaf tea in adults with mild atopic dermatitis: a double-blind, randomized, placebo-controlled preliminary trial. Nutrients. 2022;14(21):4470.
55. Fürst R, Zündorf I. Plant-derived anti-inflammatory compounds: hopes and disappointments regarding the translation of preclinical knowledge into clinical progress. Mediat Inflamm. 2014;2014:1.
56. Kim JH, Kismali G, Gupta SC. Natural products for the prevention and treatment of chronic inflammatory diseases: integrating traditional medicine into modern chronic diseases care. Evid Based Complement Alternat Med. 2018;2018:9837863.
57. Livrea MA, Tesoriere L. Antioxidative effects of cactus pear (Opuntia ficus-indica (L).) Mill. fruits from Sicily and bioavailability of betalain components in healthy humans. In: VI international congress on cactus pear and cochineal 811; 2007. p. 197–204.
58. Deshpande NS, Dwivedi PS, Revanasiddappa BC. Virtual screening, pharmacokinetics & MD simulation study of active phytoconstituents of Ficus Carica Linn. against PPAR-γ in diabetes mellitus. J Biomol Struct Dyn. 2023;1–7. https://doi.org/10.1080/07391102.2023.2279286.

Potential Prospects of Underexploited Wild Edible Fruits of Tamil Nadu

M. A. Deepa, M. Pugalenthi, S. Narthanaa, M. Pradheeba, and T. Parimelazhagan

Abstract

Underutilized Wild Edible Fruits is a treasure trove of minor, neglected and often overlooked fruits that have played significant roles in human diets for centuries. These fruits, characterized by their local, orphaned, and promising nature, have faced challenges such as low supply, poor shelf life, and limited consumer awareness, leading to their diminished importance in modern times. Despite their underutilization, wild edible fruits have been vital sources of sustenance, flavour and even medicinal properties throughout history. They offer alternative food sources during periods of scarcity, potential nutritional supplements to combat malnutrition and income opportunities for rural communities. However, with the dominance of industrial agriculture and globalization, traditional knowledge of these fruits has dwindled and commercially cultivated varieties have overshadowed them. Yet, there is a growing interest in reviving and exploring these underexploited fruits. Researchers, nutritionists, and conservationists are recognizing their nutritional richness, potential for biodiversity conservation, and contributions to sustainable development. This chapter aims to provide a comprehensive overview of underexploited wild edible fruits, delving into their nutritional composition, culinary uses, traditional knowledge, conservation status and potential for cultivation. By shedding light on these often-neglected fruits, we hope to cultivate a deeper appreciation for biodiversity and advocate for sustainable methodologies to preserve these invaluable resources for future generations.

M. A. Deepa (✉) · M. Pugalenthi · S. Narthanaa · M. Pradheeba
Department of Botany, Government Arts College (A), Coimbatore, Tamil Nadu, India

T. Parimelazhagan
Department of Botany, Bharathiar University, Coimbatore, Tamil Nadu, India

P. Thangaraj (ed.), *Pomotherapeutic Insights on Wild Edible Fruits*,
https://doi.org/10.1007/978-981-96-0543-9_8

Keywords

Underutilized fruits · *Trema orientalis* · *Berberis tinctoria* · *Ficus pumila* · Underexploited wild edible fruits

Introduction

Underutilized wild edible fruits is a treasure trove of minor, neglected and often overlooked fruits that have played significant roles in human diets for centuries. These fruits, characterized by their local, orphaned and promising nature, have faced challenges such as low supply, poor shelf life, and limited consumer awareness, leading to their diminished importance in modern times [1]. Despite their underutilization, wild edible fruits have been vital sources of sustenance, flavour and even medicinal properties throughout history. They offer alternative food sources during periods of scarcity, potential nutritional supplements to combat malnutrition and income opportunities for rural communities [2]. From South America's rainforests to Africa's deserts, indigenous cultures have long relied on the diverse array of wild fruits growing in their environments [3, 4].

However, with the dominance of industrial agriculture and globalization, traditional knowledge of these fruits has dwindled, and they have been overshadowed by commercially cultivated varieties [5]. Yet, there is a growing interest in reviving and exploring these underexploited fruits. Researchers, nutritionists, and conservationists are recognizing their nutritional richness, potential for biodiversity conservation, and contributions to sustainable development.

This chapter aims to provide a comprehensive overview of underexploited wild edible fruits, delving into their nutritional composition, culinary uses, traditional knowledge, conservation status and potential for cultivation. By shedding light on these often-neglected fruits, we hope to cultivate a deeper appreciation for biodiversity and advocate for sustainable methodologies to preserve these invaluable resources for future generations.

Nutritional Composition

Nutritional Composition

The Food and Agricultural Organization (FAO) [6] of the United Nations is currently promoting the conservation and sustainable use of biodiversity for nutrition and agriculture to increase dietary diversity. According to the World Health Organization (WHO) [7], 80% of the global population depends on traditional remedies for their medicines, which contain plant-derived compounds.

In addition to this, herbal medicine is now utilized by more than half of the global population to maintain health, and plant products have emerged as the primary source of novel medicines [8]. The resurgence of local health traditions as an essential component of local culture has been compelled by changing circumstances, particularly in the post-COVID period. In recent years, it has become more customary for individuals to consume a greater variety of plant-based natural products, which has increased the demand in the market. In particular, the post-COVID era has witnessed changes in food consumption patterns concerning local fruits and vegetables to boost immunity [9, 10].

Even though several initiatives are going on to enhance the productivity of a few domesticated crops, nutritional security remains a concern [11]. In this regard, the use of underutilized wild edible fruit species has been suggested as an alternative [12]. The challenge of ensuring food and nutrition security for a global population projected to reach 9.7 billion by 2050 is compounded by the COVID-19 pandemic, which has increased the number of people facing hunger, estimated to be between 720 and 811 million [13]. Access to adequate, healthy and nutritious food is essential for food security and micronutrient deficiencies are a hidden component of this issue [14]. The triple burden of malnutrition, undernourishment, micronutrient deficiency and overnutrition affects different strata of society in both developed and developing countries [15, 16].

Current research predominantly targets plant secondary metabolites as potential medicinal agents to provide protection and treatment against a variety of degenerative diseases. Fruits, which are a significant part of our diet, play a crucial role as essential nutrients and offer protection against numerous diseases. According to the World Health Organization [17], individuals should consume more than 400 g of fruit per day to prevent non-communicable diseases related to food. However, poor consumption is often caused by accessibility issues, high costs and lack of knowledge [18]. Wild edible fruits are the most widely used non-timber forest products [3]. Being native species that grow and reproduce without human intervention in their natural habitats, these fruits are excellent sources of nutrients and minerals that are lacking in many diets [2].

The availability of reports on the biological applications of wild edible fruit species, their use as functional foods and their nutraceutical benefits for human health has increased [19]. Reliance on a limited number of crop species for food security poses a risk. Micronutrient deficiency, including iron (Fe) deficiency, affects over two billion people worldwide and is referred to as "hidden hunger" [13]. Iron deficiency is a widespread nutritional disorder in developing countries and is also prevalent in developed countries [20]. Utilizing these untapped wild edible fruit sources can diversify our food resources and control these deficiencies, which are currently in short supply. It can be a crucial source of food and income for rural populations and malnourished children [21].

Biodiversity of Indian Underutilized Wild Edible Fruits

The rich biodiversity profile of a country reflects the presence of different varieties of taxa of flora and fauna and the consequences of their utilization and conservation [22]. India is rich in species and ecosystem diversities with 1, 03,258 species of fauna and 55,048 species of flora that have been documented in the 10 biogeographic zones in the country [23]. India boasts a rich floral diversity comprising 55,048 known plant species, among which 12,095 are endemic [24].India's diverse agroclimatic zones support a wide variety of wild edible fruits, ranging from the Himalayan foothills in the North to the Western Ghats in the South and from the arid regions of Rajasthan to the tropical forests of the Northeast [25]. These fruits exhibit a remarkable diversity in terms of flavours, textures and nutritional profiles, reflecting the country's rich ecological heritage [26].

Wild edible fruits play a significant role in the cultural traditions and dietary practices of indigenous communities across India. They are often incorporated into religious rituals, festivals and culinary traditions, reflecting the deeper connection between people and their natural environment. Furthermore, the sustainable harvesting and processing of these fruits provide livelihood opportunities for rural communities, particularly women and marginalized groups [27].

Indian traditional practitioners use wild edible fruits as folk medicine [28]. In Indian folk medicine, they are frequently incorporated into various Ayurveda formulations [29]. The vast traditional knowledge of medicinal plants in our country is currently playing a crucial role in the development of new drugs [30]. Physicians of traditional medicines such as Ayurveda, Homoeopathy and Unani use various species of medicinal plants that have been used in Hindu traditional medicine for centuries [31].

Underutilized Wild Edible Fruits of the Western Ghats

Over the past decades, numerous studies have enumerated the significant botanical wealth harboured within the Western Ghats [32]. This ecologically diverse region has emerged as a crucial source of various value-added products catering to the food, agriculture, medicine and cosmetic industries [33]. Particularly, the wild edible fruits indigenous to this region not only serve as sources of nutrition but also exhibit medicinal properties [34]. There are about 394 wild edible fruit species spanning 216 genera across 83 families in South India [35]. Habit-wise analysis revealed that 224 species (57%) were trees, 80 species (20%) were shrubs, 31 species (8%) were herbs and 58 species (15%) were climbers [36]. Among these fruit species, 359 species (91%) were native to the region, while 34 species (9%) were naturalized in South India [37]. This emphasizes the rich diversity of fruit species in the region and emphasizes the importance of understanding, conserving and utilizing these resources effectively [33].

The Western Ghats boast an impressive array of wild edible fruits with over 60 species identified across various families [38]. Recent research efforts have shed light on the abundance and diversity of wild edible plants in the Western Ghats [39]. Jadhav et al. [40] documented 159 wild edible plant species in the Northern Western Ghats, with 77 of them bearing edible fruits, predominantly consumed from January to July [40]. Similarly, Shiddamallayya and Kumar [38] reported 75 wild edible fruit species in the forest region of Hassan district in the Southern Western Ghats, belonging to 40 families and 60 genera [38]. Studies by Nazarudeen [41] have explored the nutritional composition of several wild fruits, highlighting their rich protein, fat and iron content [41].

Phytochemical Profile and Health Benefits

More than 900 different phytochemicals have been identified in foods and a single vegetable or plant food can contain over 100 different phytochemicals [42]. Many of these phytochemicals possess antioxidant properties and help protect cells against oxidative damage caused by reactive oxygen species. Antioxidants are molecules that can scavenge or inhibit the oxidation of other molecules and oxidation reactions can generate reactive oxygen species, such as oxygen free radicals, which can lead to the formation of unwanted products or cell damage, causing diseases such as cancer, arthritis, diabetes and other diseases related to humans [43]. Phytochemicals, such as polyphenols and other bioactive compounds, can prevent these chain reactions by scavenging free radicals and obstructing the oxidation of other biomolecules [44]. These phytochemicals offer endless prospects for new drug development due to their unmatched chemical variety [45]. The plant-derived food products are considered to be less toxic and free from side effects than synthetic drugs [46]. Moreover, diet-related diseases are a major contributor to disease and health inequities in rural areas. Chronic non-communicable diseases such as type-2 diabetes, cardiovascular disease, and cancer are on the rise and more prevalent among consumers of highly processed, high-fat, high-sugar convenience foods [47].

Some wild fruits possess nutritive values comparably higher than commercial fruits [48]. They also possess a wider range of phytochemicals and antioxidants than commercial fruits. The inclusion of these fruits in the daily diet can reduce the risk of diseases such as diabetes, cancer and coronary heart disease, due to their high mineral and vitamin content [42]. Furthermore, incorporating these fruits into agricultural systems can protect biodiversity, ensure food security and boost the rural economy [2]. As the global population is projected to exceed nine billion by 2050, increasing global food demand by 50% compared to 2013, it becomes essential to properly identify, document, enumerate and domesticate wild edible fruits to meet this demand [49, 50].

Ethnobotanical Review of Underutilized Wild Edible Fruits

The majority of Tribal people of Tamil Nadu principally rely on forests for their livelihood [51]. The tribal population of the state derives sustenance from forests for food, fibre, fodder as well as the traditional healthcare system's ethno-medicines [52]. Tribals of various ethnic groups like Savara, Khonds, Nayaka, Pulayans, Maratis, Irular, Uraly, Gonds, Koyas and others are predominant in the various parts of South India and follow traditional medicines using UWEFs [53]. In addition to other edible and non-edible plants in the forests, a variety of wild fruits are also present, which the tribal people have been utilizing since ancient times to treat various ailments in the traditional healthcare system [54]. However, the medicinal values of wild edible fruits have been transmitted orally from generation to generation without proper documentation [55]. In numerous villages, these fruits also provide a source of sustainable income and livelihood security [56]. At present, with the continuous growth of population and shrinking of land, as well as the utilization of forest areas for developmental and constructional purposes, these wild fruits are on the verge of extinction [57]. Although these fruits have been employed in the folk medicines of the traditional healthcare system, the traditional knowledge concerning the ethno-medicinal uses of these fruits is losing its intrinsic value due to the lack of proper documentation. Hence, the scientific community must safeguard these valuable natural resources of forests through domestication and conservation efforts [58].

The following information offers insights into several noteworthy wild fruits found in the Western Ghats, particularly in Southern India. Although the list provided may not encompass all wild edible fruit species present in South India, it offers a substantial representation of the diversity within the region, and it encompasses a diverse selection of fruits traditionally consumed by tribal and rural communities. The primary objective of this chapter was to explore the potential of these underutilized wild edible fruits in improving the socioeconomic conditions of local communities. By broadening the spectrum of available food resources, this chapter advocates for the consumption of these fruits while also assessing their phytochemical attributes and conservation status. Additionally, the study emphasizes the importance of the secondary metabolites present in these fruits, which hold promise in targeting various degenerative diseases and their underlying mechanisms through pharmaceutical interventions. This chapter discusses the following underutilized wild edible fruits in the Western Ghats, Tamil Nadu (Table 1).

Trema orientalis L.

Trema orientalis L. belongs to the family Cannabaceae, which is an evergreen tree. It is commonly called pigeon wood, hop out, charcoal tree, Indian charcoal tree, Indian nettle tree and gunpowder tree [59]. It is also known by the synonyms *Trema bracteolate* Hochst. Blume, *Trema guineensis* (Schum. and Thonn.) Ficalho, *Celtis orientalis* Linn., *Celtis guineensis* (Schum. and Thonn.) and *Sponiaorientalis* Linn. The name "*Trema*" derives from the Greek word for "hole", referencing the pitted stone of its fruit, while the common name "pigeon wood" stems from the frequent observation of pigeons nesting or roosting in these trees [60].

Description

T. orientalis is a dioecious tree. It reaches up to 15 m in height, with bark measuring 0.6 cm thick, thin and either greyish or bluish-green in colour, appearing rough and

Table 1 List of medicinal plants discussed in the article

S. No	Name of the plant	Common name	Family	Vernacular name
1	*Trema orientalis* L.	Indian charcoal tree	Cannabaceae	Yerralai
2	*Miliusa tomentosa* (Roxb.) J. Sinclair	Woolly Miliusa	Annonaceae	Periyavaay
3	*Tinospora cordifolia* (Willd.) Miers ex Hook.F. & amp	Heart-leaved moonseed	Menispermaceae	Seenthil kodi
4	*Toddalia asiatica* (L.) Lam.	Forest pepper	Rutaceae	Kattu milagu
5	*Rhodomyrtus tomentosa* (Aiton) Hassk.	Rose myrtle	Myrtaceae	Malaikoyya
6	*Bridelia retusa* L.	Spinous Kino tree	Phyllanthaceae	Mulluvengai
7	*Berberis tinctoria* Lesch.	Dye Barberry	Berberidaceae	Nilgiri Barberry
8	*Cayratia pedata* (Lam.) Gagnep	Bird foot grapevine	Vitaceae	Kattu pirandai
9	*Ficus pumila* L.	Climbing fig	Moraceae	Kodi athi
10	*Grewia subinaequalis* DC.	Phalsa	Tiliaceae	Tadachi
11	*Falcourtiamontana* Graham	Mountain sweet thorn	Salicaceae	Charalmaram
12	*Viburnum coriaceum* Blume	Cylindrical viburnum	Viburnaceae	Rashina

lenticellate. The inner bark, when exposed, shows a creamy-yellow hue streaked with other colours. Branchlets are typically scabrous or covered with adpressed pubescence. The leaves are simple and arranged alternately along the stems. Stipules are lateral and soon fall off. Petioles measure 5–10 mm in length, are slender, tomentose and grooved above. Leaf blades are ovate-lanceolate, ovate or oblong-lanceolate, measuring 7.5–15 cm long and 2.5–6 cm wide. Their bases are obliquely cordate, while the apices are acuminate. Leaf margins are serrulate, with the upper surface being scabrid and the lower surface tomentose. Leaves are chartaceous and prominently 3–5-ribbed from the base, with 3–4 pairs of lateral nerves, pinnate and prominent. Intercostae is reticulate and also prominent. The flowers are unisexual, greenish and measure 3–4 mm across. They are arranged in axillary fascicles or cymes. Male flowers are usually sessile, with four or five equal tepals, each about 2 mm long and curved, ciliate. They have five stamens and an oblong pistillode. Female flowers have unequal tepals, also ciliate, with a superior, sessile, one-celled ovary that is ovate. The style is bifid and villous, while the stigma is plumose. Fruits are drupes, measuring about 4 × 3 mm, globose, and black when mature, with a persistent stylar tip. The seed inside is globose.

Distribution

T. orientalis is a native habitat of tropical and subtropical regions and is widely distributed around three continents, especially in Africa and Asia. The tree is commonly found along stream banks and in ravine areas, particularly encountered along stream banks and in ravine areas in various regions, including Karnataka (Chikmagalur, Coorg, Hassan, Mysore, North Kanara, Shimoga, South Kanara), Kerala (all districts) and Tamil Nadu (Dindigul, Namakkal, Salem, Theni, Tiruchchirappalli, Tiruvannamalai, Viluppuram).

Conservation Status

It has been recently assessed as the least concern by the IUCN 3.1 Red List of Threatened Species [61].

Ethnobotanical Uses

T. orientalis is most popularly used to cure various ailments especially in the forms of decoctions, which include bark decoctions (used as vermifuge, to cure toothache, dysentery), leaf decoctions (used to treat malaria, venereal diseases) and the infusions of fruits, and flowers are given to children to cure pulmonary disorders and the roots are used in folk medicine to cure haematuria, blood stasis. The fruits of this tree are utilized to prepare infusions, which are administered to children as a therapeutic remedy for bronchitis, pneumonia and pleurisy [62, 63].

Scientific Validation and Economic Uses

The chemical compounds, such as Methylswertianin, decussatin, glycosides of decussatin, sweroside, scopoletin, (-)epicatechin, lupeol, p-hydroxybenzoic acid, 3,4-dihydroxybenzoic acid, adian-5-en-3-one,2a, 3a, 23-trihydroxyurs12-en-28-oic acid, 2a, 3b-dihydroxyurs-12-en-28-oic acid, b-sitosterol, 3-O-b-glucopyranosyl-b-sitosterol and hexacosanoic acid, were found in *T. orientalis*. It has exhibited several pharmacological activities, including a laxative effect, hypoglycaemic, antiplasmodial, diuretic, analgesic, anti-arthritic, anthelmintic, cardioprotective, anticancer, antioxidant and antimicrobial properties [60, 64–69].

Fig. 1 Underutilized wild edible fruits

Miliusa tomentosa (Roxb.) J. Sinclair

Miliusa tomentosa (Roxb.) J. Sinclair (Fig. 1a) belongs to the family Annonaceae. It is known by various synonyms *Saccopetalum tomentosum* (Roxb.) Hook. f. & Thomson and *Uvaria tomentosa* Roxb. It is commonly called as Umbh, Hoom, Barre duduga, Gandha palasa, Kari, Kothukalaa and Nool ilia maram.

Description

It has simple, oval, alternating, glabrous to pubescent, membrane-like leaves. Flowers are solitary, fascicled or cymes, extra-axillary or axillary, bracts oval or obovate, monoecious, dioecious, or sporadically polygamous valvate. Sepals are 3 in number, petals arranged 6 in 2 series, valvate in buds and the inner petals are larger and thinner, initially adhering by the edges, and finally freeing the outer petals, which are smaller like sepals. Anthers globose, extrorse, connectives slightly apiculate and rounded at the top, and stamens definite or indefinite. Ovules are

1–10, stigma is club-shaped, a style typically short and carpels are indeterminate and linear–oblong. Several ripe carpels that are stalked or sub-sessile, ovoid or oblong, with one to two or more seeds.

Distribution

M. tomentosa is indigenous to India. It is globally distributed in India, Nepal and Sri Lanka. In India, it is distributed in Tamil Nadu, Andhra Pradesh, Odisha, Rajasthan, Kerala, Gujarat, Madhya Pradesh and Uttar Pradesh.

Conservation Status

This plant comes under the category of not evaluated (NE) in the IUCN Red List data.

Ethnobotanical Uses

Fruits are edible and it is used to cure weakness in children, mostly used by Gond, koya, hazzar, bhil and grasia tribes of India. Unripen fruits are dried, crushed and burnt to cure post-delivery swellings. The leaves are used to cure constipation and respiratory disorders. Leaves are crushed in water and filtered with cloth, and the extract is given for relief from internal stomach swelling. The bark powder is used as a substitute for tragacanth, it is also mixed with mustard oil applied to treat paralysis and it is also used to cure dysentery. The fruits yield a pale-yellow gum known as Karee gum [70].

Scientific Validation and Economic Uses

The biological activities of *Miliusa tomentosa* fruits have not been reported so far. The primary investigations have been initiated to bring about the unrevealed therapeutic potentials of the wild edible fruits of *Miliusa tomentosa,* which exhibited promising antioxidant and nutritional profiles and the GC-MS analysis of the fruit extract revealed the presence of potential phytochemical compounds with significant therapeutic values, yet the analgesic and antimicrobial activities of *M. tomentosa* have been studied in recent years.

Tinospora cordifolia (Willd.) Hook. f. and Thoms.

Tinospora cordifolia (Fig. 1b), commonly known as Guduchi or Giloy, is indeed a large, glabrous (meaning lacking hair), deciduous climbing shrub [71]. It belongs to the family Menispermaceae. The plant is commonly known as Giloe, and it has a rich cultural and mythological significance. The name "Giloe" comes from Hindu mythology, where it refers to the heavenly elixir that bestows immortality and eternal youth upon those who consume it.

Description

They are succulent woody climbing shrubs that typically develop long aerial roots from host trees, allowing them to spread and thrive. The bark of these shrubs is corky or papery, and it tends to peel off as it dries. The leaves are simple, alternate, broadly ovate, with a slightly heart-shaped base. They have either entire or dentate margins and are not peltate. The leaves typically feature 3–7 basal veins and 1–3 pairs of lateral veins. The petioles are geniculate, meaning they have a bent or angled shape, and they are swollen near the base. The inflorescences of *Tinospora* species occur either in the axils of leaves or on old leafless branches. They are arranged in pseudo-racemes, pseudo-paniculate or thyrsoid clusters, with a few flowers per cluster, each on a peduncle. The flowers are unisexual and have pedicels. In male flowers, there are six sepals arranged in two series, with the inner series being larger than the outer series. The sepals are free, imbricate and subelliptic, with an acute apex. There are also three to six free petals that are obovate, with a cuneate apex and fleshy texture. The stamens number six, with free filaments and loculed anthers that dehisce obliquely or longitudinally. Female flowers share similarities with male flowers in terms of sepals and petals. However, there are six subulate staminodes present. The carpels number three and are ellipsoid-curved in shape, with a short, thick style and a reflexed, peltate stigma. The fruits of *Tinospora* species are drupes that are either sub-globose or ellipsoid in shape. They feature a columnar carpophore and a hard, bony endocarp [72]. The endocarp is dorsally convex and often verrucose, with deep rumination. The ventral side of the fruit typically has a central aperture. Seeds are curved or half-moon-shaped, endospermic, with flattened, leaf-like cotyledons and a short radicle.

Distribution

It is distributed in India, Sri Lanka, Bangladesh and Myanmar. It is distributed across various regions in India. Its local distribution includes states such as Arunachal Pradesh, Assam, Bihar, Delhi, Gujarat, Goa, Karnataka, Kerala, Maharashtra, Odisha, Sikkim, Tamil Nadu, Uttar Pradesh, West Bengal and likely others. These regions provide suitable habitats for the growth of *Tinospora cordifolia*, allowing it to thrive in diverse climatic and ecological conditions across the Indian subcontinent.

Conservation Status

It has been recognized as the least concern in IUCN Red List data.

Ethnobotanical Uses

Tinospora cordifolia (Guduchi) holds significant importance in the traditional medicine systems of various tribal and ethnic groups across India. Its versatile uses are documented in ethnobotanical surveys, reflecting its widespread application in folk remedies. The Baiga tribe in Uttar Pradesh uses a paste of Guduchi stem and *Solanum surattense* roots for fever, and Mumbai's tribal communities employ the whole plant for ailments like fever, jaundice and chronic diarrhoea. In North Gujarat, powdered root and stem bark are utilized for conditions like cancer, dysentery and periodic fever. Similarly, other regions like Jammu, Orissa, Bihar, Punjab, Maharashtra, Haryana and Tamil Nadu have their unique applications of *T. cordifolia* for various health issues, ranging from fever and bone fractures to general debility and cough [73–76].

Scientific Validation and Economic Uses

Tinospora cordifolia is rich in a diverse array of chemical compounds, including alkaloids, diterpenoid lactones, glycosides, steroids, sesquiterpenoids, phenolics, aliphatic compounds and polysaccharides. The leaves of this plant are particularly noteworthy for their high protein content (11.2%), as well as significant amounts of calcium and phosphorus. Among the compounds isolated from *T. cordifolia* are four new clerodane furano diterpene glucosides (amritosides A, B, C and D), whose structures were elucidated through spectroscopic studies. Additionally, a polysaccharide with various glucose and xylose linkages has been identified, exhibiting polyclonal mitogenic activity against B cells. Studies on a novel (1,4)-alpha-D-glucan (RR1) from *T. cordifolia* have explored its immunostimulating properties in macrophages. An arabinogalactan isolated from dried stems also exhibited mitogenic activity against B cells. Overall, the phytochemical characterization of *T. cordifolia* underscores its pharmacological potential and traditional medicinal uses, with compounds like tinosporaside being of particular interest due to their specified limits and potential therapeutic applications. *Tinospora cordifolia* holds a special place in Ayurvedic medicine as Medhya Rasayana, which refers to substances that enhance learning and memory. It is also noted for its efficacy in treating bhrama, or vertigo, as described in various Ayurvedic texts. Additionally, recent research suggests promising outcomes in children with moderate behaviour disorders and mental deficits, with observed improvements in IQ levels. These findings highlight the potential of *T. cordifolia* in promoting cognitive health and addressing neurological conditions, aligning with its traditional use as a brain tonic in Ayurveda [72, 77, 78].

Toddalia asiatica (L.) Lam.

Toddalia asiatica is (Fig. 1c) also known as Lopez root or Kanchanah. It indeed belongs to the Rutaceae family and is the only plant in the *Toddalia* genus. Its aliases include kanch and mullakasinda, among others. The interesting aspect of *Toddalia asiatica* is its seed dispersal mechanism, facilitated by birds and monkeys that consume its fruits. The scaly breasted appearance, in particular, shows a preference for nesting in these trees, which suggests a potential ecological relationship between the plant and the bird species. This mutualistic interaction aids in the dispersal of seeds and potentially contributes to the plant's distribution and survival in its habitat.

Description

It is a woody liana characterized by its corky, thorny stems that can climb trees, often reaching lengths of up to 10 m. It has shiny green leaves with a lemon scent, and its flowers are creamish white in colour. The sepals of the plant measure between 0.3 and 0.5 mm, while the petals are ovate to elliptic and range from 1 to 3.5 mm in size. In male flowers, the stamens are 3–4 mm long, while in female flowers, they are much smaller, measuring only 0.2–0.8 mm and ligulate in shape. The disk of the flower is typically 0.2–0.5 mm. The gynoecium, or female reproductive organ, in female flowers, is ovoid to ellipsoid and measures 1.5–2.5 mm, whereas in male flowers, it is sub-cylindrical and approximately 1–2 mm in size. The fruits of the *Toddalia asiatica* are small, orange in colour and about half a centimetre wide. They have a taste reminiscent of orange peel. Overall, it is a visually striking plant with aromatic foliage and fruit.

Distribution

It is commonly found along shola borders and forest thickets at altitudes above 1000 m. Its distribution spans from India to China and Malaysia, including countries in Indo-Malaysia and Africa. In India, it is widely distributed. In the state of Kerala, it is found in all districts. Additionally, it is present in Assam, Karnataka and Odisha. In Assam specifically, the Orange Climber is distributed throughout the state.

Conservation Status

It comes under the category of not evaluated (NE) in IUCN Red List data.

Ethnobotanical Uses

Traditionally, *Toddalia asiatica* serves multiple purposes, with significant importance placed on its medicinal properties. Among various uses, the fruit is employed to treat malaria and coughs, while the roots are utilized for addressing indigestion and influenza. Additionally, the leaves are utilized in the treatment of lung diseases and rheumatism. Beyond its medicinal applications, it also serves as a browse for goats and is utilized as a hedge among communities such as the Maasai and Kipsigis. In recent studied by Jakob et al. 2020, the plant is reported to possess Immunosuppressant activity [79].

Scientific Validation and Economic Uses

T. asiatica is rich in bioactive compounds with over 165 reported so far. These include 69 coumarins, 69 alkaloids, 8 terpenoids, 5 flavonoids and 14 other compositions, such as lipids, alcohols, phenolic acids, lignans, steroids and fatty acids. Coumarins and alkaloids are particularly characteristic of *T. asiatica*, reflecting its pharmacological potential and traditional medicinal uses. These compounds contribute to the plant's diverse therapeutic properties and its significance in traditional medicine systems. Research on the pharmacological activities of *T. asiatica* has revealed various biological effects, including anti-inflammatory and analgesic properties, haemostatic coagulation, antibacterial activity, antioxidant effects, and antitumour effects [80]. It is reported to possess 15 compounds, out of which toddalic acid and two enantiomers of toddanolic acid. Among them, nitidine chloride inhibited the proliferation of primary human T cells with IC50 value of 0.4 μm [81].

Rhodomyrtus tomentosa (Aiton) Hassk.

Rhodomyrtus tomentosa (Aiton) Hassk. (Fig. 1d) belongs to the family Myrtaceae. It is also known by the synonym *Myrtus tomentosa* Aiton. They are commonly called rose myrtle, hill Gooseberry, Thavittukoya and Malai koyya.

Description

It is an evergreen shrub, a stout shrub, densely covered with fine hairs (tomentose); branchlets are cylindrical (terete). Leaves measure up to 5 × 2.5 cm, elliptic-obovate, with three prominent veins, and densely covered with white hairs underneath;

petioles are 5 mm long. Cymes are broad, measuring 15 cm across, and are terminal or located in the upper axils of the branches. Flowers are pedicelled, with a globose calyx tube that is 6 mm long and pink, densely covered with fine hairs, with five orbicular lobes. Petals number five, white, orbicular and hairy on the outer surface. Stamens are numerous, with filaments measuring 8 mm long and pink in colour. The ovary has 2–3 cells, and the style is stellate, with a few hairy ovules. The capsule is 10 mm across, globose and densely covered with hairs.

Distribution

R. tomentosa is native to Southeastern and Southern Asia, Indo-Malaysia and China. In India, it is distributed in Tamil Nadu, particularly in Coimbatore and Nilgiris. In Kerala, the plant is reported in Idukki, Palakkad and Malapuram.

Conservation Status

This plant is assessed as Not Evaluated (NE) in IUCN Red List data.

Ethnobotanical Uses

R. tomentosa has been used in traditional Chinese, Vietnamese and Malaysian medicine from immemorial. The berries of *R. tomentosa* are edible; in Malaysia and Vietnam, the fruits are used to cure dysentery and diarrhoea. The aerial parts are used as a medicine to treat postpartum women. Traditionally, they are used to cure urinary tract infections and wounds and are also used as an antidote.

Scientific Validation and Economic Uses

The fruits of *R. tomentosa* are reported to possess high nutritional value with high amounts of proteins and total dietary fibre, and the main amino acid was found to be tryptophan. The analysis of *R. tomentosa* fruit has shown a clear observation regarding minerals and vitamins. It contains different minerals with high level of potassium (221.76 mg/150 g fruit), calcium (73.65 mg/150 g fruit), manganese (3.23 mg/150 g fruit), iron (1.54 mg/150 g fruit), zinc (0.61 mg/150 g fruit), and copper (0.40 mg/150 g fruit). Meanwhile, the vitamin E level (3.89 mg/150 g fruit) was higher than that of mango and avocado [82]. Numerous phytochemical compounds have been isolated and the principal components were found to be phenols, flavonoids, meroterpenoids and triterpenoids (Lupeol, botulin, botulin, barnyrin, beta-amphenol). An acylphlroglucinol compound rhodomyrtone is considered a more effective antibiotic agent and exhibits potential pharmacological activities. Rhodomyrtone has also been reported to be a potential antiproliferative and apoptosis, inducing agent in HeLa T keratinocyte cells. A skin cosmetic agent piceatannol 4′- O-β- D-glucopyranoside was isolated from the leaves, and combretol, a flavonoid compound, was isolated from the bark. From the fruits of *R. tomentosa*, a familiar fermented drink Ruou sim and many processed products, such as jam, jellies, wines and canned syrup, are being prepared in Vietnam and China for consumption [83].

Bridelia retusa (L.) A. Juss

Bridelia retusa (L.) A. Juss (Fig. 1e), commonly known as Kasi, is a tree belonging to the Phyllanthaceae family. It is a wild tree [84]. It is typically found at altitudes up to 1000 m, excluding very dry regions. The tree is known for its edible fruits. It is commonly cultivated in dry evergreen and deciduous forests, as well as open land with sandy, loamy, granite and basalt-derived soils.

Description

Bridelia retusa is characterized by its self-supporting growth form by presenting as a tree species. The leaves are simple and broad, arranged alternately, with stipules measuring 7 mm in length, which are lanceolate and lateral in position. The tree can reach heights of up to 15 m. The bark of *Bridelia retusa* appears blackish-brown externally and yellowish-brown internally. It has an astringent taste and odourless and feels rough to the touch. Flowers of *Bridelia retusa* are creamy-white and arranged in terminal panicles, which can be either erect or drooping. The fruits are loculicidal capsules, smooth on the enclosed perianth and ripen from a greenish-purple to wrinkled texture. Flowering and fruiting typically occur between August and December [85–87].

Distribution

Bridelia retusa is primarily found in the Indo-Malayan region. In India, *Bridelia retusa* is distributed throughout the country, with a significant presence in Kerala, spanning all districts and in Tamil Nadu, particularly in Nilgiris.

Conservation Status

It is categorized as the least concern by IUCN Red List data.

Ethnobotanical Status

Bridelia retusa is highly esteemed in South India for its medicinal properties, particularly in the treatment of rheumatic pain. The leaves of this plant are utilized for various ailments, including wound healing, urinary tract infections, dysentery and diarrhoea. The roots of *Bridelia retusa* are employed as a remedy for gonorrhoea, and they possess astringent and purgative properties. Furthermore, the fruits and seeds are utilized to alleviate earaches and indigestion. The bark of *Bridelia retusa* finds application in more specialized treatments, such as inducing sterility in women and serving as a contraceptive. Additionally, it is utilized as an antidote against snakebites [84]. The wood is hard and close-grained and is used locally for house construction, carts, agricultural implements, joinery and carving. The mottled silver grain makes the wood suitable for decorative purposes. It yields a good quality fodder, which has a high protein content and is commonly used at the foot of the Siwalik Range, Nepal [88].

Scientific Validation and Economic Uses

The bark of *Bridelia retusa* contains a variety of phytochemicals with reported biological activity, including steroids, triterpenoids and tannins. Additionally, minerals such as calcium, copper, iron, manganese, magnesium, phosphorous and zinc have been identified in this species [89]. *Bridelia retusa* is characterized by its pungent and bitter taste and is traditionally considered useful in balancing the Vata dosha. Furthermore, it is utilized in the treatment of jaundice, often in combination with other plants. Studies have shown that the ethanol extract of *Bridelia retusa* leaves contains chitosan flavonoid, which exhibits significant anti-inflammatory and analgesic activity [90]. These findings underscore the potential therapeutic value of *Bridelia retusa* in managing inflammatory conditions and pain. The fruits are edible and used for medicinal purposes. Tribal communities of Odisha and other states collect them and use them to sell in local markets.

Berberis tinctoria **Lesch.**

Berberis tinctoria Lesch., commonly referred to as Nilgiri Barberry, is an evergreen shrub belonging to the Berberidaceae family. This species is endemic to the Nilgiris and Palani Hills of Tamil Nadu, India. It is commonly known as Oosikala in Nilgiris.

Description

Berberis species are characterized as deciduous or evergreen shrubs, with some being spiny while others are glabrous. The stems can be angular, terete or sulcate, with spines typically occurring in groups of 3 or 5. The leaves are generally simple, often jointed at the junction of the lamina and petiole. They are arranged in whorls or fascicles, attached to the node and protected by spines. Long shoots may be transformed into spines, while short shoot axils bear clusters of green leaves. The leaves are alternate, lanceolate or ovate, with entire or dentate margins, and are usually petiolate. Inflorescences are solitary or fascicled flowers arranged axillary in cymes. The flowers are bisexual, with typically six sepals arranged in two to four imbricate series. Bracts are usually 2–3, appressed and caducous. Six petals are arranged in two series, free and imbricate, often with two basal glands towards the inner side. Stamens range from 3 to 6, with anther locules opening by recurved valves. The ovary is simple and club-shaped, containing 1–12 sub-basal erect ovules. The style is short or sessile, with a stigma. Fruits are few-seeded berries, varying in colour from blood red to black and in shape from ellipsoid to globose or ovoid. Seeds typically number 2–10 and are reddish-brown or black in colour.

Distribution

It is found in the Shola forests of Nilgiris and Palani Hills, altitude of about 1800 m. In Kerala, *Berberis tinctoria* is found in the Idukki district. In Tamil Nadu, it is found in the Coimbatore, Dindigul and Nilgiri districts.

Conservation Status

It comes under not evaluated (NE) category in IUCN Red List data book.

Ethnobotanical Uses

The genus *Berberis* indeed holds a significant place in traditional medicine across various cultures, including Ayurveda and homeopathy. The plant has been utilized by many tribal communities for its medicinal properties, as documented ethnobotanically. Among the tribal peoples like the Toda, Kota and Irula, *Berberis* plants are valued for their laxative and tonic properties, and they have been used for various ailments such as ulcers, asthma, toothaches, swelling and blood purifier. The availability of the plant in local environments makes it a convenient resource for these communities. Additionally, *Berberis* is utilized for preparing mosquito coils and acts as a natural mosquito repellent, aiding in the prevention of malaria and other mosquito-borne diseases. This dual purpose of medicinal use and pest control underscores the versatility and importance of *Berberis* in traditional practices. Furthermore, *Berberis* is applied in the treatment of skin diseases, wound healing, inflammatory conditions, menorrhagia, diarrhoea, jaundice and eye infections. Its efficacy as a remedy for snakebites, particularly through the application of leaf juice, highlights its potential in emergency situations.

Scientific Validation and Economic Uses

Over the past 50 years, a wide range of phytocompounds have been isolated and identified from various Berberis plant species. These compounds include terpenoids, anthocyanins, alkaloids, lipids, flavonoids, vitamins, sterols, lignans, proteins and carotenoids, among others. Among these compounds, alkaloids stand out as the preeminent bioactive chemical constituents of *Berberis* species. Studies have reported that alkaloids extracted from *B. tinctoria* have been quantified using high-performance thin-layer chromatography (HPTLC) densitometric methods. It was found that 3.36% berberine is present in the form of berberine hydrochloride. Berberine is one of the most well-known and studied alkaloids found in *Berberis* plants. It possesses various pharmacological properties, including antimicrobial, anti-inflammatory, antioxidant and antidiabetic effects. The presence of berberine and other alkaloids in *Berberis* species contributes significantly to their medicinal properties and traditional uses in various cultures [91].

Cayratia pedata (Lam.) Gagnep

Cayratia pedata (Lam.) Gagnep commonly known as "Kattuppirandai", belongs to the family Vitaceae. It is an endemic and endangered species found in Thaishola, Nilgiris South Division, within the Western Ghats region.

Description

It is a woody liana with terete branchlets bearing longitudinal ridges and sparse pilosity. The tendrils are bifurcate. The leaves are pedately 5-foliolate, with brownish ovate-lanceolate stipules measuring 3–4 × 1.5–3 mm. The petiole is 5.5–16 cm long, with the central petiolule measuring 1.5–5 cm and lateral petiolules 2–4 cm. The petiolules of the lateral leaflet complex range from 1–3.5 cm and are sparsely pilose. The leaflets are pilose on the abaxial surface and have appressed hairs on the adaxial surface. They have 6–11 pairs of lateral veins, with inconspicuously raised veinlets. The central leaflet is obovate-elliptic, while the lateral leaflets are ovate-elliptic, measuring 5–22 × 2.5–9 cm, with a nearly truncate, rounded or slightly cordate base, irregularly toothed margin and caudate-acuminate apex. The inflorescence is a corymbose polychasium arising from the axils, with a pilose peduncle measuring 15–16 cm. Pedicels are 2–3 mm long and densely pubescent. The buds are oval-shaped, measuring 2–2.5 mm, with a rounded apex. The calyx is saucer-shaped, pubescent and entire, while the petals are ovate-elliptic, measuring 1.7–2.2 mm and nearly glabrous. The anthers are ovoid-elliptic and poorly developed in female flowers. The disk is well-developed with an undulate margin. The lower part of the ovary is adnate to the disk, and the style is slender, with a slightly expanded stigma or cylindrical in male flowers. The berry is elliptic, measuring 1.2–1.5 cm in diameter and containing two or three seeds. The seeds are semi-globose, with a rounded base and subrounded apex, an inconspicuous rostrum, a narrow upper raphe, a smooth surface and nearly rounded ventral holes measuring 3–5 mm in diameter.

Distribution

It is primarily found growing in the Kodanad and Thaishola border areas, where it often scrambles over hedges and trees. Additionally, this species has been reported in other regions beyond India, including Sri Lanka and Malaysia.

Conservation Status

It is listed as an endangered plant in IUCN Red List data book.

Ethnobotanical Uses

Traditionally, the leaves of this plant have been used as a dietary ingredient in the treatment of ulcers and diarrhoea. In Ayurveda, the extract from *C. pedata* is used to prepare formulations prescribed to treat microbial infections, ulcers, inflammations and arthritis. The leaf of this plant is traditionally used as a medicine for the treatment of ulcers, diarrhoea and inflammation. The paste of *C. pedata* has been applied by tribals as a near cure for wounds and migraine.

Scientific Validation and Economic Uses

Analysis of the essential oil extracted from *C. pedata* revealed the presence of various phytoconstituents, including type acids, esters, alcohols, ethers, hydrocarbons, alkane and fatty acids. Fifteen compounds were identified. The major compounds detected in the essential oil include Heptadecane (17.36%), silane, trichlorooctadecyl (17.29%), Eicosane (15.44%), Heptane, 2,2,3,3,5,6,6-heptamethyl (14.71%), n-tetradecyltrichlorosilane (13.64%), pentadecane (10.55%), iron, tricarbonyl [N-(phenyl-2-pyridinylmethylene) benzenamine-N, N′] (10.91%) and octadecane (8.93%). The predominant compounds have the property of antioxidant, antifungal, antibacterial, anti-allergic, chemopreventive and anticancer activity [92].

Ficus pumila L.

Ficus pumila, of the family Moraceae commonly referred to as creeping fig or climbing fig, belongs to the mulberry family and is native to East Asia, including regions such as China, Japan and Vietnam. While it thrives naturally in these areas, it has also become naturalized in certain parts of the southeastern and south-central United States.

Description

It is a shrubby, climbing or scandent growth habit. The branchlets that root is sterile. Stipules are lanceolate and adorned with yellow-brown silk-like hairs. The leaves are arranged distichously, meaning they are in two opposite rows along the stem. On fertile branchlets, the leaf blades differ in shape from those on sterile branches, typically ovate-cordate, ovate-elliptic or oblong-ovate, measuring 5–12 × 2–5 cm. The abaxial (lower) surface of the leaves is pubescent, while the base is rounded to slightly cordate, and the margin is entire. The apex may be obtuse, acute or acuminate. The veins are conspicuous and have a honeycomb-like pattern. Basal lateral veins are elongated, with three or four secondary veins on each side of the midvein, prominently visible on the abaxial surface and slightly impressed on the adaxial surface. The figs are axillary, arising from normal leafy branches, and are typically solitary. They range in colour from yellowish-green to pale red when mature and have a pear-shaped to ± globose or cylindrical shape, measuring 4–8 × 3–5 cm. When young, they are shortly yellow pubescent. The figs have a short stalk at the base and an apical pore that may be truncate, navel-like or acuminate. The peduncle is thick, reaching about 1 cm in length, and the involucral bracts are triangular-ovate, densely covered with long pubescence and persistent. Male flowers are numerous and arranged in several rows near the apical pore, each pedicellate, with two linear calyx lobes and two stamens with short filaments. Gall flowers are pedicellate, with three or four linear calyx lobes and a short lateral style. Female flowers have a long pedicel, with four or five calyx lobes, and develop into ± globose achenes with adherent liquid.

Distribution

It has a global distribution across Asia, specifically in China, India, Japan and Vietnam. In India, it is found in several states, including Bihar, Himachal Pradesh, Madhya Pradesh, Maharashtra, Rajasthan, Tamil Nadu and Uttar Pradesh. In Tamil Nadu, this species can be found across all districts of the state.

Conservation Status

This plant is not evaluated (NE) by the IUCN Red List data.

Ethnobotanical Uses

The leaves of this plant are utilized in Japan for preparing beverages aimed at treating diabetes and high blood pressure. Additionally, a decoction made from fruits, stems and leaves is employed to alleviate joint pain associated with conditions such as arthritis, rheumatism and sprains. Moreover, it is believed that consuming fruits can promote lactation in nursing mothers and may also have potential benefits for addressing issues related to impotency.

Scientific Validation and Economic Uses

Pectin and various fatty acids have been applied in the food industry. It also contains many bioactive components, such as phenolic acids, flavonoids, polysaccharides, terpenoids, alcohols and steroids. *Ficus pumila* L. extracts and their bioactive compounds exhibit multiple therapeutic activities, including antioxidant, anti-inflammatory, antibacterial, antitumor, hypoglycaemic and cardiovascular protective activities [93].

Grewia subinaequalis DC.

Grewia subinaequalis DC. (syn. *Grewia asiatica* L.) belongs to the family Tiliaceae. It is commonly called as Phalsa, dhaman, tadachi and tadasala.

Description

The shrub grows to a height of 4–5 m. The leaves are large and range in length from 5 to 18 cm. The leaves are simple, alternate and broadly cordate with pointed tips with petiole. The inflorescence is cymes, and the individual flowers are yellowish, with five large (12 mm) sepals and five smaller (4–5 mm) petals. Fruit is a fibrous, drupe that turns greyish-purple colour when matures, seeds are one or two with a pointed end and the seed coat is hard and stony with two-chambered oily endosperms.

Distribution

It is native to the Indian subcontinent and South Asia and distributed in India, Sri Lanka, Bangladesh and Myanmar.

Conservation Status

It is categorized as not evaluated (NE) in Red List of IUCN.

Ethnobotanical Uses

The edible fruit of *G. subinaequalis* is used to cure intestinal infections, jaundice, rheumatism, skin disorders and anaemia. The fruits and roots are together used in the treatment of osteoporosis and wound healing.

Scientific Validation and Economic Uses

The fruits of *G. subinaequalis* have high nutritional value (vitamins, fibres and minerals) with low calorific value with threonine, methionine, serine and taurine as the predominant essential amino acids. Fruits are reported to contain quercetin, tannins, pelargonidin 3,5-diglucoside and cyanidin-3-glucoside components. Notable pharmacological properties, such as radioprotective, antiemetic, antimalarial, antihyperglycaemic and antioxidant antipyretic activities, are reported so far. Ripen fruits of *G. subinaequalis* is used to prepare squash and soft drink during summer. The bark powder is used as substitute for soap, and the mucilage extracted from bark is used in the process of clarifying sugar. The wood from *G. subinaequalis* is specially used to make arch, poles and shingles to carry heavy loads [94].

Falcourtiamontana Graham

Falcourtiamontana (Syn. *Falcourtia inermis* Miq. ex Hook. fil. & Thoms) (Fig. 1f) is an indigenous plant species of Western ghats, categorized under the Salicaceae family. It is commonly called as mountain sweet thorn, Gajale and charal-maram.

Description

They are thorny trees that can reach heights of up to 8 m. They bear thorns approximately 5 cm long and have grey bark, which is thin and smooth. The leaves are simple, alternate and spiral in arrangement. The petioles are stout, measuring between 5 and 10 mm, and are glabrous. Leaf blades are ovate or elliptic, ranging from 12 to 22 cm in length and 6–8 cm in width. The base of the leaf can be acute or rounded, while the apex is acute or acuminate. Leaf margins are crenate-serrate, and the texture is coriaceous. The upper surface of the leaves is glabrous, except for the midrib below, giving a shining appearance. There are typically three to five prominent nerves originating from the base of the leaf, with four to six pairs of pinnate lateral veins that are also prominent. Intercostal veins are scalariform and slender, standing out prominently. The flowers are unisexual, small and arranged in axillary congested pubescent cymes. Sepals, usually four or five in number, are tomentose, small and imbricate. Petals are absent. Stamens are numerous, with versatile anthers. The ovary is superior, urn-shaped, glabrous and incompletely two to five locular, with two ovules in each cell. The styles, numbering 5, are reflexed and notched at the tip. The fruit is a berry, globose in shape, obtusely ribbed and measures 1–1.5 cm across. When ripe, it is bright red and possesses an agreeable acid flavour. Each fruit contains a few reddish seeds.

Distribution

It is endemic to the Western Ghats, moist deciduous to semi-evergreen forests of southern India. In India, it is distributed in Tamil Nadu, Karnataka, Kerala and Maharashtra.

Conservation Status

Categorized as not evaluated (NE) by the IUCN Red List.

Ethnobotanical Uses

The fruits of *F. montana* are edible and used as an appetizer, and fruit juice is used as a digestive tonic and diuretic for jaundice. The root is used in nephritic colic, and the bark is used to treat intermittent fever.

Scientific Validation and Economic Uses

The fruits tend to be sweet and edible and contain rich sources of calcium, sugar, magnesium and secondary metabolites. GC-MS profiling of *F. montana* fruit revealed the presence of 36 volatile constituents; the major compounds were found to be alcohol, alkanes and esters. 1-Undecanol was found to be the predominant volatile compound responsible for the floral, citrus-like odour of *F. montana* fruit. Many compounds with different aromatic profiles were detected, which include phenylethyl alcohol and n-hexyl-2-butenoate. The leaves of *F. montana* are reported to have anti-inflammatory, hepatoprotective and antioxidant properties and also exhibited antihyperglycaemic activity in streptozotocin-induced diabetic rats [95].

Viburnum coriaceum Blume

Viburnum coriaceum Blume (Fig. 1g) has synonym *Viburnum cylindricum* Buch-Ham. Ex D. Don belongs to the family Viburnaceae. It is commonly called Mottumookan in Malayalam.

Description

This evergreen tree reaches up to 7 m in height with grey bark. Its leaves are simple and opposite, lacking stipules. The petioles measure between 10 and 40 mm in length, slender, with a groove on the upper side, and glabrous. Leaf blades range from 6.4 to 17.5 cm long and 2.5–5.5 cm wide, varying in shape from elliptic to obovate, ovate-lanceolate or elliptic-lanceolate. The base of the blade can be acute or obtuse, while the apex is acuminate. Leaf margins are either entire or slightly serrate, and the texture is chartaceous. There are typically 4–6 pairs of prominent pinnate lateral nerves, with well-defined reticulate intercostal veins. Domatia are present. The bisexual flowers are white, approximately 3 mm in diameter and arranged in terminal or subterminal umbellate corymbs. The deciduous bracteoles surround the flowers. The calyx tube measures 2 mm in length, is pubescent and consists of five persistent obtuse lobes. The corolla tube is tubular, 4 mm long and glabrous, with five short, ovate-obtuse, erect lobes that are ciliate. There are five stamens inserted into the corolla tube with short filaments. The ovary is inferior, oblong and one-celled with a single pendulous ovule. The style is short, and the stigma is three-lobed. The fruit is a drupe, measuring 8 × 5 mm, with an elliptic, compressed, ribbed and beaked structure, appearing reddish-brown when ripe. Each fruit contains a single seed, elliptic in shape, measuring 6 × 4 mm.

Distribution

It is globally distributed in India, Myanmar, Indonesia and Bhutan. In India, it is present in Kerala, Tamil Nadu and Karnataka and especially found in the margins of Shola forests.

Conservation Status

Categorized as not evaluated (NE) by the IUCN Red Rist.

Ethnobotanical Uses

The fruits are edible, eaten by the tribes and reported to have folkloric values of antispasmodic, uterine relaxant and anticonvulsive activity, but still it is the fruits of *V. coriaceum* that is one of the least scientifically explored edible fruits.

Scientific Validation and Economic Uses

The phytochemical analysis revealed the presence of triterpenoids, carbohydrates, glycosides and phytosterols and also evinced significant antioxidant and anti-inflammatory activity. The ayurvedic fermented formulation of *V. coriaceum* has proved to exhibit anticonvulsant properties against tonic-clonic seizures and is being used in the field of medicine [96].

Conclusion

The exploration of underexploited wild edible fruits reveals a significant nutritional and phytochemical potential that holds promise for various applications. These fruits, often overlooked or underutilized, offer a wealth of essential nutrients, including vitamins, minerals and dietary fibres, contributing to a balanced and healthy diet. Moreover, their rich phytochemical composition, comprising antioxidants, polyphenols, flavonoids and other bioactive compounds, endows them with diverse health-promoting properties, such as anti-inflammatory, antioxidant, anticancer and antimicrobial activities. The utilization of these underexploited wild edible fruits presents opportunities for enhancing food security, promoting biodiversity conservation and supporting sustainable agricultural practices. Incorporating these fruits into local diets and culinary traditions can diversify food options, improve nutrition and contribute to the preservation of traditional knowledge and cultural heritage. Furthermore, the development of value-added products, such as jams, juices and dietary supplements, derived from these fruits can create economic opportunities for rural communities and small-scale farmers. Research and innovation in processing technologies and product development are essential to unlock the full potential of these underutilized resources and promote their integration into mainstream food systems.

Overall, recognizing the nutritional and phytochemical potential of underexploited wild edible fruits underscores the importance of their conservation, sustainable management and promotion for human health and well-being, as well as for the preservation of biodiversity and cultural diversity. This chapter serves as a call to action for further exploration, utilization and valorization of these valuable natural resources to address current and future challenges in food security, nutrition and public health.

References

1. Tomar A, Kumar A, Dubey K. Underutilized wild edible fruits of nutritional and medicinal value. J Res Educ Indian Med. 2015;21(1):3–70.
2. Bharucha Z, Pretty J. The roles and values of wild foods in agricultural systems. Philos Trans R Soc B: Biol Sci. 2010;365(1554):2913–26.
3. Shackleton C, Shackleton S. The importance of non-timber forest products in rural livelihood security and as safety nets: a review of evidence from South Africa. S Afr J Sci. 2004;100(11):658–64.
4. Etkin NL, editor. Eating on the wild side: The pharmacologic, ecologic and social implications of using noncultigens. University of Arizona Press; 2000.

5. Cunningham AB. Applied ethnobotany: people, wild plant use and conservation. Routledge; 2014.
6. FAO and Traditional Knowledge. The linkages with sustainability, food security and climate change impacts. Rome: Food and Agriculture Organization (FAO); 2009.
7. World Health Organization. Traditional medicine: fact sheet, no. 134. Geneva: World Health Organization; 2008.
8. Newman DJ, Cragg GM. Natural products as sources of new drugs over the nearly four decades from 01/1981 to 09/2019. J Nat Prod. 2020;83(3):770–803.
9. Rasul A, Saleem Khan M, Ali M, Li J. Post-COVID-19 pandemic: healthcare, social and economic restoration measures. World J Clin Cases. 2021;9(16):3856–65.
10. Kawabata T, Nguyen MT, Yamamoto S. Changes in food consumption patterns in the COVID-19 pandemic: analysis of consumer survey data from Vietnam. Nutrients. 2022;14(1):67.
11. Fanzo J, Hawkes C, Udomkesmalee E, Afshin A, Allemandi L, Assery O, Baker P, Battersby J, Bhutta Z, Chen K, Corvalan C. 2018 global nutrition report: Shining a light to spur action on nutrition. Bristol: Development Initiatives; 2018. p. 88149.
12. Grivetti LE, Ogle BM. Value of traditional foods in meeting macro-and micronutrient needs: the wild plant connection. Nutr Res Rev. 2000;13(1):31–46.
13. World Health Organization. The state of food security and nutrition in the world 2021: transforming food systems for food security, improved nutrition and affordable healthy diets for all. Food & Agriculture Org; 2021.
14. FAO. The state of food and agriculture 2013: food systems for better nutrition. FAO; 2013.
15. FAO. The State of Food Security and Nutrition in the World 2020: transforming food systems for affordable healthy diets. FAO; 2020.
16. WHO. The double burden of malnutrition: policy brief. WHO; 2016.
17. World Health Organization. Diet, nutrition, and the prevention of chronic diseases: report of a joint WHO/FAO expert consultation. World Health Organization; 2003.
18. Lock K, Pomerleau J, Causer L, Altmann DR, McKee M. The global burden of disease attributable to low consumption of fruit and vegetables: implications for the global strategy on diet. Bull World Health Organ. 2005;83:100–8.
19. Kausar H, Saeed S, Ahmad MM, Salam A, Shah MA. Wild edible fruits: a source of nutraceuticals and functional foods. J Food Sci Nutr. 2019;7(1):25–33.
20. McLean E, Cogswell M, Egli I, Wojdyla D, De Benoist B. Worldwide prevalence of anaemia, WHO vitamin and mineral nutrition information system, 1993–2005. Public Health Nutr. 2009;12(4):444–54.
21. Smith GC, Eyzaguirre PB. African leafy vegetables: their role in the World Health Organization's global fruit and vegetable initiative. Afr J Food Agric Nutr Dev. 2007;7(3):1–17.
22. Myers N, Mittermeier RA, Mittermeier CG, Da Fonseca GA, Kent J. Biodiversity hotspots for conservation priorities. Nature. 2000;403(6772):853–8.
23. Ministry of Environment, Forest and Climate Change. India's National Biodiversity Action Plan. Government of India; 2019.
24. Botanical Survey of India (BSI). Plant Discoveries 2018. Government of India; 2019.
25. National Biodiversity Authority (NBA). Biodiversity of India. Government of India; 2020.
26. Arora RK, Pandey A. Wild edible plants of India diversity, conservation and use. New Delhi: National Bureau of Plant Genetic Resources; 1996.
27. Sinha S, Singh RK. Role of non-timber forest products in the economy of tribal people in India. J Trop For Sci. 2002;14(2):146–59.
28. Gupta AK, Tandon N. Reviews on Indian medicinal plants. Indian Council of Medical Research; 2004.
29. Singh V, Govil JN. Recent progress in medicinal plants. Ethnomedicine and pharmacognosy II. 11Houston: Studium Press LLC; 2006.
30. Patwardhan B, Mashelkar RA. Traditional medicine-inspired approaches to drug discovery: can Ayurveda show the way forward? Drug Discov Today. 2009;14(15–16):804–11.
31. Kapoor LD. Handbook of Ayurvedic medicinal plants: herbal reference library. Routledge; 2017.

32. Reddy CS, Jha CS. Assessment and conservation prioritization of the ecosystems in the Western Ghats. J Biodivers Endanger Species. 2012;1(4):1–8.
33. Ravikanth G, Ganeshaiah KN. Conservation value of the Western Ghats. India J For Res. 2011;22(3):357–67.
34. Chandrasekara A, Shahidi F. Anti-diabetic potential of phytochemicals in the Western Ghats region. J Funct Foods. 2011;3(4):332–8.
35. Reddy CS, Ugle P. Species diversity and distribution of wild edible fruit species in southern India. Biodivers Conserv. 2008;17(12):3091–101.
36. Singh A, Reddy CS. Ecological analysis of the life forms and biological spectrum of wild edible plants in South India. J Environ Biol. 2009;30(3):361–6.
37. Kumar S, Bhatt V. Native and naturalized wild edible fruit species in the Western Ghats, India: Diversity and utilization. Trop Ecol. 2010;51(1):121–33.
38. Shiddamallayya N, Kumar S. Wild edible fruits of Western Ghats: a review. J Appl Nat Sci. 2016;8(2):830–7.
39. Nayar TS, Beegam AR. Diversity and conservation of wild edible plants in the Western Ghats. J Plant Sci Res. 2014;30(3):255–60.
40. Jadhav SN, Patil RB, Mule SB. Wild edible plants of Northern Western Ghats, India. Int J Biodivers Conserv. 2015;7(2):78–87.
41. Nazarudeen A. Nutritional composition of some lesser-known fruits used by the ethnic communities and local folks of Kerala. Indian J Tradit Knowl. 2010;9(2):398–402.
42. Liu RH. Potential synergy of phytochemicals in cancer prevention: mechanism of action. J Nutr. 2004;134(12):3479S–85S.
43. Lobo V, Patil A, Phatak A, Chandra N. Free radicals, antioxidants and functional foods: Impact on human health. Pharmacogn Rev. 2010;4(8):118.
44. Pandey KB, Rizvi SI. Plant polyphenols as dietary antioxidants in human health and disease. Oxidative Med Cell Longev. 2009;2(5):270–8.
45. Atanasov AG, Waltenberger B, Pferschy-Wenzig EM, Linder T, Wawrosch C, Uhrin P, Temml V, Wang L, Schwaiger S, Heiss EH, Rollinger JM. Discovery and resupply of pharmacologically active plant-derived natural products: a review. Biotechnol Adv. 2015;33(8):1582–614.
46. Balunas MJ, Kinghorn AD. Drug discovery from medicinal plants. Life Sci. 2005;78(5):431–41.
47. Popkin BM, Gordon-Larsen P. The nutrition transition: worldwide obesity dynamics and their determinants. Int J Obes. 2004;28(3):S2–9.
48. Moyo M, Aremu AO, Van Staden J. Medicinal plants: an invaluable, dwindling resource in sub-Saharan Africa. J Ethnopharmacol. 2015;174:595–606.
49. United Nations. World population prospects 2019: highlights. Department of Economic and Social Affairs, Population Division; 2019.
50. FAO. The future of food and agriculture—Trends and challenges. FAO; 2017.
51. Singh H, Kumar M. Livelihood dependence of tribal communities on forests in Tamil Nadu. J Hum Ecol. 2012;38(3):165–72.
52. Ravikumar K, Ved DK. Ethnomedicinal practices among the tribes of Tamil Nadu. Indian J Tradit Knowl. 2000;1(1):69–75.
53. Jain A, Sharma A. Ethnobotanical studies on the tribals of South India. Ethnobot Res Appl. 2012;10:189–204.
54. Nithya N, Anuradha R. Traditional uses of wild edible fruits in Tamil Nadu. Indian J Nat Prod Resour. 2013;4(4):411–8.
55. Prasad S, Pandey BL. Oral traditions and the need for documentation of indigenous knowledge. J Ethnopharmacol. 2010;128(1):119–24.
56. Bhattacharya P, Hayat SA. Sustainable livelihood through NTFP management: a case from Tamil Nadu. Int For Rev. 2005;7(2):124–9.
57. Kumar A, Bhatt VP. Impact of anthropogenic activities on the forest and tribal livelihood: a case study from Tamil Nadu. J Hum Ecol. 2008;24(2):107–15.
58. Pullaiah T, Naidu KC. Conservation of medicinal plants. J Med Plants Res. 2003;8(7):341–7.
59. Sankara Rao K, Swamy RK, Kumar D, Arun Singh R, Bhat KG. Flora of Peninsular India. 2019. Available in http://flora-peninsula-indica.ces.iisc.ac.in/herbsheet.php

60. Abiodun O, Gbotosho G, Ajaiyeoba E, Happi T, Falade M, Wittlin S, Sowunmi A, Brun R, Oduola A. In vitro antiplasmodial activity and toxicity assessment of some plants from Nigerian ethnomedicine. Pharm Biol. 2011;49(1):9–14.
61. Barstow M. Trema orientalis. The IUCN Red List of Threatened Species 2018: e.T61988133A61988136. 2018.
62. Watt JM, Breyer-Brandwijk MG. The medicinal and poisonous plants of Southern and Eastern Africa being an account of their medicinal and other uses, chemical composition, pharmacological effects and toxicology in man and animal. E. & S. Livingstone.
63. Rasoanaivo P, Petitjean A, Ratsimamanga-Urverg S, Rakoto-Ratsimamanga A. Medicinal plants used to treat malaria in Madagascar. J Ethnopharmacol. 1992;37(2):117–27.
64. Rajalakshmi P. Evaluation of nutritional value and anti obesity property of wild fruits syzygium Jambos L Alston Trema Orientalis L Blume and Rubus Ellipticus Smith Collected From the Nilgiris South India. Thesis submitted to Bharathiar University.
65. Barbera R, Trovato A, Rapisarda A, Ragusa S. Analgesic and antiinflammatory activity in acute and chronic conditions of *Trema guineense* (Schum. et Thonn.) *ficalho* and *Trema micrantha* blume extracts in rodents. Phytother Res. 1992;6:146–8.
66. Perlmann P, Troye-Blomberg M. Malaria blood-stage infection and its control by the immune system. Folia Biol. 2000;46(6):210–8.
67. Diehl MS, Atindehou KK, Téré H, Betschart B. Prospect for anthelminthic plants in the Ivory Coast using ethnobotanical criteria. J Ethnopharmacol. 2004;95(2–3):277–84.
68. Dimo T, Ngueguim FT, Kamtchouing P, Dongo E, Tan PV. Glucose lowering efficacy of the aqueous stem bark extract of Trema orientalis (Linn) Blume in normal and streptozotocin diabetic rats. Die Pharmazie-An Int J Pharm Sci. 2006;61(3):233–6.
69. Uddin SN, Uddin KM, Ahmed F. Analgesic and antidiarrhoeal activities of Trema orientalis Linn. in mice. Orient Pharm Exp Med. 2008;8:187–91.
70. Anonymous. The Wealth of India (Raw material). New Delhi: Publication of Information Directorate, Council of Scientific and Industrial Research; 1991. p. 377–8.
71. Upadhyay AK, Kumar K, Kumar A, Mishra HS. Tinospora cordifolia (Willd.) Hook. f. and Thoms.(Guduchi)–validation of the Ayurvedic pharmacology through experimental and clinical studies. Int J Ayurveda Res. 2010;1(2):112.
72. Chorachoo J, Saeloh D, Srichana T, Amnuaikit T, Musthafa KS, Sretrirutchai S, Voravuthikunchai SP. Rhodomyrtone as a potential anti-proliferative and apoptosis inducing agent in HaCaT keratinocyte cells. Eur J Pharmacol. 2016;772:144–51.
73. Jain SK. Dictionary of folk medicine and ethnobotany. New Delhi: Deep Publishers; 1991. p. 179–80.
74. Singh V, Pandey RP. Ethnobotany of Rajasthan India. Jodhpur: Scientific publisher (India); 1998. p. 203–31.
75. Singh J, Sinha K, Sharma A, Mishra NP, Khanuja SP. Traditional uses of *Tinospora cordifolia* (Guduchi). J Med Aromat Plant Sci. 2003;25:748–51.
76. Sood SK, Parmar S, Lakhanpal TN. Ethnic plants of India used in cancer cure. Dehradun: Bishen Singh Mahendra Pal Singh; 2005. p. 235.
77. Khosa RL, Prasad S. Pharmacognostical studies on guduchi (Tinospora cordifolia Miers). J Res Ind Med. 1971;6:261–9.
78. Shama P, Dwivedee BP, Bisht D, Dash AK, Kumar D. The chemical constituents and diverse pharmacological importance of Tinospora cordifolia. Heliyon. 2019;5(9):e02437.
79. Reinhardt JK, Zimmermann-Klemd AM, Danton O, Smieško M, Gründemann C, Hamburger M. Compounds from Toddalia asiatica: immunosuppressant activity and absolute configurations. J Nat Prod. 2020;83(10):3012–20.
80. Alagaraj P, Muthukrishnan S. *Toddalia asiatica* L. – a rich source of phytoconstituents with potential pharmacological actions, an appropriate plant for recent global Arena. Cardiovasc Hematol Agents Med Chem. 2020;18(2):104–10.
81. Zeng Z, Tian R, Feng J, Yang NA, Yuan L. A systematic review on traditional medicine *Toddalia asiatica* (L.) Lam.: Chemistry and medicinal potential. Saudi Pharm J. 2021;29(8):781–98.

82. Lai TN, André C, Rogez H, Mignolet E, Nguyen TB, Larondelle Y. Nutritional composition and antioxidant properties of the sim fruit (Rhodomyrtus tomentosa). Food Chem. 2015;168:410–6.
83. Zhao Z, Wu L, Xie J, Feng Y, Tian J, He X, Li B, Wang L, Wang X, Zhang Y, Wu S. Rhodomyrtus tomentosa (Aiton.): a review of phytochemistry, pharmacology and industrial applications research progress. Food Chem. 2020;309:125715.
84. Tekameren W, Mahanaik T, Dash N, Kumar S. Kasi Bridelia retusa (L.): a medicinal tree of India. In: Medico Bio wealth of India, volume XI. ISBN 978-81-965138-8-7. ; 2023.
85. Nguyen VT, Nguyen VD. Rhodomyrtus tomentosa: a source of bioactive compounds. Phytochem Rev. 2019;18(3):671–92.
86. Kumar S, Mishra S, Mishra AK, Kumar SN. Floral diversity of Koira and Barsuan ranges of Bonai Forest Division, Odisha. Odisha: Bonai Forest Division and Ambika Prasad Research Foundation; 2022. p. 1–480.
87. Nayak SU, Kumar SA. Medicinal plants used by tribals of Odisha. In: APRF and SMPB. Odisha: APRF Publishers; 2023. p. 1–665.
88. Sosef MS, Hong LT, Prawirohatmodjo S. Plant resources of South-East Asia. SV Kerkwerve, Backhuys; 1998.
89. Banerjee SK, Bonde CG. Total phenolic content and antioxidant activity of extracts of Bridelia retusa Spreng Bark: Impact of dielectric constant and geographical location. J Med Plants Res. 2011;5(5):817–22.
90. Patil PS, Jadhav VD. Bioactive potential of Bridelia retusa (L.) A. Juss. plant: a wild edible plant. J Drug Res Ayurvedic Sci. 2022;7(1):55–61.
91. Vignesh A, Sivalingam R, Selvakumar S, Vasanth K. A review on ethnomedicinal and phytopharmacological potential of traditionally wild and endemic plant Berberis tinctoria Lesch. Thai J Pharm Sci. 2022;46(2):137–48.
92. Aswathy TR, Gayathri E, Praveen J, Nair AS, Sugunan VS. Phytoprofiling of medicinal plant Cayratia pedata by qualitative and quantitative method. J Pharmacogn Phytochem. 2019;8(2):1637–42.
93. Qi ZY, Zhao JY, Lin FJ, Zhou WL, Gan RY. Bioactive compounds, therapeutic activities, and applications of Ficus pumila L. Agronomy. 2021;11(1):89.
94. Sinha J, Purwar S, Chuhan SK, Rai G. Nutritional and medicinal potential of Grewia subinaequalis DC.(syn. G. asiatica.)(Phalsa). J Med Plants Res. 2015;9(19):594–612.
95. Mundaragi AC, Thangadurai D, Sangeetha J, Byahatti SD, Appaiah KA. Chemical analysis of volatile constituents of Flacourtia montana J. Graham (Salicaceae): an edible and wild fruit of Western Ghats of India. Plant Archives. 2017;17(1):506–8.
96. Vijaytha V, Anupama RV, Haridas M. Profiling, and anti-oxidant, anti-bacterial, and anti-inflammatory properties of *Viburnum coriaceum* Blume. Future J Pharm Sci. 2020;6(1):84.

Curative Potentialities of *Garcinia gummi-gutta* (L.) Roxb.: A Less-Exploited Tropical Fruit Tree

B. Vishnu, R. M. Deepthy, Parimelazhagan Thangaraj, and E. A. Siril

Abstract

The genus *Garcinia* is composed of around 200 species distributed in the tropics and subtropics. Out of 36 species from India, *Garcinia gummi-gutta* (L.) Roxb. is an important non-timber forest product (NTFP) from the southwestern ghats of India. It is a small- or medium-sized fruit tree valued for its fruits as a condiment. Moreover, the plant is a better source of phytochemicals such as xanthones, benzophenones, organic acids, and amino acids. Among these phytochemicals, hydroxycitric acid (HCA), an important organic acid enriched in fruit rinds, has the mechanism for regulation of lipid metabolism and obesity control. HCA helps to regulate serotonin synthesis and thereby suppresses food intake in our body. Also, it competitively inhibits the ATP-citrate lyase meant for the conversion of citrate to oxaloacetate and acetyl coenzyme A, a building block of fatty acid synthesis. Furthermore, an array of pharmacological effects such as antioxidant, antitumor, anti-allergic, anti-inflammatory, antibacterial, antifungal, antiviral, neuroprotective, anti-HIV, enzyme inhibition, and hypoglycemic effects has been well elucidated based on in vitro as well as in vivo experimental models. A generous portion of phytoconstituents with therapeutic potentialities has been contributed by the fruits, followed by leaf and bark. An array of health, wellness, and nutraceutical supplements are being produced from the plant by reputed pharma companies all over the world. To fulfil the needs of pharmaceutical industries, breeding approaches could be enhanced to improve the genome of *G. gummi-gutta*. Thus, underutilized plants, in turn, may contribute significantly

B. Vishnu · R. M. Deepthy · E. A. Siril (✉)
Department of Botany, University of Kerala, Kariavattom, Thiruvananthapuram, Kerala, India
e-mail: easiril@keralauniversity.ac.in

P. Thangaraj
Department of Botany, Bharathiar University, Coimbatore, Tamil Nadu, India
e-mail: drparimel@buc.edu.in

P. Thangaraj (ed.), *Pomotherapeutic Insights on Wild Edible Fruits*,
https://doi.org/10.1007/978-981-96-0543-9_9

to plant-based industry, thus benefiting the rural people who engage in the collection, processing, and marketing of such resources.

Keywords

Condiment · Hydroxycitric acid · Non-timber forest product (NTFP) · Pharmacological effects · Therapeutic potentialities · Underutilized plants

Introduction

The therapeutic potentials of natural products from several plants have been recognized from ancient period. In the past several clinically approved herbal medicines have developed. Simultaneously, the demand for Ayurvedic medicines is increasing at an alarming rate. Several formulations are the combinations of different plant parts such as fruits, leaves, roots, seeds, stems, and bark. Leaves are the most frequently used parts, followed by roots, bark, and fruits [1]. Epidemiological surveys confirmed that fruits are rich source of nutrients and antioxidants. The fruits of *Garcinia gummi-gutta* (L.) Roxb. (syn. *G. cambogia* (Gaertn.) Desr.) is coming under this category. It is a multipurpose tropical fruit tree cultivated or maintained in homesteads for its fruit rind in south India, especially in Kerala and southern districts of Karnataka [2]. It is a medium-sized to large fruit tree, commonly known as brindle berry or Malabar tamarind. The fruits are extremely sour and difficult to eat even after ripening. The sourness is due to organic acids, mainly hydroxycitric acid (HCA) [3]. The genus *Garcinia* belongs to the family Clusiaceae and is a well-recognized source of several non-timber forest products (NTFPs), distributed in the tropics and subtropics of the world [4, 5], and different species are distributed in the wild as well as semidomesticated territories across India [3, 6, 7]. The genus has an immense role in the flora of the south Western Ghats, and it was named in honor of the Dutch army doctor and naturalist Laurentius Garcin [8]. About 30 species of *Garcinia* are in cultivation out of the 200 existing worldwide [3]. From India, about 36 *Garcinia* species have been reported, and 11 are endemic to the south Western Ghats.

Geographical Distribution and Botanical Description

The species *G. gummi-gutta* is believed to have originated from Southeast Asia, whereas the Southwestern Ghats of India are considered as its secondary center of origin [7, 9]. It is introduced to other tropical and subtropical regions of Asia, including China, Malaysia, and the Philippines [10]. Also, it is endemic to the Western Ghats and Sri Lanka. The species strives prolifically in Peninsular India and western Sri Lanka, common along streams at lower altitudes of 300–750 masl. The species is extensively grown in the coastal region of Gujarat, the Konkan region of Maharashtra, Goa, Karnataka, and wild and domestic habitats of Kerala and

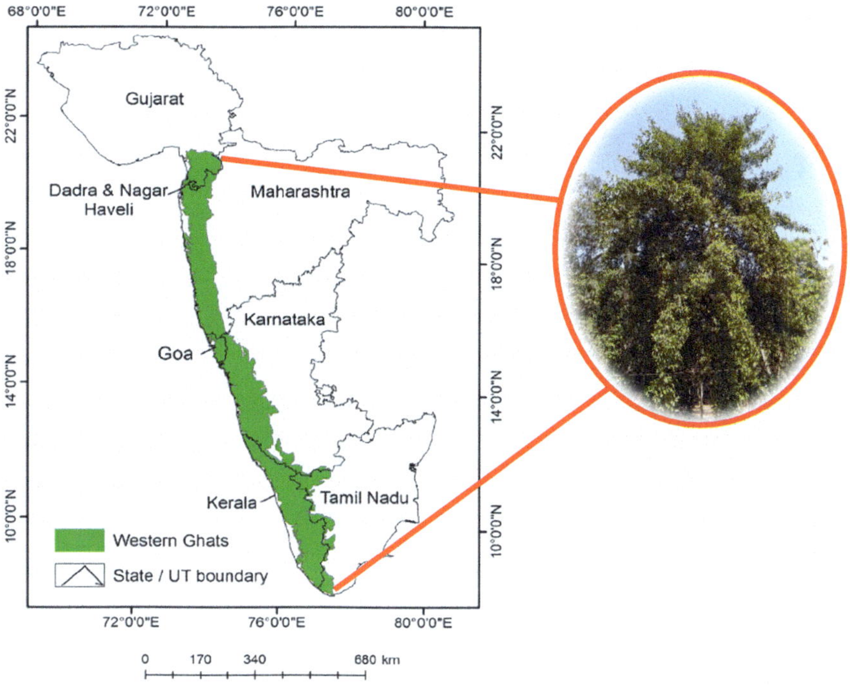

Fig. 1 Distribution of *G. gummi-gutta* across south Western Ghats of India

Tamil Nadu (Fig. 1). This fruit tree is maintained in the homesteads of Kerala mainly for its fruits [3]. *Garcinia gummi-gutta* grows up to a height of 20 m with a dense canopy. The species is a polygamodioecious with separate male and hermaphrodite trees (Fig. 2). The trunk has dark brown, smooth bark and exudes yellow gum. The branching pattern is dichotomous with horizontal or drooping branchlets, which form conical or rounded crowns. Leaves are simple, oblong, or elliptic to lanceolate, and entire with a thick, shiny appearance; mature leaves are dark green and pale green tender leaves.

The flowers are either androecious or bisexual, so the plant has a polygamodioecious nature. Male flowers are seen as auxiliary or terminal clusters of 4–8 having 4 obovate, unequal sepals and petals and 15–20 stamens, inserted on the prominent receptacles with bilocular anthers, basifixed and dehiscing vertically. Hermaphrodite flowers may be solitary or clusters of 2–3 seen as auxiliary or terminal clusters. They have numerous staminodes and ovoid or subglobose ovary and stigma rays spreading and free near the base. Recently, scientists have identified that hermaphrodite flowers do possess viable pollen and the flowers are reproductively fit for self-pollination [11]. Fruit is a berry with depressed globose shape with green mature fruits and ripened yellow fruits. The pericarp consists of 6–8 grooves, and the seeds are encapsulated in white or pale brown succulent aril (Fig. 3) [6, 9, 12].

Fig. 2 Habitat of *G. gummi-gutta*. (**a**) Male tree. (**b**) Hermaphrodite tree

Fig. 3 (**a**) A branch bearing fruits, (**b**) open fruit showing rind and seed encapsulate aril, (**c**) fruit bearing prominent nipple, (**d**) seeds encapsulated in succulent aril, and (**e**) de-coated seeds

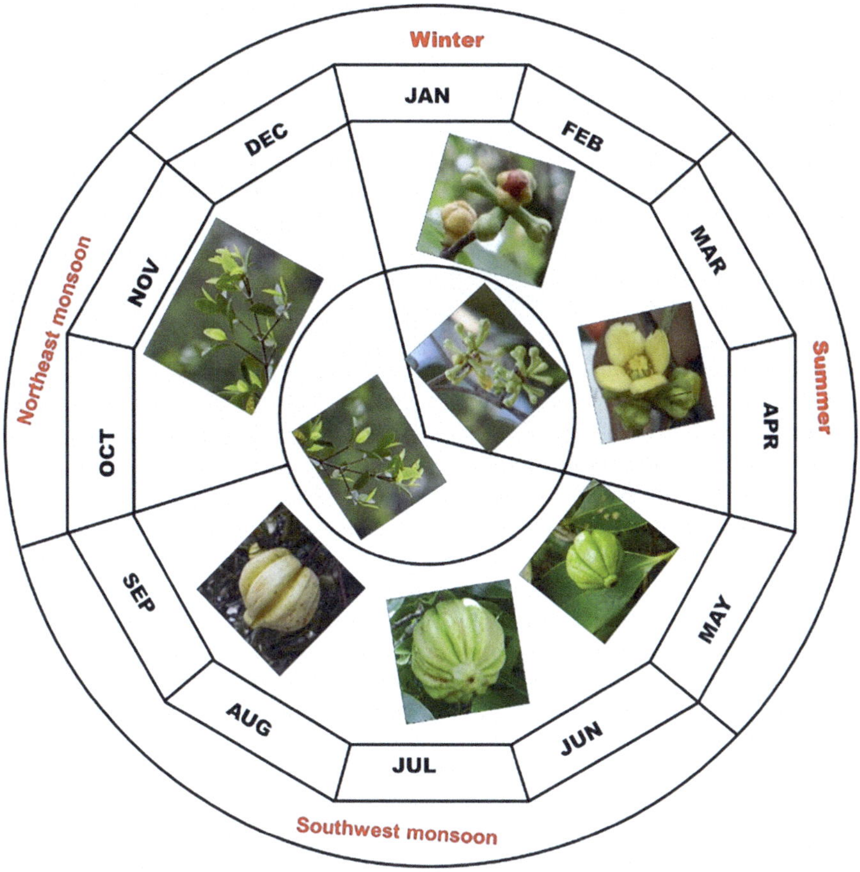

Fig. 4 Phenology calendar showing the flowering and fruiting events male (inside the circle) and hermaphrodite (inside the dodecagon) of *G. gummi-gutta*

The embryo is an elongated hypocotyl containing vascular bundles, which fill up the seeds. The primary root and shoot emerge from the opposite sides of the seed, known as the "garcinia-type" germination [13, 14]. The rind may or may not contain protrusions, known as nipples. It is an evergreen fruit crop, and in male and hermaphrodite individuals the flowering starts during January and extends up to April. The fruits emerged from April and ripened during June, which may extend beyond August. The harvesting period of *G. gummi-gutta* will be during the southwest monsoon season in Kerala (Fig. 4). Based on fruit morphology, three varieties of the species were recognized in India, viz., *G. gummi-gutta* var. *gummi-gutta*, *G. gummi-gutta* var. *papilla* (Wight) N. P. Sing., and *G. gummi-gutta* var. *conicarpa* (Wight) N. P. Sing [9, 15, 16]. As per the International Union for Conservation of Nature (IUCN), the species is included in the Least Concern category of the IUCN Red List of Threatened Species in 2020 [17].

A Gold Mine of Future Therapeutics

Garcinia atroviridis, *Garcinia cowa*, *Garcinia dulcis*, *Garcinia forbesii*, *G. gummi-gutta*, *Garcinia hombroniana*, *Garcinia humilis*, *Garcinia indica*, *Garcinia intermedia*, *Garcinia livingstonei*, *Garcinia macrophylla*, *Garcinia madruno*, *Garcinia malaccensis*, *Garcinia mangostana*, *Garcinia nervosa*, *Garcinia nitida Pierre*, *Garcinia parvifolia*, *Garcinia prainiana*, *Garcinia schomburgkiana*, *Garcinia spicata*, and *Garcinia xanthochymus* are the other important species of genus *Garcinia* that contribute diverse non-timber forest products (NTFPs). An extremely different application was found in the case of *G. lucida*, where tree bark is being used for medicinal, food, and incense purposes in South Cameroon [18]. In India, *G. mangostana* and *G. indica* are widely cultivated species for their fruits [19]. The fruit rind of *G. indica* has an imperative role in the spice industry of south India and is extensively cultivated in states like, Goa, Karnataka, and Maharashtra [20]. But *G. gummi-gutta* and *G. xanthochymus* are the semidomesticated ones across south India, especially in Kerala and Karnataka [3, 21]. Due to the presence of many socioeconomically valuable species, the genus is acclaimed as "a gold mine of future therapeutics" [8]. Especially the species *G. gummi-gutta* is being traditionally used by Adiyan tribal community in Wayanad district of Kerala, India, as a food and oral medicine for various diseases, and it is extensively used in different regions of India. The fruits are nutrient rich and contain more than 30 essential minerals [22]. Genetic improvement in this crop may significantly influence fruit production, and it could enhance the surplus availability of raw material for preparations and formulations.

Crop Improvement

In the recent past, various approaches were initiated to accomplish the improvement of this valuable fruit crop. Researchers recognized that marker-assisted breeding approaches can be served as an effective tool in manipulating the genome of this species. A morphotype of *G. gummi-gutta* with higher fruit weight and rind recovery was identified, which could be used for crop improvement programs [6]. Better performing candidate plus trees could be screened based on species-specific fruit mass gain and metric traits that could be used for the genetic improvement [23]. The estimation of genetic variability based on molecular markers could be applied to evaluate the efficiency of the selection procedure. An efficient DNA isolation procedure was developed for this plant, which could act as the starting material for every molecular experiment [24] (Table 1).

Species-specific simple sequence repeats (SSR) markers were successfully isolated and characterized in this plant [25]. It could be effectively transferred for assessing the morphogenetic diversity among 35 candidate plus trees (CPTs) of *G. gummi-gutta*. Molecular variability analysis in this species was started with the random amplified polymorphic DNA (RAPD) analysis of 12 accessions from the National Bureau of Plant Genetic Resources (NBPGR) Regional station, Thrissur,

Table 1 Evidences of genetic evaluation and crop improvement, propagation, and conservation of *G. gummi-gutta*

Genetic evaluation for crop improvement				
Types of analysis	**No. of individuals studied**	**Study area**	**Major outcomes**	**References**
Gender-specific SCAR marker	44	Western Ghats region of Kerala and Karnataka, India	Male-specific SCAR marker was developed for selection at seedling stage	[31]
Isolation and characterization of microsatellite markers	–	College of Forestry, Sirsi (University of Agricultural Sciences, Dharwad), India	Total genomic DNA of *G. gummi-gutta* has 241,141,804 bp length 32 SSR primer pairs were standardized for the species	[25]
Morphological and biochemical evaluation	2 morphotypes (GG-02 and GG-05)	Division of Horticulture and Forestry of ICAR-Central Island Agricultural Research Institute, Port Blair, ANI, India	Significant variations observed in morphological and physiological characters A morphotype with higher fruit weight and rind recovery was identified	[6]
Molecular identification of '*Candidatus Phytoplasma oryzae*'	–	Kumarakom, Kottayam District of Kerala, India,	Based on 16S rDNA sequencing *Candidatus, Phytoplasma oryzae* associated little leaf and witches' broom disease of *G. gummi-gutta* was identified from Kerala	[30]
Exploration genetic diversity using RAPD markers	12	National Bureau of Plant Genetic Resources (NBPGR) Regional station, Thrissur, Kerala, India	Existence of high genetic diversity Average polymorphism 97.7%	[26]
Morphogenetic analysis using species-specific morphological and SSR markers	35	Kerala state, India	Total no. of fruits per tree and no. of fruits/Kg are economically potential traits Morphogenetic similarities between high yielding individuals	[28]

(continued)

Table 1 (continued)

Genetic evaluation for crop improvement				
Types of analysis	**No. of individuals studied**	**Study area**	**Major outcomes**	**References**
Genetic evaluation using ISSR markers	50	Kerala state, India	Moderate degree of molecular variability Possibility of eight assumed subpopulations	[28]
Selection of promising individuals based on morphological traits	35	Kerala, India	Individual with 40-fold higher fruit gaining per year than the average of 180 trees was identified	[23]
Establishment of efficient DNA isolation protocol	–	Kerala state, India	DNA mini preparation method using silica-based spin column Fast, efficient and using toxic at its minimum	[24]
Association analysis between conc. of HCA and SCoT markers	35	Kerala state, India	Strong general linear model (GLM) association between HCA content and SCoT marker allele	[29]
Development of DNA barcoding	10 species of *Garcinia*	Western Ghats, India	DNA fingerprinting technique was developed for discriminating different species of Garcnia, including *G. gummi-gutta*	[32]
Selection and improvement through morphological traits	–	Indian Council of Agricultural Research-National Bureau of Plant Genetic Resources (ICAR-NBPGR), Regional Research Centre, Thrissur, Kerala, India	Two early maturing (6-7 years juvenility) varieties, i.e., IC244100-2 (INGR No.04061) and IC244111-1 (INGR No. 04062) were released	[33]
Conservation measures				
Experimental pattern	**Specific treatments**		**Presumptions**	**References**
Enhancement of seed germination	De-coating of seeds		Improvement of germination from 50% to 90%	[13, 14, 36]

(continued)

Table 1 (continued)

Genetic evaluation for crop improvement				
Types of analysis	**No. of individuals studied**	**Study area**	**Major outcomes**	**References**
Vegetative propagation using juvenile shoot cuttings	IBA		54% rooting by 4000ppm IBA	[22]
In vitro development of plantlets from seeds	MS medium supplemented with various concentrations of 6-Benzylaminopurine (BAP), Thidiazuron (TDZ), or Kinetin		Best adventitious bud development in 12.5μM BAP supplemented media and rooting response at 10μM IBA with 0.2% activated charcoal	[21]
In vitro development of plantlets from seeds	MS medium supplemented with varying conc. of BAP and NAA		Multiple shoot initiation in 2.5mg/l BAP and faster shoot elongation in 2 mg/l of NAA	[38]

Kerala, India. The study revealed high genetic divergence among the accessions [26]. Similarly, the highest percentage of genetic diversity was noticed among 35 CPTs based on SSR markers, and they showed some morphogenetic relationships [27]. Moderate levels of genetic variability and a probability of eight assumed subpopulations were observed among 50 individuals of *G. gummi-gutta* from Kerala while conducting the diversity study using 21 ISSR makers. Correspondingly, there exists a possibility of horizontal gene flow between the individuals, preferably through their ancestors [28]. Recently, start codon targeted (SCoT) amplification markers were also applied to analyze the genetic diversity between 35 CPTs, and a strong general linear model association of SCoT marker alleles with HCA content was reported [29]. Based on rDNA sequencing, *Candidatus, Phytoplasma oryzae* associated little leaf, and witches' broom disease of *G. gummi-gutta* were identified from Kerala [30]. The finding could also be extended for resistance breeding programs in the plant. In addition to this, male-specific sequence characterized amplified region (SCAR) markers have already been developed for this plant [31]. It is highly applicable for identifying the sex of the saplings because of their long juvenility period. At the same time, DNA fingerprinting technique has been developed for discriminating different species from the genus [32], which could be effectively utilized for the exploitation of different species of *Garcinia* or even other genera. Apart from that, the Indian Council of Agricultural Research-National Bureau of Plant Genetic Resources (ICAR-NBPGR), Regional Research Centre, Thrissur, have released two early maturing (6–7 years juvenility) varieties of Malabar tamarind, i.e., IC244100–2 (INGR No.04061) and IC244111-1 (INGR No. 04062), for the farmers [33]. It could be used for further breeding and improvement programs in *G. gummi-gutta* (Table 1).

Propagation and Conservation

The species is polygamodioecious, with separate male and hermaphrodite trees with a long juvenility period. The trees are not evenly distributed in the woodland as well as homestead habitats, and the seedling percentage is too low [34]. The climatic change analysis of the Western Ghats based on maximum entropy modeling (MaxEnt) revealed that the suitable environment for *G. gummi-gutta* in the biodiversity hotspot will be drastically reduced by 2050 and 2070 [35]. So there exists an urgent need to develop conservation measures for this plant via in vitro or ex vitro system, and propagation through seeds can also be used for the species. However, the seeds with intact coats require almost 24 weeks to germinate and are typically tropical dormant recalcitrant. Moreover, they are difficult to germinate due to their non-deep physiological dormancy and sensitivity to chilling and desiccation [14]. In complement to this, the de-coated seeds showed a cumulative germination from 50% to 90% [13, 14, 36]. Also, researchers found that each fragment of the seed is capable of germination as it is an elongated hypocotyl with vascular tissues and embryoid [37]. On the other hand, vegetative propagation succeeded in this plant with 54% rooting by treating it with indole-3-butyric acid (IBA) [22]. The totipotent property of the seed fragments has been efficaciously exploited by in vitro techniques. Murashige-Skoog (MS) medium supplemented with IBA [21] and 1-naphthaleneacetic acid (NAA) [38] were effective for the in vitro experiments (Table 1). Promising fruiting trees could be easily multiplied through this approach, and this could also be extended for the creation of *G. gummi-gutta* plantations.

Multipurpose Applications

The sun-dried and smoke-dried fruit rind is traditionally used as a souring agent for various dishes in India and Sri Lanka [3, 7, 39]. Salted, dried rinds are normally used to cure fish, due to its bacteriostatic properties [40]. In connection with that, the shelf life of fresh *Rastrelliger kanagurta* was increased by up to 24 days when stored in an icing medium combined with the aqueous extracts of *G. gummi-gutta* and *G. indica* fruits [41]. Traditionally, the fruit rinds are used to polish gold and silver ornaments [3]. Recently it was discovered that the leaf extract combined with potassium iodide salt drastically inhibited the corrosion in mild steel [42, 43]. Similarly, aqueous and alcoholic extracts of fruit rind showed a corrosion inhibition efficiency at room temperature in an acidic medium [44]. The seeds are a rich source of edible oil, and about 33.5% yield was obtained while conducting extraction using hexane. The isolated oil was rich in stearic acid and oleic acid, and the presence of former stable fatty acid proves its suitability for value-added food products such as chocolates [45]. Using the transesterification process, seed oil could be converted into biodiesel [46–52]. The seed oil-based biodiesel can be blended with ethanol to improve fuel properties and engine performance and reduce carbon monoxide emission [47]. A combination of 20% *G. gummi-gutta* biodiesel and 80% diesel (B20 blend) fortified with titanium oxide ($TiO2$) nano additives created a greater impact

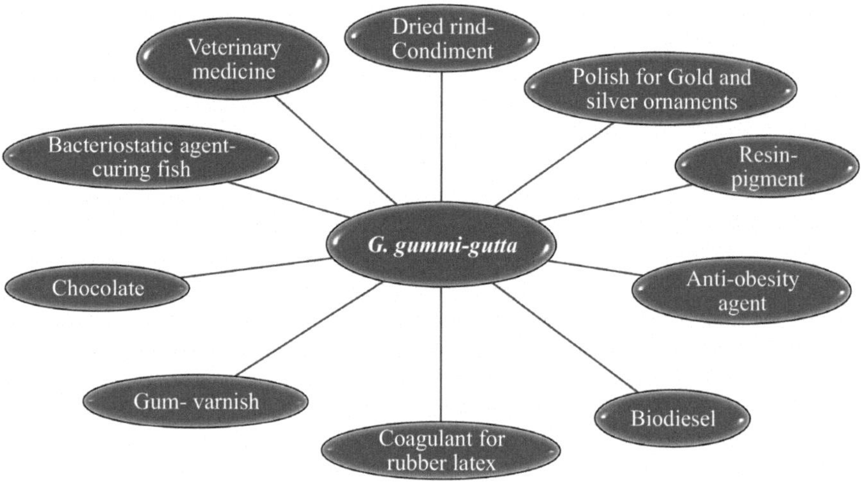

Fig. 5 Multipurpose uses of *G. gummi-gutta*

and low emission of unburnt hydrocarbons, carbon monoxide, and smoke [53, 54] (Fig. 5). In addition to this, the pharmaceutical potential of *G. gummi-gutta* is yet to be discussed.

Phytochemical Profiling

A brief analysis of previous literature indicates that in the field of phytochemical profiling, *G. gummi-gutta* is less exploited. Based on certain indigenous knowledge and experiences, advanced studies were mainly focused on the fruit, which led to the identification of almost all the active principles in the fruit, whereas leaves, bark, root, and stem remained less explored. Xanthones, benzophenones, organic acids, and amino acids are the major class phytochemicals isolated from the plant (Table 2) (Fig. 6). Also, certain pharmaceutically active ingredients have been identified from *G. gummi-gutta*.

Therapeutic Potentials of Major Phytochemicals

Xanthones

A new xanthone, garbogiol, was isolated from the root of *G. gummi-gutta*; in addition, another known xanthone (rheediaxanthone A) and two known benzophenones (garcinol and isogarcinol) were obtained from the bark by high-resolution electrospray ionization mass spectrometry (HREIMS) and nuclear magnetic resonance (NMR) spectroscopy [55]. *Garcinia hanburyi* [56] and *G. mangostana* [57] are

Table 2 Details of phytoconstituents identified from *G. gummi-gutta*

Plant tissue	Compounds	Extract	Method adopted	References
Fruit	(-)-hydroxycitric acid	Aqueous	Chromatography, IR Spectrometry, X-Ray crystallography	[66–68]
Root	Xanthone (Garbogiol)	Methanol	HREIMS, ^{1}H NMR, ^{13}C NMR	[55]
Bark	Xanthone (Rheediaxanthone A); Benzophenones (Garcinol and Isogarcinol)	Benzene, Acetone	HREIMS, ^{1}H NMR, ^{13}C NMR	[55]
Fruit	Xanthochymol	Methanol	LC–MS/MS	[59]
Fruit	Amino acids (Arginine, Asparagine, Glutamine, Threonine, Glycine, Proline, γ-aminobutyric acid, Leucine, Isoleucine, Ornithine, and Lysine)	–	HPLC	[90]
Fruit	Guttiferone (I, J, and K); Polyisoprenylated benzophenones (Guttiferone M, Guttiferone N, Oxy-Guttiferone K)	Diethyl ether	RP-HPLC	[91]
Fruit	Luteolin 7-O-glucuronide; Kaempferol 3-O-(6-O-acetyl) glycoside; Dicaffeoylquinic acid; Apigenein-6-C-pentosyl-8-C-hexoside; p-Coumarylquinic acid	Hydro alcohol	LC–MS/MS	[91]
Fruit	Isoxanthochymol; Camboginol	Methanol	HPLC and LCMS/MS	[92]
Fruit	Garcinol; Guttiferone K	Ethanol	HPLC/DAD	[93]
Leaf	Organic acid (Hydroxy citric acid; Garcicnic acid); Phenolic acid (Protocatechic acid); Flavaonoids (Apigenin, Vitexin, Isovitexin, Kaempferol); Prenylated xanthones (Gambogic acid); Triterpenoids (Betulinic acid, Ursolic acid); Polyisoprenylated benzophenones (Garcinol)	Methanol	UPHLC–QqQLIT–MS/MS	[94]
Fruit	Luteolin 7-O-glucuronide; Kaempferol 3-O-(6-O-acetyl) glycoside; Dicaffeoylquinic acid; Apigenein-6-C-pentosyl-8-C-hexoside; p-coumarylquinic acid	Ethanol	LC-MS	[95]
Fruit	Malic acid; Tartaric acid; Protocatechic acid 4-glucoside; Caffeoyl glucose; Hydroxy citric acid (HCA); Quercetin 3-glucoside; Quercetin 3-rhamanoside	Methanol	LC–MS	[96]
Fruit	4,8-Epi-uralione; 4,8-Epi-uralione G; Uralione S; Coccinone J; 6-Epi-coccinone C; Coccinone I; 36-Hydroxy-guttiferone J; Multiflorone I; Garciniagifolone F; 36-Hydroxy-garciniagifolone F	Ethanol	HPLC	[97]

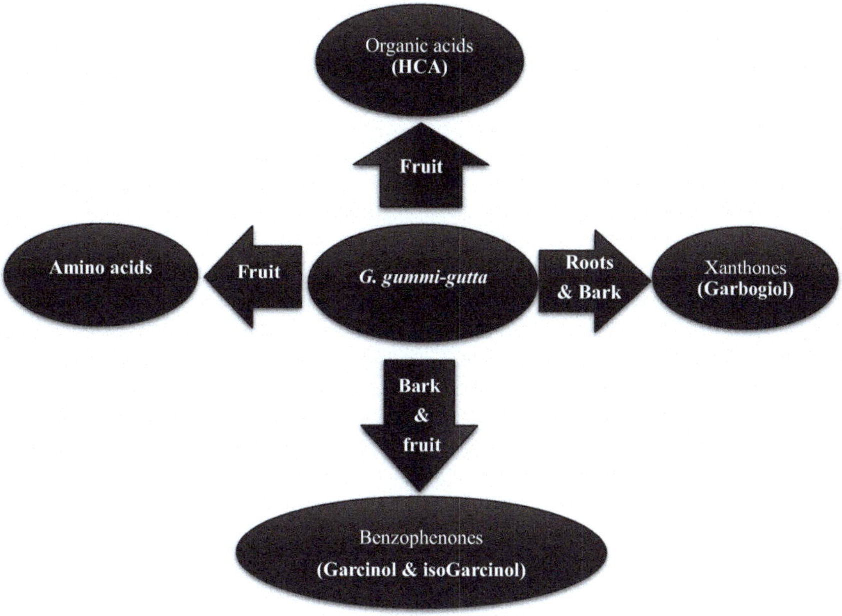

Fig. 6 Phytochemical profiling of *G. gummi-gutta*

other potent sources of this compound. Xanthone has the molecular formula $C_{13}H_8O_2$, having two benzene rings fused to a γ-pyrone ring. Most of the biological activities of xanthones are confined to their radical scavenging abilities. Their glycosides exhibit broad-spectrum pharmaceutical properties, such as antioxidant, antitumor, anti-allergic, anti-inflammatory, antibacterial, antifungal, antiviral, neuroprotective, anti-HIV, enzyme inhibition, and hypoglycemic activities [58].

Xanthochymol and Isoxanthochymol

Xanthochymol and isoxanthochymol are the two polyisoprenylated benzophenones; they were identified in the extracts of fruit rinds, stem bark, seed pericarps, and leaves of *G. indica* and in the fruit rinds of *G. gummi-gutta* using liquid chromatography/electrospray ionization tandem mass spectrometry (LC/ESI-MS/MS) [59]. Based on LC-MS/MS analysis, the therapeutic effects of these compounds from *G. indica* have evaluated. A combination of xanthochymol and isoxanthochymol at 1:2 ratio showed maximum cytotoxic activity against three human cancer cell lines, colon (COLO-320-DM), breast (MCF-7), and liver (WRL-68), as determined by 3-(4,5-dimethylthiazol-2-yl)-2,5 diphenyl tetrazolium bromide assay (MTT) [60].

Garcinol

Garcinol (camboginol) is an isoprenylated benzophenone first isolated from *G. gummi-gutta* [61]. Garcinol extracted from *G. dulcis* proved to be a potent anti-inflammatory potentials by its reduction of the pro-inflammatory cytokines and mediators in mRNA and protein expression as well as secretion levels via the modulation of the NF-κB signaling pathway. It helped in decreasing the production of pro-inflammatory molecules by alteration in the nuclear factor NF-κB signaling pathway, normally organizing cellular resistance to invading pathogens. The modification of NF-κB reduced the translocation of proNF-κB from the cytosol into the nucleus, which subsequently decreased the production of pro-inflammatory molecules such as pIKKα/β, pIκBα, and pNF-κB [62].

Guttiferones

They are prenylated benzophenone identified from *G. gummi-gutta* [55], *G. humilis* [63] and *G. multiflora* [64]. Guttiferone F from *G. multiflora* exhibited anti-hepatocellular carcinoma activity, and it showed cytotoxic effect against liver cancer cell lines like PLC/PRF/5, HepG2, and Hep3B. Likewise, it induced apoptosis in PLC/PRF/5 by upregulating cleaved pro-caspase-3 and poly (ADP-ribose) polymerase (PARP) expressions, i.e., enzymes which modulate apoptosis [64]. Guttiferone I isolated from *G. humilis* increased cholesterol efflux, lowered low-density lipoprotein (LDL), and raised high-density lipoprotein (HDL) levels. Similarly, it was proved to be a potent antagonist of liver X receptors (LXR), meant for cholesterol homeostasis [63].

Hydroxycitric Acid (HCA)

Hydroxycitric acid (HCA) is a derivative of citric acid and is an α-, β-dihydroxy tricarboxylic acid [65]. The acid was reported to be present in the fruit rinds of *G. gummi-gutta*, *G. indica*, and *G. atroviridis*, in the calyxes of *Hibiscus cannabinus* and *Hibiscus sabdariffa* in higher quantities [66–68], and also in the fruits of *G. cowa*, *Garcinia oblongifolia*, *G. dulcis* [69], *G. pedunculata* [70], *G. atroviridis* [71], and *Garcinia binucao* [72]. The fruit rind of *G. gummi-gutta* contains 10–60% HCA and is the major natural source of this organic acid [29, 73, 74]. Structurally, HCA has two asymmetric centers, hence two pairs of diastereoisomers or four isomers, and it can easily be crystalized in the lactone form [67] or as mineral salt of sodium, potassium, and calcium [68]. Reports on the isolation and characterization of chemical and physical properties of HCA are available [75–77]. Recently, researchers have identified superior *G. gummi-gutta* individuals with annual fruit yield, which is 40 times higher than the average of 250 individuals from Kerala, India [23]. This type of germplasm could significantly contribute to the extraction of HCA at an industrially feasible level.

Role in Anti-obesity and Weight Management

HCA helps in weight management and is a natural anti-obesity agent identified so far. It reduces the synthesis of fatty acids by blocking the enzyme adenosine triphosphate (ATP)-citrate lyase, which is meant for the extramitochondrial conversion of citrate to oxaloacetate and acetyl-coenzyme A (acetyl-CoA), thereby reducing the availability of subunit for fat synthesis (lipogenesis) (Fig. 7). Consequently, the consumed carbon source will be converted to glycogen in the liver; due to this metabolic alteration an impulse will be given by the brain to elevate the level of serotonin to suppress appetite [68]. Similarly, HCA reduces the rate of de novo lipogenesis from noncarbohydrate sources. The reduction in de novo lipogenesis and serotonin regulation will lead to weight management by the administration of HCA [78] (Fig. 7). Likewise, during exercise carbohydrate oxidation will be increased in the presence of HCA, thus generating ATP. This will result in the unavailability of carbohydrates for fatty acid synthesis without affecting adipose conversion. During the body exercise, the oxidation of stored fat will occur in the presence of HCA [79].

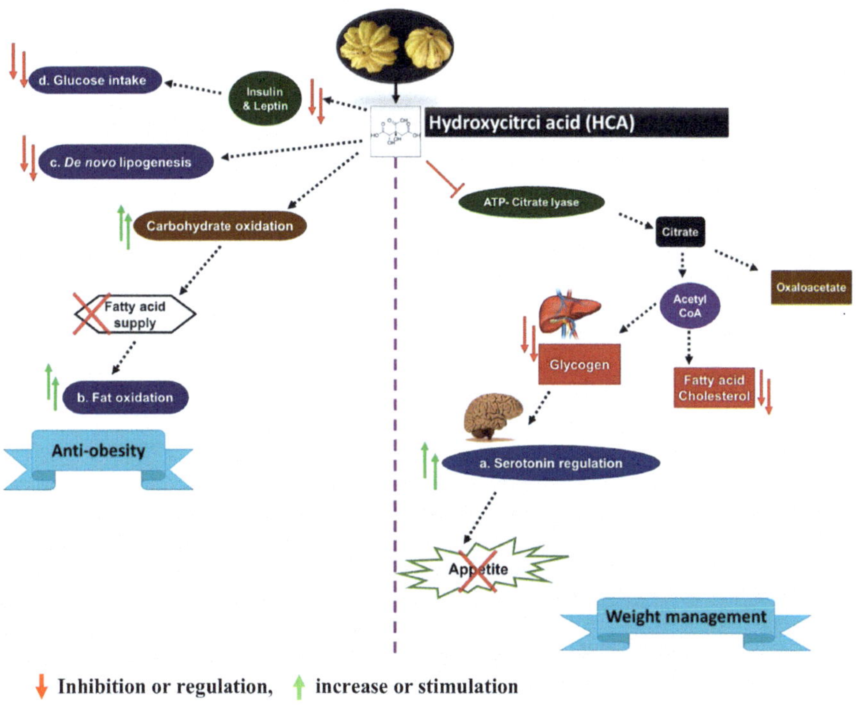

Fig. 7 Possible multiple mechanisms for weight management and anti-obesity effects of HCA, (**a**) summary of appetite suppression, (**b**) summary of stimulation of fat oxidation to synthesis carbohydrates, (**c**) summary of inhibition of de novo lipogenesis, and (**d**) summary of suppression of glucose intake by HCA, (**a** and **c**) contributed to weight management and (**b** and **c**) contributed to anti-obesity properties of HCA

Similarly, HCA reduces the concentration of leptin, thereby suppressing the excretion of insulin which will result in the reduced uptake of glucose [80, 81]. Increased oxidation of stored fat and reduced glucose uptake will contribute to the anti-obesity effect by the consumption of *G. gummi-gutta*, i.e., HCA [78] (Fig. 7). Thus, HCA increases the energy expenditure and subsequently reduces body fat accumulation and weight gain, as evident from the studies on animal models [82, 83].

Commercialized Products and Their Features

HCA is the major supplement in medications due to its anorexigenic effect, increased fat oxidation, and regulation of endogenous lipid biosynthesis. The pharmacological potentials of HCA have been proven, in connection with the formulation of nutraceuticals and wellness products of *G. gummi-gutta* [78, 84]. Among various commercial products, Super CitriMax® (HCA-SX) and Citrin® K are still marketed globally as health and wellness products. The safety and efficacy of these two weight loss supplements have ben evidenced through their achieved patents, placebo-controlled trials, and clinical and preclinical studies (Table 3). Adulteration in these types of natural products will lead to their debasement. In order to overcome such discrepancies, gas chromatography-mass spectrometry (GC-MS)-based analytical techniques are now available especially for supplements from *G. gummi-gutta* [74]. This could, in turn, improve the quality of weight loss supplements from the plant.

Table 3 Details of commercialized weight loss supplements of HCA

Sl. No.	Brand name	Manufacturer	Concentration of HCA	Usage	References	Patents
1.	Super CitriMax® (HCA-SX) (capsule)	Lonza Capsules & Health Ingredients, Morristown, USA	40-64% HCA in free form and 1% HCA in lactone form	900mg HCA per day for females and 2800mg per day for males up to 2 weeks with 2000kcal diet with exercise	[79, 87, 98–101]	[102–106]
2.	Citrin®K (capsule)	Sabinsa Corporation, East Windsor, USA	50 to 60% of hydroxycitric acid (HCA) and 28-35% of potassium	500 mg of extract (250 mg HCA) 30 min to 1 hr before meals—3 times daily	[107]	[108, 109]

Biological Activities and Safety of Extracts

Several extracts and supplements from different parts of *G. gummi-gutta* showed biological activities with in vitro and in vivo experimental models. Extracts prepared from fruit or fruit rind and leaf were the most experimented ones, where aqueous and methanolic extracts showed maximum biological activities (Table 4). Evidences revealed that *G. gummi-gutta* extracts and its supplements are safe and do not have any toxic effects [85–87]. *Garcinia gummi-gutta* containing formulations are safe up to 2800 mg/day toward spermatogenesis, and it is in the "no

Table 4 Details of biological activities of various extracts of *G. gummi-gutta*

Plant part	Extract type	Experimental model	Bioactivity	References
Fruit	Petroleum ether	Male Sprague-Dawley rats	Lipid lowering	[110]
Fruit	Commercially available fruit extract	Male Wistar rats	Antiulcer	[111]
Fruit	Aqueous	Female mice	Antidiabetic and body fat lowering	[80]
Fruit	*G. gummi-gutta* powder	Male Zucker obese rats	suppression of epididymal fat accumulation	[112]
Fruit	Aqueous	Male, Wistar rats and female, Swiss mice	Anti-inflammatory	[113]
Dried fruit	Ethanol	Human blood platelets and plasma	Antioxidant	[93]
–	*G. gummi-gutta* capsule containing 500mg HCA	Male Wistar rats	Effect against obesity-induced nephropathy	[83]
Fruit	Aqueous	Female Sprague–Dawley rats	Mineral balancing	[114]
Fruit and leaf	Aqueous and methanol	Swiss albino mice	Geno-protective and Cyto-protective	[115]
Fruit rind	Methanol	*Pheretima posthuma* *Pseudomonas aeruginosa* and *Staphylococcus aureus*	Anthelmintic Antibacterial	[116]
Fruit rind	Aqueous, Ethyl acetate	HeLa cell line CaCo2 cell line	Antimicrobial and Anticancer	[117]
Fruit	Methanol	Human liver microsomes	Inhibitory effects on CYP2B6 activity (reduces cholesterol synthesis)	[118]

(continued)

Table 4 (continued)

Plant part	Extract type	Experimental model	Bioactivity	References
Fruit	Commercially available fruit extract	Male Wistar rats	Anti-obesity	[119]
Fruit rind	Aqueous	Adult male Wistar rats	Reducing the side effects of allopathic drug by interacting with *G. gummi-gutta* extract	[120]
Leaves	Essential oil	Gram-positive and Gram-negative bacteria	Antibacterial and cytotoxic against human breast cancer	[121]
Fruit rind	Zinc oxide nanoparticles	HeLa cell line MCF-7 cell line	Antibacterial, antifungal, antioxidant, and cytotoxic	[122]
Fruit	*G. gummi-gutta* extract	Mice	Arousal-inducing Effect	[123]
G. gummi-gutta fruit powder	-	Male mice	Amelioration of nonalcoholic fatty liver Disease	[124]
Garcinia herbal supplement, Garcinol, and HCA Fruit	Methanol	murine macrophage cell line	Anti-inflammatory	[125]
Hydroxycitric acid	Aqueous	Chronic myelogenous leukemia cell lines (K562)	Inhibition of chronic myelogenous leukemia	[126]
Fruit	Aqueous	Male mice	Alleviation of insulin resistance and hepatic lipid accumulation	[127]

observed adverse effect level" (NOAEL) [85]. Super CitriMax (HCA-SX), a calcium/potassium salt of (−)-HCA, showed non-toxicity in rats as the oral LD_{50} dose was determined to be greater than 5000 mg/kg [88]. An intraperitoneal injection of HCA-SX, at doses of up to 12,500 μmol/kg, does not cause genotoxicity [64, 89]. Based on the mechanism of action, and pharmacological and wellness properties of HCA, several weight loss supplements of the plant have been developed and marketed in India by reputed pharmaceutical companies.

Apart from HCA in fruits, *G. gummi-gutta* contains several other active principles in various plant parts. However, phytopharmacology has not yet been comprehensively evaluated. Numerous biological properties of *G. gummi-gutta* such as anti-obesity, appetite suppressant, anticancer, lipid-lowering, antimicrobial, antioxidant, and anti-inflammatory have been well elucidated by applying in vitro as well as in vivo experimental models. Clinical trial reports revealed that extracts and supplements from *G. gummi-gutta* are beneficial to obese individuals in many cases. Besides, they do not cause any health problems up to the recommended level.

Conclusion

Garcinia gummi-gutta is an important component of the Southwestern Ghats of India. Several approaches were performed for the identification and authentication of pharmaceutical potential of this plant. A generous portion of phytoconstituents has been identified from fruits, followed by leaf and bark. Based on the anti-obesity properties of hydroxycitric acid from the fruits of *G. gummi-gutta*, several health, wellness, or nutraceutical supplements are being formulated and marketed in India. Due to unscientific harvesting and processing, large quantities of fruit resource remain unprocessed and wasted every year. At the same time, pharma companies demand large quantities of resources for their needs. In the current scenario, fruit dealers could not fulfill the requirements of resources for medicinal industries. The availability of raw materials for medicinal companies could be enhanced by improving the quality and productivity of *G. gummi-gutta* through breeding approaches. HCA is the major phytochemical identified from the species and is an organic acid that imparts extreme sourness to fruits. Xanthones and benzophenones such as xanthochymol, isoxanthochymol, garcinol, and guttiferone are the other major phytoconstituents identified with medicinal applications. Individual or synergetic effects of active principles have been evaluated in light of in vitro as well as in vivo models with sufficient clinical trials. Phytoconstituent-based nutraceuticals and preparations from *G. gummi-gutta* are safe at its prescribed dosage. Thus, a proportional relationship is existing between the botany and pharmacology of natural products from the plant. However, more research is to be focused on the exploration of pharmaceutical potentials of non-fruit tissues and the genome level elucidation of mechanisms behind the action of diverse phytochemicals.

References

1. Tugume P, Nyakoojo C. Ethno-pharmacological survey of herbal remedies used in the treatment of paediatric diseases in Buhunga parish, Rukungiri District, Uganda. BMC Complement Altern Med. 2019;19:1–10. https://doi.org/10.1186/s12906-019-2763-6.
2. Chandrashekara UM. Tree species yielding edible fruit in the coffee-based homegardens of Kerala, India: their diversity, uses and management. Food Secur. 2009;1:361–70. https://doi.org/10.1007/s12571-009-0032-0.
3. Abraham Z, Malik SK, Rao GE, Narayanan SL, Biju S. Collection and characterisation of Malabar tamarind (Garcinia cambogia (Gaertn.) Desr.). Genet Resour Crop Evol. 2006;53:401–6. https://doi.org/10.1007/s10722-004-0584-y.
4. Baas P. Mabberley's Plant-book – a portable dictionary of plants, their classification and uses. In: Mabberley DJ, editor. 4th ed. Cambridge University Press; 2017. p. 1102. ISBN

978-1-107-11502-6. Price: EUR 69.00 or GBP 59.99 (hardback)., IAWA J. 38 (2017) 573. 10.1163/22941932-20170192.
5. Kusters EK, Belcher B. Forest products, livelihoods and conservation, n.d.
6. Bohra P, Waman AA. Morphological and biochemical studies in Garcinia gummi-gutta (L.) Roxb. Erwerbs-obstbau. 2019;61:217–23. https://doi.org/10.1007/s10341-019-00419-3.
7. Parthsarathy U, Nandakishore OP. A study on nutrient and medicinal compositions of selected Indian Garcinia species. Curr Bioact Compd. 2014;10:55–61. https://doi.org/10.2174/1573407210011140725001152.
8. Angami T, Wangchu L, Debnath P, Sarma P, Singh B, Singh AK, Singh S, Hazarika BN, Singh MC, Aochen C. Lungmuana, Garcinia L.: a gold mine of future therapeutics. Genet Resour Crop Evol. 2021;68:11–24. https://doi.org/10.1007/s10722-020-01057-5.
9. Parthasarathy U, Nandakishore OP. Morphological characterisation of some important Indian Garcinia species. Dataset Pap Sci. 2014;2014:1–4. https://doi.org/10.1155/2014/823705.
10. Orwa C, Mutua A, Kindt R, Jamnadass R, Anthony S. Agroforestree database: a tree reference and selection guide version 4.0. World Agroforestry Centre, Kenya; 2009. http://worldagroforestry.org/output/agroforestree-database.
11. Remya Ramachandran SS, Jayan B, Vishnu AN, Gangaprasad AP, Siril EA. Floral Polychromatism and female fitness in hermaphrodites of Garcinia gummi-gutta (L.) Roxb. Biol Bull. 2023;50:S351–62. https://doi.org/10.1134/S1062359023600332.
12. T.K. Lim, Garcinia gummi-gutta, in: Edible Med. non-medicinal plants, Springer Dordrecht, 2012: pp. 45–55. https://doi.org/10.1007/978-94-007-1764-0_7.
13. Anilkumar C, Babu KP, Krishnan PN. Dormancy mechanism and effects of treatments on the germination of Garcinia gummi-gutta (Clusiaceae) seeds. J Trop For Sci. 2002;14:322–8.
14. Joshi G, Phartyal SS, Arunkumar AN. Non-deep physiological dormancy, desiccation and low-temperature sensitivity in seeds of Garcinia gummi-gutta (Clusiaceae): a tropical evergreen recalcitrant species. Trop Ecol. 2017;58:241–50.
15. Shameer PS, Sabu T, Mohanan NN. Taxonomic reinstatement of Garcinia conicarpa Wight (Clusiaceae). Phytotaxa. 2021;490:191–6. https://doi.org/10.11646/phytotaxa.490.2.5.
16. Singh MK, Bindhu OS. Plant latex: A rich source of haemostatic proteases. Herbal Med India. 2019. https://doi.org/10.1007/978-981-13-7248-3_10.
17. Deepu S, Geethakumary MP. Garcinia gummi-gutta, 8235 (2020).
18. Guedje NM, Zuidema PA, During H, Foahom B, Lejoly J. Tree bark as a non-timber forest product: the effect of bark collection on population structure and dynamics of Garcinia lucida Vesque. For Ecol Manag. 2007;240:1–12. https://doi.org/10.1016/j.foreco.2006.09.029.
19. Lim TK. Edible medicinal and non-medicinal plants. 2016. https://doi.org/10.1007/978-94-017-7276-1.
20. Palkar RS, Janarthanam MK, Sellappan K. Prediction of potential distribution and climatic factors influencing Garcinia indica in the Western Ghats of India using ecological niche modeling. Natl Acad Sci Lett. 2020;43:585–91. https://doi.org/10.1007/s40009-020-00918-y.
21. Kalia RK, Malik SK, Chaudhury R. In vitro morphogenetic studies on three apomictic species of Garcinia. J Plant Biochem Biotechnol. 2012;21:279–85. https://doi.org/10.1007/s13562-011-0085-y.
22. Nair KKN, Mohanan C, Mathew G. Preliminary observations on seed and plantation technology for garcinia gummi-gutta (Guttiferae): an indigenous and economic tree crop of The Indian Peninsula. For Trees Livelihoods. 2005;15:75–87. https://doi.org/10.1080/14728028.2005.9752508.
23. Ramachandran VBR, Gangaprasad A, Siril EA. Selection of promising candidate of Malabar tamarind (Garcinia gummi-gutta (L.) Roxb.)- a multipurpose fruit tree. Int J Fruit Sci. 2022;22:664–74. https://doi.org/10.1080/15538362.2022.2095552.
24. Ramachandran R, Vishnu B, Gangaprasad A, Siril EA. Effective and reliable protocol for DNA isolation from polyphenolic leaves and fruits of Garcinia gummi-gutta (L.) robs. Proc Natl Acad Sci India Sect B Biol Sci. 2022;92:825–31. https://doi.org/10.1007/s40011-022-01356-0.

25. Ravishankar KV, Vasudeva R, Hemanth B, Sandya BS, Sthapit BR, Parthasarathy VA, Rao VR. Isolation and characterization of microsatellite markers in Garcinia gummi-gutta by next-generation sequencing and cross-species amplification. J Genet. 2017;96:213–8. https://doi.org/10.1007/s12041-017-0756-0.
26. Tharachand C, Immanuel Selvaraj C, Abraham Z. Molecular insights into the genetic diversity of Garcinia cambogia germplasm accessions. Braz Arch Biol Technol. 2015;58:765–72. https://doi.org/10.1590/S1516-89132015050197.
27. Babu V, Ramachandran R, Gangaprasad A, Siril EA. Assessment of morphogenetic diversity in Garcinia gummi-gutta (L.) Roxb. Using species-specific morphological and SSR markers. Ecol Genet Genomics. 2021;18. https://doi.org/10.1016/j.egg.2021.100081.
28. Ramachandran VBR, Aswanilal L, Appukuttannair G, Siril EA. Genetic evaluation of Garcinia gummi-gutta L. (Roxb.) accessions based on inter simple sequence repeat markers. Ecol Genet Genomics. 2022;24:100130. https://doi.org/10.1016/j.egg.2022.100130.
29. Vishnu B, Ramachandran R, Gangaprasad A, Siril EA. Start codon targeted (SCoT) variability analysis and its association with hydroxy citric acid (HCA) in Garcinia gummi-gutta (L.) Roxb. Plant Gene. 2023;34:100415. https://doi.org/10.1016/j.plgene.2023.100415.
30. Mitra S, Subramanian RN, Mall S, Rao GP. Molecular identification of 'Candidatus Phytoplasma oryzae' in Garcinia gummi-gutta in India. Phytopathogenic Mollicutes. 2019;9:5. https://doi.org/10.5958/2249-4677.2019.00003.3.
31. Joseph KS, Murthy HN, Ravishankar KV. Development of SCAR marker for sex identification in dioecious Garcinia gummi-gutta. Trees. 2014;28:1645–51. https://doi.org/10.1007/s00468-014-1074-2.
32. Anerao J, Sharangi G, Jha V, Pardhi V, Chavan S, Desai N, Mangaonkar K. Molecular analysis and phytochemical profiling to explore the diversity of Garcinia species collected from Western Ghats of India. Acta Bot Hung. 2021;63:1–30. https://doi.org/10.1556/034.63.2021.1-2.1.
33. H.N. Murthy, V.S. Dandin, D. Dalawai, S.Y. Park, K.Y. Paek, Breeding of garcinia spp, Advances in plant breeding strategies: fruits 2018. https://doi.org/10.1007/978-3-319-91944-7_19.
34. S.R. Suja, R.R. Nair, S. Rajasekharan, R. Prakashkumar, Relevance of Ethnopharmacological research related to threatened medicinal plants associated with traditional knowledge, in: P.E. Rajasekharan, S.H. Wani (Eds.), Conservation and sustainable uses of medicinal plants, Springer International Publishing, Cham, 2020: pp. 463–487. https://doi.org/10.1007/978-3-030-39793-7_16.
35. Pramanik M, Diwakar AK, Dash P, Szabo S, Pal I. Conservation planning of cash crops species (Garcinia gummi-gutta) under current and future climate in the Western Ghats, India. Environ Dev Sustain. 2021;23:5345–70. https://doi.org/10.1007/s10668-020-00819-6.
36. Chacko KC, Chandrasekhara Pillai PK. Seed characteristics and germination of garcinia gummi-gutta (L.) Robs. Indian For. 1997;123:123–6.
37. Joshi G, Arun Kumar AN, Gowda B, Srinivasa YB. Production of supernumerary plants from seed fragments in Garcinia gummi-gutta: evolutionary implications of mammalian frugivory. Curr Sci. 2006;91:372–6.
38. Mohan S, Parthasarathy U, Nirmal Babu K. In vitro and in vivo adventitious bud differentiation from mature seeds of three Garcinia spp. Indian J Nat Prod Resour. 2012;3:65–72.
39. Rai N. The socio economic and ecological impact of Garcinia gummi-gutta fruit harvest in the Western Ghats, India. In: Forest products, livelihoods and conservation: Case studies non-timber forest product systems Vol. 1 – Asia; 2004. p. 23–41.
40. T.K. Lim, Edible medicinal and non-medicinal plants: Volume 2, fruits (2012) 1–1088. https://doi.org/10.1007/978-94-007-1764-0.
41. Apang T, Xavier KAM, Lekshmi M, Kannuchamy N, Layana P, Balange AK. Garcinia spp. extract incorporated icing medium as a natural preservative for shelf life enhancement

of chilled Indian mackerel (Rastrelliger kanagurta). LWT. 2020;133:110086. https://doi.org/10.1016/j.lwt.2020.110086.
42. Jeeja Rani AT, Kuruvilla M, Joseph A. Computational modelling and theoretical calculations on hydroxy citric acid and mangostine and comparing the corrosion inhibition effect of aqueous and alcoholic extracts Garcinia cambogia leaves for mild steel in hydrochloric acid. J Mol Struct. 2023;1274:134302. https://doi.org/10.1016/j.molstruc.2022.134302.
43. Shamnamol GK, Rugma P, John S, Mary Jacob J. Unraveling the synergistic effect of cationic and anionic salt on the corrosion inhibition performance of Garcinia gummi-gutta leaf extract against mild steel in HCl medium. Results Chem. 2023;5:100728. https://doi.org/10.1016/j.rechem.2022.100728.
44. Rani ATJ, Thomas A, Arshad M, Joseph A. The influence of aqueous and alcoholic extracts of Garcinia cambogia fruit rind in the management of mild steel corrosion in hydrochloric acid: theoretical and electroanalytical studies. J Mol Liq. 2022;346:117873. https://doi.org/10.1016/j.molliq.2021.117873.
45. Priya Rani M, Gokul Raj MR, Bahuleyan Rameshkumar K. Garcinia gummi-gutta seeds: a novel source of edible oils. J Sci Food Agric. 2022;102:3475–9. https://doi.org/10.1002/jsfa.11671.
46. Ajith BS, Math MC, Manjunath Patel GC. Evaluation of performance and emission characteristics of diesel engine Fuelled with Garcinia gummi Gutta biodiesel and diethyl ether blends. Eur J Sustain Dev Res. 2020;4:em0134. https://doi.org/10.29333/ejosdr/8353.
47. Ajith BS, Math MC, Manjunath Patel GC, Parappagoudar MB. Engine performance and exhaust emissions of Garcinia gummi-gutta based biodiesel–diesel and ethanol blends. SN Appl Sci. 2021;3:1–11. https://doi.org/10.1007/s42452-021-04537-0.
48. Ajith BS, Math MC, Patel MGC, Parappagoudar MB. Analysis and optimisation of transesterification parameters for high-yield Garcinia Gummi-Gutta biodiesel using RSM and TLBO. Aust J Mech Eng. 2022;20:1463–78. https://doi.org/10.1080/14484846.2020.1815998.
49. Arun SB, Suresh R, Yatish KV, Omkaresh BR, Channa Keshava Naik N. Use of CaO and Na_3PO_4 catalysts in the synthesis of biodiesel and investigation of fuel properties. Mater Today Proc. 2017;4:11111–7. https://doi.org/10.1016/j.matpr.2017.08.074.
50. Lingesan S, Annamalai K, Parthasarathy M, Ramalingam KM, Dhinesh B, Lalvani JIJ. Production of Garcinia gummi-gutta methyl Ester (GGME) as a potential alternative feedstock for existing unmodified DI diesel engine: combustion, performance, and emission characteristics. J Test Eval. 2018;46:20170246. https://doi.org/10.1520/JTE20170246.
51. Rajagopalachar S, Joshi SS, Reddy RP. Biodiesel synthesis from Garcinia gummi-gutta (L. Robson) seed oil: fuel feasibility evaluation of a novel feedstock by homogeneous and heterogeneous transesterification. Biofuels. 2019;10:403–10. https://doi.org/10.1080/17597269.2017.1306681.
52. Yatish KV, Lalithamba HS, Suresh R, Omkaresh BR. Synthesis of biodiesel from Garcinia gummi-gutta, Terminalia belerica and Aegle marmelos seed oil and investigation of fuel properties. Biofuels. 2018;9:121–8. https://doi.org/10.1080/17597269.2016.1259524.
53. Janakiraman S, Lakshmanan T, Chandran V, Subramani L. Comparative behavior of various nano additives in a DIESEL engine powered by novel Garcinia gummi-gutta biodiesel. J Clean Prod. 2020;245:118940. https://doi.org/10.1016/j.jclepro.2019.118940.
54. L. Subramani, M. Parthasarathy, D. Balasubramanian, K.M. Ramalingam, Novel Garcinia gummi-gutta methyl ester (GGME) as a potential alternative feedstock for existing unmodified DI diesel engine, Elsevier B.V., 2018. https://doi.org/10.1016/j.renene.2018.02.134.
55. Iinuma M, Ito T, Miyake R, Tosa H, Tanaka T, Chelladurai V. A xanthone from Garcinia cambogia. Phytochemistry. 1998;47:1169–70. https://doi.org/10.1016/S0031-9422(98)80096-4.
56. Jia B, Li S, Hu X, Zhu G, Chen W. Recent research on bioactive Xanthones from natural medicine: Garcinia hanburyi. AAPS PharmSciTech. 2015;16:742–58. https://doi.org/10.1208/s12249-015-0339-4.

57. Nauman MC, Johnson JJ. The purple mangosteen (Garcinia mangostana): defining the anticancer potential of selected xanthones. Pharmacol Res. 2022;175:106032. https://doi.org/10.1016/j.phrs.2021.106032.
58. Shan T, Ma Q, Guo K, Liu J, Li W, Wang F, Wu E. Xanthones from Mangosteen extracts as natural Chemopreventive agents: potential anticancer drugs. Curr Mol Med. 2011;11:666–77. https://doi.org/10.2174/156652411797536679.
59. Chattopadhyay S, Kumar S. Identification and quantification of two biologically active polyisoprenylated benzophenones xanthochymol and isoxanthochymol in Garcinia species using liquid chromatography–tandem mass spectrometry. J Chromatogr B. 2006;844:67–83. https://doi.org/10.1016/j.jchromb.2006.07.045.
60. Kumar S, Chattopadhyay S, Darokar M, Garg A, Khanuja S. Cytotoxic activities of Xanthochymol and Isoxanthochymol substantiated by LC-MS/MS. Planta Med. 2007;73:1452–6. https://doi.org/10.1055/s-2007-990255.
61. Rama Rao AV, Venkatswamy G, Pendse D. Camboginol and Cambogin. Tetrahedron Lett. 1980;21:1975–8. https://doi.org/10.1016/S0040-4039(00)93661-X.
62. Chantree P, Martviset P, Thongsepee N, Sangpairoj K, Sornchuer P. Anti-inflammatory effect of Garcinol extracted from Garcinia dulcis via modulating NF-κB signaling pathway. Nutrients. 2023;15. https://doi.org/10.3390/nu15030575.
63. Herath K, Jayasuriya H, Ondeyka JG, Guan Z, Borris RP, Stijfhoorn E, Stevenson D, Wang J, Sharma N, MacNaul K, Menke JG, Ali A, Schulman MJ, Singh SB. Guttiferone I, a new prenylated benzophenone from Garcinia humilis as a liver X receptor ligand. J Nat Prod. 2005;68:617–9. https://doi.org/10.1021/np050045j.
64. Lee L-T, Tsai H-P, Wang C-C, Chang C-N, Liu W-C, Hsu H-C, Hsieh C-T, Chen Y-C, Tseng H-W, Gau R-J, Liu S-H, Chen I-S, Iinuma M. Guttiferone F from the fruit of Garcinia multiflora and its anti-hepatocellular carcinoma activity. Biomed Prev Nutr. 2013;3:247–52. https://doi.org/10.1016/j.bionut.2012.10.007.
65. Antony MEB, Varghese W. Preparation and evaluation of hydroxy citric acid from garcinia cambogia extract using RP-Amide HPLC. Indian J Pharm Sci. 2004;66:208–11. https://www.ijpsonline.com, https://www.ijpsonline.com/abstract/preparation-and-evaluation-of-hydroxy-citric-acid-from-garcinia-cambogia-extract-using-rpamide-hplc-641.html.
66. Lewis YS, Neelakantan S. (−)-Hydroxycitric acid—the principal acid in the fruits of Garcinia cambogia desr. Phytochemistry. 1965;4:619–25. https://doi.org/10.1016/S0031-9422(00)86224-X.
67. Lewis YS. Isolation and properties of hydroxycitric acid. Methods Enzymol. 1969;613–9. https://doi.org/10.1016/0076-6879(69)13085-2.
68. Lowenstein JM, Brunengraber H. (36) Hydroxycitrate. Methods Enzymol. 1981;486–97. https://doi.org/10.1016/S0076-6879(81)72038-X.
69. Stiyani H, Irwanto RR. The potential of garcinia dulcis (Roxb.) kurz. as a source of hydroxycitric acid and its habitat conditions. Malaysian Appl Biol. 2020;49:111–4. https://doi.org/10.55230/mabjournal.v49i2.1532.
70. Bhattacharjee S, Devi R. A comprehensive review of Garcinia pedunculata Roxb. and its therapeutic potential, mini-reviews. Med Chem. 2021;21:3113–43. https://doi.org/10.2174/1389557521666210217094152.
71. Niyomdecha A, Wattanachant C, Rattanawut J, Plodpai P, Ngampongsai W. Antibacterial activity of ethanol extract of asam gelugur (Garcinia atroviridis) fruits from Southern Thailand, Thai. J Pharm Sci. 2022;46:346–52.
72. Bainto LC, Dizon EI, Castillo-Israel KAT. Effects of various processing methods on hydroxycitric acid content of "batuan" (Garcinia binucao (Blanco) Choisy) fruits. Int Food Res J. 2018;25:S13–9.
73. Jayaprakasha GK, Sakariah KK. Determination of (-) hydroxycitric acid in commercial samples of Garcinia cambogia extract by liquid chromatography with ultraviolet detection. J Liq Chromatogr Relat Technol. 2000;23:915–23. https://doi.org/10.1081/JLC-100101498.
74. Mena-García A, Sanz ML, Díez-Municio M, Soria AC, Ruiz-Matute AI. Development of a new method using gas chromatography coupled to mass spectrometry to deter-

mine (−)-hydroxycitric acid in Garcinia gummi-gutta food supplements. Microchem J. 2022;181:107811. https://doi.org/10.1016/j.microc.2022.107811.

75. Jena BS, Jayaprakasha GK, Singh RP, Sakariah KK. Chemistry and biochemistry of (−)-Hydroxycitric acid from Garcinia. J Agric Food Chem. 2002;50:10–22. https://doi.org/10.1021/jf010753k.
76. H.N. Murthy, V.S. Dandin, D. Dalawai, S.-Y. Park, K.-Y. Paek, Bioactive compounds from Garcinia fruits of high economic value for food and health, Springer 2019: pp. 1643–1670. https://doi.org/10.1007/978-3-319-78030-6_65.
77. Yamada T, Hida H, Yamada Y. Chemistry, physiological properties, and microbial production of hydroxycitric acid. Appl Microbiol Biotechnol. 2007;75:977–82. https://doi.org/10.1007/s00253-007-0962-4.
78. Chuah LO, Ho WY, Beh BK, Yeap SK. Updates on antiobesity effect of garcinia Origin (-)-HCA. Evid Based Complement Altern Med. 2013;2013. https://doi.org/10.1155/2013/751658.
79. Preuss HG, Bagchi D, Bagchi M, Rao CVS, Satyanarayana S, Dey DK. Efficacy of a novel, natural extract of (-)-hydroxycitric acid (HCA-SX) and a combination of HCA-SX, niacin-bound chromium and Gymnema sylvestre extract in weight management in human volunteers: a pilot study. Nutr Res. 2004;24:45–58. https://doi.org/10.1016/j.nutres.2003.09.007.
80. Hayamizu K, Hirakawa H, Oikawa D, Nakanishi T, Takagi T, Tachibana T, Furuse M. Effect of Garcinia cambogia extract on serum leptin and insulin in mice. Fitoterapia. 2003;74:267–73. https://doi.org/10.1016/S0367-326X(03)00036-4.
81. Lustig RH. Childhood obesity: behavioral aberration or biochemical drive? Reinterpreting the First Law of Thermodynamics. Nat Clin Pract Endocrinol Metab. 2006;2:447–58. https://doi.org/10.1038/ncpendmet0220.
82. Kim YJ, Kim K-Y, Kim MS, Lee JH, Lee KP, Park T. A mixture of the aqueous extract of Garcinia cambogia, soy peptide and l-carnitine reduces the accumulation of visceral fat mass in rats rendered obese by a high fat diet. Genes Nutr. 2008;2:353–8. https://doi.org/10.1007/s12263-007-0070-1.
83. Amin KA, Kamel HH, Abd Eltawab MA. Protective effect of Garcinia against renal oxidative stress and biomarkers induced by high fat and sucrose diet. Lipids Health Dis. 2011;10:1–13. https://doi.org/10.1186/1476-511X-10-6.
84. Márquez F, Babio N, Bulló M, Salas-Salvadó J. Evaluation of the safety and efficacy of Hydroxycitric acid or Garcinia cambogia extracts in humans. Crit Rev Food Sci Nutr. 2012;52:585–94. https://doi.org/10.1080/10408398.2010.500551.
85. Chuah LO, Yeap SK, Ho WY, Beh BK, Alitheen NB. In vitro and in vivo toxicity of Garcinia or Hydroxycitric acid: a review. Evid Based Complement Altern Med. 2012;2012:1–12. https://doi.org/10.1155/2012/197920.
86. Pittler MH, Schmidt K, Ernst E. Adverse events of herbal food supplements for body weight reduction: systematic review*. Obes Rev. 2005;6:93–111. https://doi.org/10.1111/j.1467-789X.2005.00169.x.
87. Stohs SJ, Lau FC, Kim D, Kim SU, Bagchi M, Bagchi D. Safety assessment of a calcium-potassium salt of (−)-hydroxycitric acid. Toxicol Mech Methods. 2010;20:515–25. https://doi.org/10.3109/15376516.2010.521207.
88. Ohia SE, Opere CA, LeDay AM, Bagchi M, Bagchi D, Stohs SJ. Safety and mechanism of appetite suppression by a novel hydroxycitric acid extract (HCA-SX). Mol Cell Biochem. 2002;238:89–103. https://doi.org/10.1023/a:1019911205672.
89. Deshmukh NS, Bagchi M, Yasmin T, Bagchi D. Safety of a novel calcium/potassium Saltof (-)-Hydroxycitric acid (HCA-SX): II.Developmental toxicity study in rats. Toxicol Mech Methods. 2008;18:443–51. https://doi.org/10.1080/15376510802055022.
90. Carratù B, Boniglia C, Giammarioli S, Mosca M, Sanzini E. Free amino acids in botanicals and botanical preparations. J Food Sci. 2008;73:323–8. https://doi.org/10.1111/j.1750-3841.2008.00767.x.

91. Masullo M, Bassarello C, Suzuki H, Pizza C, Piacente S. Polyisoprenylated benzophenones and an unusual polyisoprenylated tetracyclic xanthone from the fruits of Garcinia cambogia. J Agric Food Chem. 2008;56:5205–10. https://doi.org/10.1021/jf800416j.
92. Kumar S, Sharma S, Chattopadhyay SK. High-performance liquid chromatography and LC-ESI-MS method for identification and quantification of two isomeric polyisoprenylated benzophenones isoxanthochymol and camboginol in different extracts of Garcinia species. Biomed Chromatogr. 2009;23:888–907. https://doi.org/10.1002/bmc.1202.
93. Kolodziejczyk J, Masullo M, Olas B, Piacente S, Wachowicz B. Effects of garcinol and guttiferone K isolated from Garcinia cambogia on oxidative/nitrative modifications in blood platelets and plasma. Platelets. 2009;20:487–92. https://doi.org/10.3109/09537100903165182.
94. Pandey R, Chandra P, Kumar B, Srivastva M, Anu Aravind AP, Shameer PS, Rameshkumar KB. Simultaneous determination of multi-class bioactive constituents for quality assessment of Garcinia species using UHPLC-QqQLIT-MS/MS. Ind Crop Prod. 2015;77:861–72. https://doi.org/10.1016/j.indcrop.2015.09.041.
95. Sulaiman CT, Balachandran I. LC/MS characterization of phenolic antioxidants of Brindle berry (Garcinia gummi-gutta (L.) Robson). Nat Prod Res. 2017;31:1191–4. https://doi.org/10.1080/14786419.2016.1224871.
96. Sulaiman CT, George S, Balachandran I. Characterization of phenolic compounds in fruit extracts of three medicinal plants using liquid chromatography-negative electrospray ionization tandem mass spectrometry. Anal Chem Lett. 2013;3:322–9. https://doi.org/10.1080/22297928.2013.861167.
97. Feng Z, Chen J, Feng L, Chen C, Ye Y, Lin L. Polyisoprenylated benzophenone derivatives from Garcinia cambogia and their anti-inflammatory activities. Food Funct. 2021;12:6432–41. https://doi.org/10.1039/d1fo00972a.
98. Preuss HG, Rao CVS, Garis R, Bramble JD, Ohia SE, Bagchi M, Bagchi D. An overview of the safety and efficacy of a novel, natural(-)-hydroxycitric acid extract (HCA-SX) for weight management. J Med. 2004;35:33–48. http://www.ncbi.nlm.nih.gov/pubmed/18084863.
99. Preuss HG, Bagchi D, Bagchi M, Rao CVS, Dey DK, Satyanarayana S. Effects of a natural extract of (-)-hydroxycitric acid (HCA-SX) and a combination of HCA-SX plus niacin-bound chromium and Gymnema sylvestre extract on weight loss. Diabetes Obes Metab. 2004;6:171–80. https://doi.org/10.1111/j.1462-8902.2004.00328.x.
100. Soni MG, Burdock GA, Preuss HG, Stohs SJ, Ohia SE, Bagchi D. Safety assessment of (−)-hydroxycitric acid and Super CitriMax®, a novel calcium/potassium salt. Food Chem Toxicol. 2004;42:1513–29. https://doi.org/10.1016/j.fct.2004.04.014.
101. Downs BW, Bagchi M, Subbaraju GV, Shara MA, Preuss HG, Bagchi D. Bioefficacy of a novel calcium–potassium salt of (−)-hydroxycitric acid. Mutat Res Mol Mech Mutagen. 2005;579:149–62. https://doi.org/10.1016/j.mrfmmm.2005.02.021.
102. Morency M.(-)-Hydroxycitric acid for controlling inflammation, US 8,394,856 B2, 2002.
103. Gokaraju RR, Pratha S. Process for preparing highly water soluble double salts of Hydroxycitric acid particularly alkali and alkaline earth metal double salts, US 6,875,891 B2. 2005.
104. Raju GG. Hydroxycitric acid compositions, pharmaceutical and dietary supplements and food products made therefrom, and methods for their use in reducing body weight, US 7,858,128 B2. 2005.
105. Ganga RG. Hydroxycitric acid compositions, pharmaceutical and dietary supplements and food products made therefrom, and methods for their use in reducing body weight, US 7,943,186 B2. 2011.
106. Bagchi D, Preuss HG, Ohia SE. Compositions incorporating (-)-hydroxycitric acid and related methods for promoting fat oxidation, US 7,335,651 B2. 2008.
107. Louter-van de Haar J, Wielinga PY, Scheurink AJ, Nieuwenhuizen AG. Comparison of the effects of three different (-)-hydroxycitric acid preparations on food intake in rats. Nutr Metab (Lond). 2005;2:23. https://doi.org/10.1186/1743-7075-2-23.

108. Majeed Muhammed RR, Vladimer B. Potassium hydroxycitrate for the suppression of appetite and induction of weight loss, USOO5783.603A. 1998.
109. Badmaey V, Lewis YS, Examiner P, Seaman DM, Examiner A, Oh TV. Process for the production of potassium hydroxy citric acid, and compositions containing the potassium hydroxy citric acid, US 6,770,782 B1. 2004.
110. Koshy AS, Vijayalakshmi NR. Impact of certain flavonoids on lipid profiles potential action of Garcinia cambogia flavonoids. Phyther Res. 2001;15:395–400. https://doi.org/10.1002/ptr.725.
111. Mahendran P, Vanisree AJ, Shyamala Devi CS. The antiulcer activity of Garcinia cambogia extract against indomethacin induced gastric ulcer in rats. Phyther Res. 2002;16:80–3. https://doi.org/10.1002/ptr.946.
112. Saito M, Ueno M, Ogino S, Kubo K, Nagata J, Takeuchi M. High dose of Garcinia cambogia is effective in suppressing fat accumulation in developing male Zucker obese rats, but highly toxic to the testis. Food Chem Toxicol. 2005;43:411–9. https://doi.org/10.1016/j.fct.2004.11.008.
113. dos Reis SB, de Oliveira CC, Acedo SC, da Conceição Miranda DD, Ribeiro ML, Pedrazzoli J, Gambero A. Attenuation of colitis injury in rats using Garcinia cambogia extract. Phyther Res. 2009;23:324–9. https://doi.org/10.1002/ptr.2626.
114. Gürsel FE, Ateş A, Bilal T, Altiner A. Effect of dietary garcinia cambogia extract on serum essential minerals (Calcium, Phosphorus, Magnesium) and trace elements (Iron, Copper, Zinc) in rats fed with high-lipid diet. Biol Trace Elem Res. 2012;148:378–82. https://doi.org/10.1007/s12011-012-9385-x.
115. Iqbal E, Salim KA, Lim LBL. Phytochemical screening, total phenolics and antioxidant activities of bark and leaf extracts of Goniothalamus velutinus (Airy Shaw) from Brunei Darussalam. J King Saud Univ Sci. 2015;27:224–32. https://doi.org/10.1016/j.jksus.2015.02.003.
116. Tharachand C, Immanuel Selvaraj C, Abraham Z. Comparative evaluation of anthelmintic and antibacterial activities in leaves and fruits of Garcinia cambogia (gaertn.) desr. And Garcinia indica (dupetit-thouars) choisy. Br Arch Biol Technol. 2015;58:379–86. https://doi.org/10.1590/S1516-8913201500062.
117. Hart C, Cock IE. An examination of the antimicrobial and anticancer properties of Garcinia cambogia fruit pericarp extracts. Biol Eng Med Sci Rep. 2016;2:55–63. https://doi.org/10.5530/BEMS.2016.2.10.
118. Yu JS, Choi MS, Park JS, Rehman SU, Nakamura K, Yoo HH. Inhibitory effects of Garcinia cambogia extract on CYP2B6 enzyme activity. Planta Med. 2017;83:895–900. https://doi.org/10.1055/s-0043-104934.
119. Guillén-Enríquez C, López-Teros V, Martín-Orozco U, López-Díaz JA, Del Hierro-Ochoa J, Ramos-Jiménez A, Astiazarán-García H, del Martinez-Ruiz NR, Wall-Medrano A. Selected physiological effects of a Garcinia gummi-Gutta extract in rats fed with different hypercaloric diets. Nutrients. 2018;10:1–13. https://doi.org/10.3390/nu10050565.
120. Ventura S, Rodrigues M, Falcão A, Alves G. Short-term effects of Garcinia cambogia extract on the pharmacokinetics of lamotrigine given as a single-dose in Wistar rats. Food Chem Toxicol. 2019;128:61–7. https://doi.org/10.1016/j.fct.2019.03.051.
121. Tan WN, Tong WY, Leong CR, Nik Mohamed Kamal NNS, Muhamad M, Lim JW, Khairuddean M, Her Man MB. Chemical composition of essential oil of Garcinia gummi-gutta and its antimicrobial and cytotoxic activities. J Essent Oil-Bearing Plants. 2020;23:832–42. https://doi.org/10.1080/0972060X.2020.1828179.
122. Mannarmannan M, Biswas K. Biological activity of ZnO nanoparticles synthesized from the dried rinds of Garcinia gummi Gutta. ChemistrySelect. 2019;4:12739–42. https://doi.org/10.1002/slct.201903159.

123. Kim D, Kim J, Kim S, Yoon M, Um M, Kim D, Kwon S, Cho S. Arousal-inducing effect of garcinia cambogia peel extract in pentobarbital-induced sleep test and electroencephalographic analysis. Nutrients. 2021;13. https://doi.org/10.3390/nu13082845.
124. Han JH, Park MH, Myung CS. Garcinia cambogia ameliorates non-alcoholic fatty liver disease by inhibiting oxidative stress-mediated steatosis and apoptosis through nrf2-are activation. Antioxidants. 2021;10. https://doi.org/10.3390/antiox10081226.
125. Kalita A, Das M, Das B, Baro MR. Molecular docking prediction and in vitro studies elucidate anti-inflammatory effect of Garcinia extract against inducible nitric oxide synthase and cyclooxygenase-2 targets. Beni-Suef Univ J Basic Appl Sci. 2022;11. https://doi.org/10.1186/s43088-022-00214-2.
126. Verrelli D, Dallera L, Stendardo M, Monzani S, Pasqualato S, Giorgio M, Pallavi R. Hydroxycitric acid inhibits chronic Myelogenous leukemia growth through activation of AMPK and mTOR pathway. Nutrients. 2022;14:1–13. https://doi.org/10.3390/nu14132669.
127. Dong J, Li W, Du X, He X, Deng B, Zheng H, Tian Y, Sheng J, Fang C. Garcinia cambogia water extract alleviates insulin resistance and hepatic lipid accumulation in mice fed a high-fat diet. Food Nutr Res. 2023;67:211–8. https://doi.org/10.29219/fnr.v67.8977.

Psidium guajava L. (Poor Man's Fruit) with Anticancer Potential: A Comprehensive Review

Anju Rani George, Kavimani Thangasamy, and Natesan Geetha

Abstract

Psidium guajava is profusely grown in several tropical and subtropical countries for its fruit. The fruit has been used for food and medicinal purpose, and therefore it is called in the nickname as the poor man's apple. Up-and-coming research has found that various parts of guava fruit such as peel, pulp, and seed and their bioactive phytochemicals exert anti-tumorigenic effects against various human cancers through multiple mechanisms. The majority earlier reviews of *P. guajava* present a broad overview of ethnopharmacological uses, phytochemical compositions, pharmacological and biological properties, and therapeutic potential in the treatment of numerous diseases. So far, critical review of the available research data on cancer prevention and therapeutic potential of different parts of guava fruit such as fruit peel, pulp, and seed has not been published to the best of our knowledge. Therefore, the intention of this review is to collect the available literature and to provide comprehensive information of the cancer preventive and therapeutic potential of various parts of guava fruit and their isolated bioactive compounds with mechanism of action. To present this review in systematic manner, a complete literature survey was made to collect and compile available research papers on varied parts of guava fruit and their isolated bioactive compounds to reveal their usage in the prevention and treatment of cancer through in vitro and in vivo studies. Different electronic bases such as PubMed, Science Direct, and Google Scholar, etc., were used to collect information. The literature survey showed that guava fruit possesses various medicinal properties which have been reported from across the globe in the form of ethnobotanical or ethnopharmacological analysis, laboratory explorations, and clinical trials. In addition, this review highlights the various patents and commercial products that use

A. R. George · K. Thangasamy · N. Geetha (✉)
Department of Botany, Bharathiar University, Coimbatore, Tamil Nadu, India

P. Thangaraj (ed.), *Pomotherapeutic Insights on Wild Edible Fruits*,
https://doi.org/10.1007/978-981-96-0543-9_10

guava fruit as raw material. This review could be a good reference for researchers who are enthusiastic to undertake further research on drug development using bioactive compounds of guava fruit for prevention and treatment of cancer.

Keywords

Psidium guajava · Anticancer · Ethnopharmacological · Pharmacological therapeutic potential

Introduction

Cancer is the second most common cause of death globally. By the year 2035, the World Health Organization (WHO) predicts that there will be 24 million new cancer cases and 14.5 million cancer-associated deaths per year [1]. In order to fight this growing health problem, it is very imperative that novel therapeutic strategies are required. Approximately, 30–40% of cancers are curable by following suitable dietary and nutritional measures, physical activity, and maintenance of a healthy body weight [2]. Various studies have shown that oxidative stress and human pathophysiological disorders are fundamentally related. It is well known that oxidative stress affects the DNA molecule and changes signaling pathways and finally leads to the development of a variety of malignancies including those of the breast, lung, liver, colon, prostate, ovary, and brain [3]. Antioxidants are any substance that can minimize the damage caused by free radicals through donating some of their electrons to balance free radicals [4]. Phytochemicals are a nonnutritive compounds that are present in plants and significantly possess various pharmacological properties. These biologically dynamic phytocompounds are classified into five major groups, namely, phenolics, carotenoids, organosulfur compounds, nitrogen-containing compounds, and alkaloids, which are known for their potential health benefits in the hindrance of various diseases including cancer [5]. Till today, numerous researches have been undertaken on the usage of fruits with high antioxidant and anticancer properties for the prevention of various types of cancer.

The *Psidium* genus has around 150 species of shrubs, of which *Psidium guajava* L. (guava) is the best-known shrub (10 m tall) belonging to the Myrtaceae family and dispersed worldwide [6]. Even though guava is a native to America, it can grow in tropical environments around the world [7]. The plant flowers mainly in spring, and fruiting can emerge throughout the year mostly during the summer months [8]. The economic importance of this plant lies mainly in guava fruit which is round, oval, or pear shaped (size ranges from 1 to 48 ounces), and when ripe it releases strong sweet and musky odor. The outer peel is green or yellow, and the flesh has a variety of color, including white, yellow, pink, salmon, and red with several small hard white seeds [9, 10]. The guava worldwide production is anticipated to be around 40 million tons. Even if fresh guava trade is restricted internationally, the marketing of an extensive range of products resultant from its fruit such as preserves, jams, jellies, and syrups is becoming more common [11]. India, Thailand,

Brazil, and Mexico are the highest regions for production of guava [12]. Although guava is grown in almost all states of India, Andhra Pradesh, Assam, Bihar, Maharashtra, Uttar Pradesh, and West Bengal are the important cultivators of this plant [13].

The majority prior reviews of *P. guajava* provide a broad overview of ethnopharmacological uses, phytochemistry, pharmacological and biological properties, and curative potential of guava plant in the treatment of various diseases. Different parts of guava plant including leaves, shoots, barks, fruits (ripe and unripe), flower buds, and whole plants and their phytochemical compositions and pharmacological properties and their usage in various countries are reviewed [13]. How guava plant is effective against gastrointestinal infections, malaria, respiratory infections, oral/dental infections, skin infections, diabetes, cardiovascular/hypertension, cancer, malnutrition, women problems, pain, fever, liver diseases, and kidney problems which are common among rural communities of India is also reviewed [14]. The health effects of guava leaves against diabetes mellitus, cardiovascular diseases, cancer, and parasitic infections are indicated [15]. Up-to-date information was given on various bioactive phytochemicals present in various parts of guava plant and their effects on different diseases [16]. The potential of agro-industrial waste pulp of guava fruit as an antimicrobial agent was discussed [17]. One of the early studies [18] revealed leaf and fruit extracts of guava may have anticancer activity, but they came into this conclusion only on the basis of very limited number of both in vitro and in vivo studies. The anticarcinogenic activity of leaf extracts of *P. guajava* against various cancer cell lines was reviewed [19]. Complete analysis of the cancer-preventive and anticancer therapeutic potential of isolated bioactive components and various plant extracts of guava with a focus on the cellular and molecular mechanisms of action against different cancer cell lines through in vitro and in vivo studies was reviewed [20]. The efficiency of different extracts of guava plant and bioactive molecules derived from them to treat various human cancers was also reviewed [21]. All the reviews contain some information about the anticancer potential of guava fruit, but they did not focus on the in vitro and in vivo anticancer potential of various parts of guava fruit such as peel, pulp, and seed and their isolated bioactive compounds with their mechanisms of action against cancer disease. Therefore, the purpose of this review is to carry out a critical analysis of the existing literature and to provide a complete evaluation of the cancer-preventive and therapeutic potential of the various parts of guava fruit.

Review Methodology

The data collected for this review include the main guava-producing states of India, various health benefits of guava fruit, in vitro and in vivo anticancer potential of the different parts of guava fruit such as peel, pulp, skin, and entire fruit extracts and their isolated phytocompounds, mechanism of action, details of patents, and list of value-added products derived from guava fruit. Different electronic bases or search engines such as PubMed, Science Direct, Google Scholar, Elsevier, Springer, and

MDPI were used to collect reviews and articles related to *P. guajava* information. Documents from 2010 to 2023 were collected using keywords like *P. guajava*, distribution, fruit peel, pulp, seed, in vitro and in vivo anticancer potential, bioactive compounds, patents, and value-added products.

Distribution and Varieties Grown

Guava (*Psidium guajava* L.) is an important fruit crop of India and is considered the poor man's fruit or "apple of tropics" as reported by Nakasone and Paull [22]. It is the fifth most imperative fruit crop of India, occupying 3.38% of the total area under fruit farming. The area under guava is 2.05 lakh/ha with a production of 24.62 lakh tons. Maharashtra, Bihar, Orissa, Uttar Pradesh, and West Bengal are the major states growing guava. Productivity is maximum in Karnataka (21.65 t/ha) followed by Gujarat (15.2 t/ha), Andhra Pradesh (14.0 t/ha), and Uttar Pradesh (12.0 t/ha) as reported by Singh [23].

In India, guava improvement program was initiated during 1907 at Ganeshkhind Fruit Experimental Station, Pune, with the objectives to develop dwarf as well as high-yielding, improved postharvest characteristics in fruits and to evolve resistant varieties to various biotic and abiotic stresses [24]. As a result, many varieties and hybrids of guava with agronomic and commercial importance have been evolved through selection and hybridization. The systematic breeding work done for the improvement in guava was reviewed by Subramanyan and Lyer [25]. In India, various states cultivate different varieties of guava suited to their climates and preferences. Some important guava varieties cultivated in different states of India are as follows:

- *Allahabad Safeda*: It is cultivated in Uttar Pradesh, particularly in the Allahabad region. It is known for its sweet flavor and large size.
- *L-49*: It is also known as Lucknow-49, and it is another popular variety cultivated in Uttar Pradesh. It is favored for its high yield and sweet taste.
- *Sardar*: It is grown in the state of Punjab for its distinct flavor and aroma.
- *Harijha*: It is cultivated in Bihar for its large size and sweet taste.
- *Baruipur Local*: This variety is popular in West Bengal for its unique flavor and aroma.
- *Arka Mridula*: This variety is grown in Karnataka for its disease resistance and high yield.
- *Banarasi*: It is cultivated in the state of Bihar for its sweet taste and aroma.
- *Lalit*: It is grown in Maharashtra for its high yield and excellent taste.
- *Safed Jam*: It is cultivated in the state of Bihar for its juicy and sweet taste.
- *Taiwan Guava*: This variety is grown in states like Maharashtra, Karnataka, and Tamil Nadu for its crunchy texture and sweet taste.

Other than the above varieties, some of the popular varieties existing today in various states of India were developed from 600 seedlings selected from different

Table 1 Prominent guava varieties grown across India's states

State			Varieties grown
1.	Andhra Pradesh	:	Anakapalli, Chittidar, and Hafshi
2.	Madhya Pradesh	:	Gwalior-27, Hafshi, and Seedless Chittidar
3.	Jharkhand	:	L-49 and Allahabad Safeda
4.	Karnataka	:	Arka Amulya, Bangalore, and Dharwar
5.	Tamil Nadu	:	Anakapalli, Banarasi, Bangalore, Chittidar, Hafshi, Nagpur Seedless, and Smooth Green
6.	West Bengal	:	Dudhe Khaja, Gole Khaja, Kabli, Baruipur, Chittidar, Harijha, and Sardar
7.	Maharashtra and Gujarat	:	Nagpur Seedless, Dharwar, Dholka, Kothrud, L-24, L-49, Nasik, and Sindh
8.	Northeastern States	:	Allahabad Safeda, Sardar, and Red Fleshed
9.	Uttar Pradesh	:	Lucknow Safeda, Apple Color, Chittidar, Red Fleshed, Allahabad Surkha, Sardar, Mirzapuri Seedless, CISH-G-1, CISH-G-2, and CISH-G-3
10.	Bihar and Jharkhand	:	Apple Color, Chittidar, Hafshi, and Selection-8
11.	Haryana	:	L-49, Hisar Safeda, Hisar Surkha, and Allahabad Safeda
12.	Assam	:	Am Sophri, Madhuri Am, and Safrior Payere

regions of the country (Table 1). In India, 8–10-year-old trees raised from the selected seedlings may produce 400–500 fruits per year, while grafted trees at the same age may produce 1000–2000 fruits. It could be harvested throughout the year [26].

Health Benefits and Nutritional Composition of Guava Fruit

Guava fruits offer a variety of health benefits due to their rich nutritional profile. Raw fruit peels and aqueous unripe fruit peel extracts of *P. guajava* showed hypoglycemic effect in streptozotocin-induced diabetic rats [27, 28]. Various scientific studies have reported the antioxidant activities of guava fruit [29–33]. Lin and Lin [34] studied the anticancer activity of guava seed polysaccharides on MCF-7 cells. Findings of the study showed MCF-7 cell viability was considerably inhibited in a dose-dependent manner. Intake of guava fruits helps in menstrual problems because their juice increases the erythrocytes and avoids anemia in adolescents [35]. The fruit is also helpful for treating male infertility (due to high vitamin content) caused by sperm clumping, adhesion, and other abnormalities [36]. The various health advantages [81–90], attributed to guava fruit consumption are illustrated in Fig. 1.

Guava fruits are used for both fresh consumption and processing. It excels most other fruit trees in productivity, hardness, adaptability, and vitamin C content. It contains low to moderate level of protein, fat, carbohydrate, calcium, iron, sodium, zinc, selenium, thiamin, riboflavin, niacin, vitamin B6, folate, vitamin B12, vitamin A, vitamin C, vitamin E, and vitamin K [18, 19], and it contains 85% water [14]. It

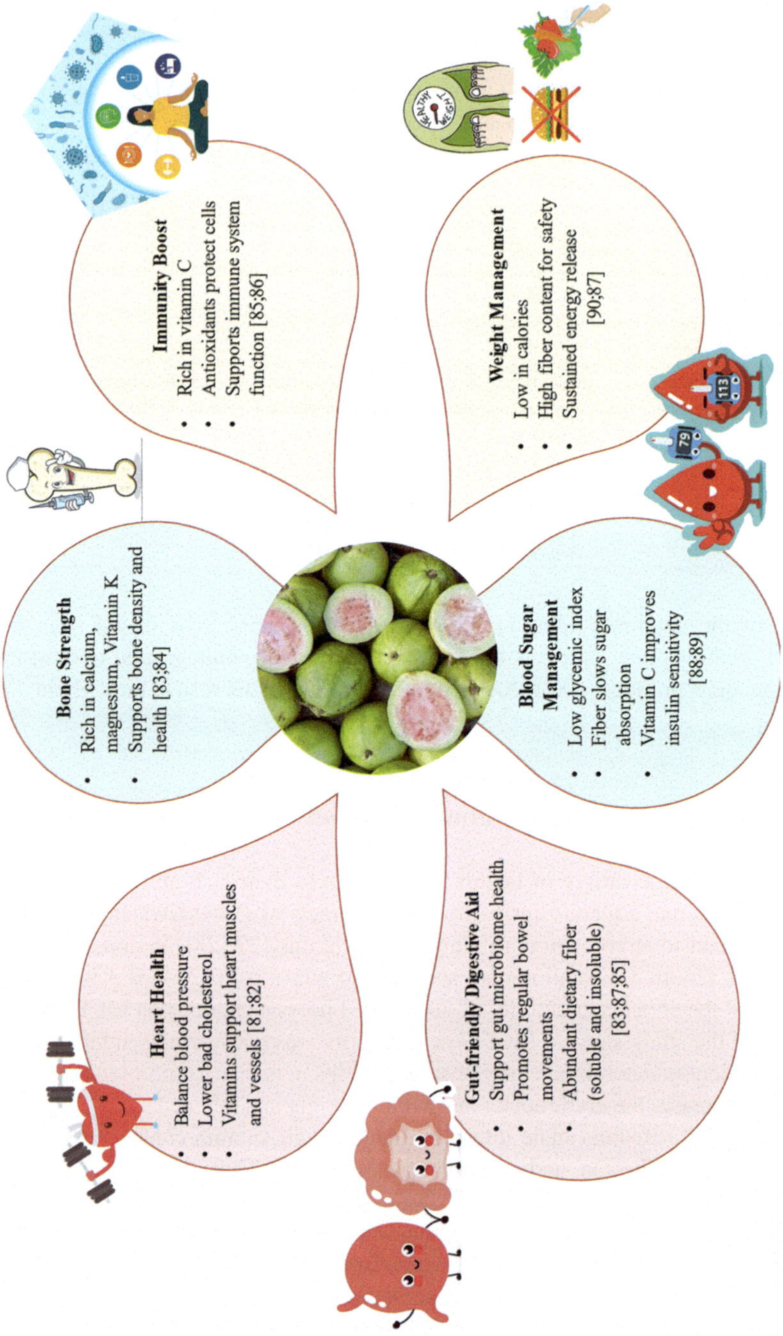

Fig. 1 Health advantages attributed to guava fruit consumption

Table 2 The nutritional breakdown of guava fruit

Sl. No	Components	Amount	References
Macronutrients			
1.	Calories	77–86 kcal	[38]
2.	Moisture	80–86 g	[39]
3.	Crude fiber	0.9–1 g	[38]
4.	Protein	0.1–2.55 g	
5.	Fat	0.43–0.7 g	
6.	Total sugar	8.92 g	[39]
7.	Carbohydrates	9.1–17 g	[38]
8.	Ash	9.5–10 g	
Micronutrients minerals			
1.	Iron	0.26 mg	[40]
2.	Magnesium	20 mg	
3.	Manganese	0.15 mg	
4.	Phosphorous	40 mg	
5.	Potassium	417 mg	
6.	Sodium	2 mg	
7.	Zinc	0.23 mg	
8.	Lycopene	5204 μg	
Vitamins			
1.	Ascorbic acids	56–500 mg	[31]
2.	Riboflavin	0.04 mg	[39]
3.	Niacin	1.084 mg	
4.	Thiamin	0.067 mg	
5.	Vitamin B6	0.11 mg	

is a good source of fiber and potassium and contains four times the vitamin C compared to orange fruit, and most of the vitamin C is found in the fruit's skin. The pulp of the guava fruits contains 30% sugar, 20% total protein, 0.5% fat, and a high level of crude fiber [37]. The fruit gives 68 calories per 100 g [18, 19]. Table 2 shows the nutritional breakdown of guava fruit.

Phytochemicals of Guava Fruit with Anticancer Activity

Guava fruit is a rich source of various phytocomponents, such as polyphenols, flavonoids, isoflavonoids, tannins, phenolic acids, terpenoids, saponins, lectins, essential oils, and fatty acids [18, 19, 37, 41, 42]. Numerous volatile compounds are detected in guava fruit, and the major components are esters followed by terpenes, aldehydes, alcohols, acids, ketones, and furans. The most typical aroma active compounds of guava fruits are (E)-b-ionone, ethyl hexanoate, ethyl butanoate, hexanal, (Z)-3-hexenal, hexyl acetate, (E)-2-hexenal, and limonene [43]. Glycoside, carotenoids, and phenolic compounds are found in guava seeds [42].

Figure 2 shows various anticancer phytochemicals present in the guava fruit. The compound lycopene was extracted from guava fruit directly, and its in vitro anticancer potential against breast cancer cells was studied [44]. Psiguajavadial A and psiguajavadial B, guajadial, guajadial B, guajadial D, and guajadial F are meroterpenoids which show cytotoxic effect against five human cancer cell lines [45]. Delphinidin-3-o-glucoside is an anthocyanin glucoside which shows an antiproliferative activity on human colon adenocarcinoma cells [46]. Betulinic acid is a

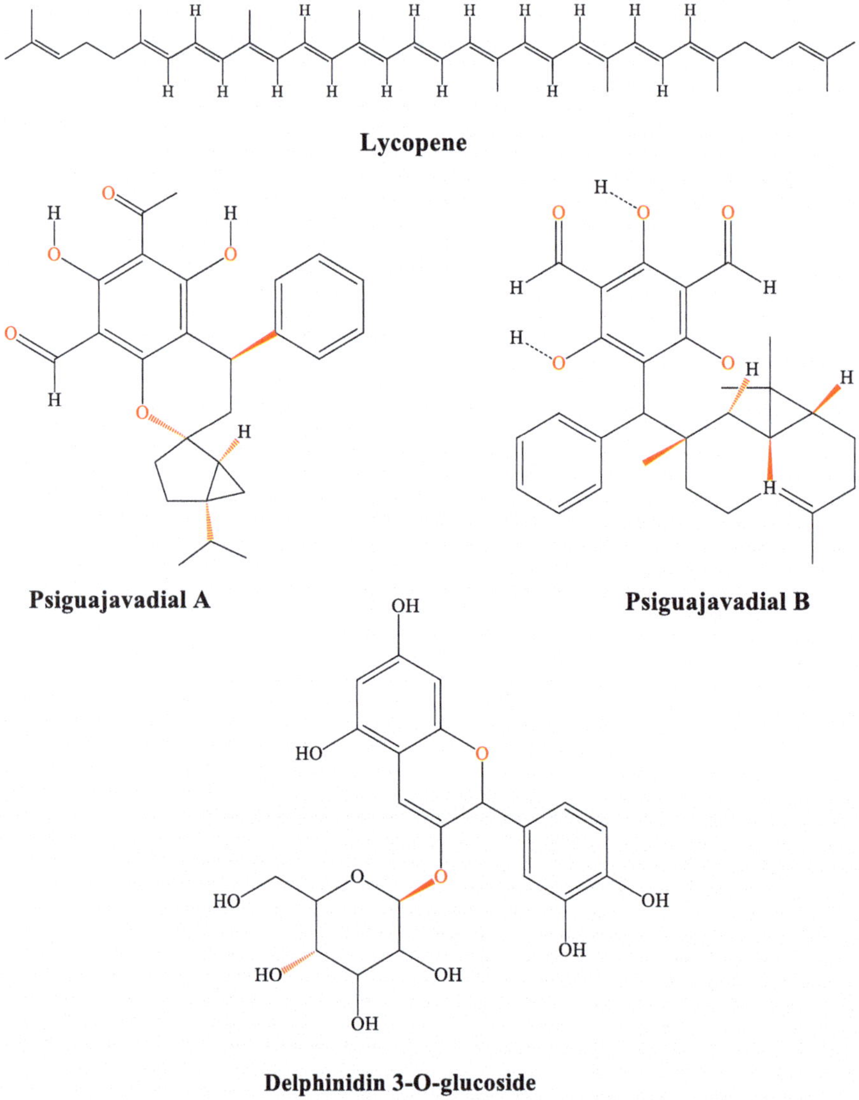

Fig. 2 Chemical structure of isolated phytocompounds of guava fruit with anticancer potential

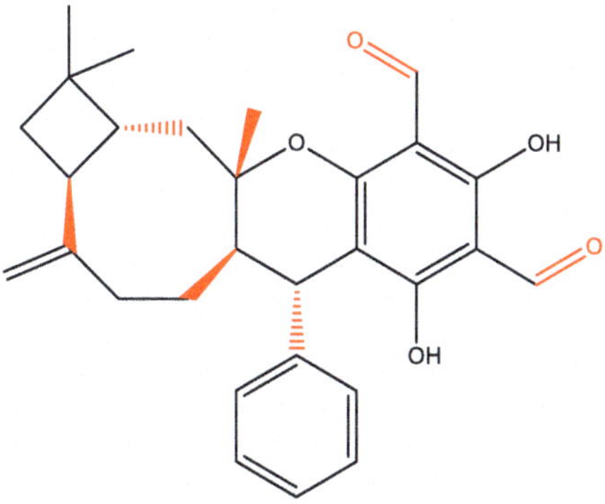

4,5-Diepipsidial A

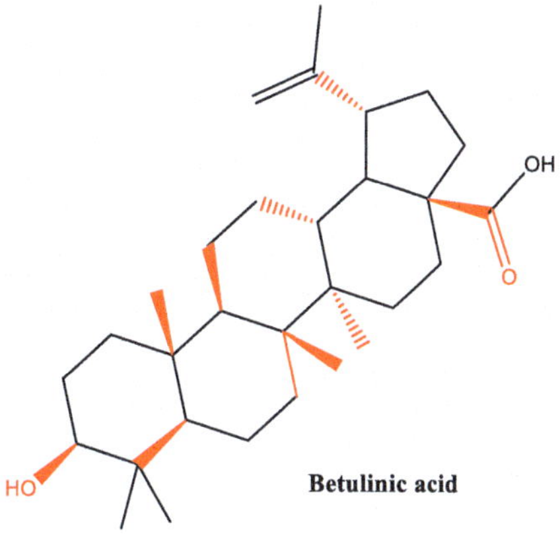

Betulinic acid

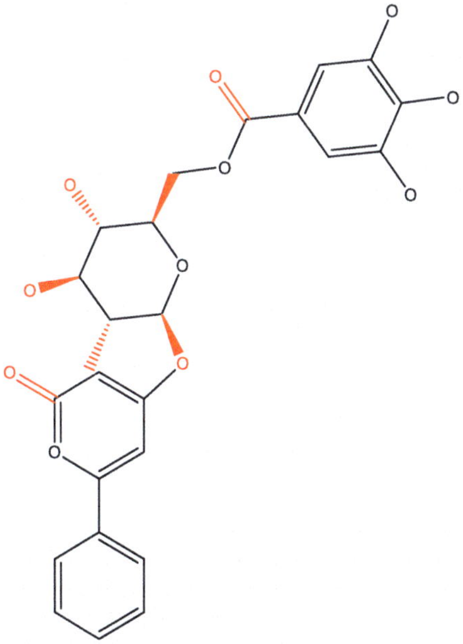

Guavinoside B

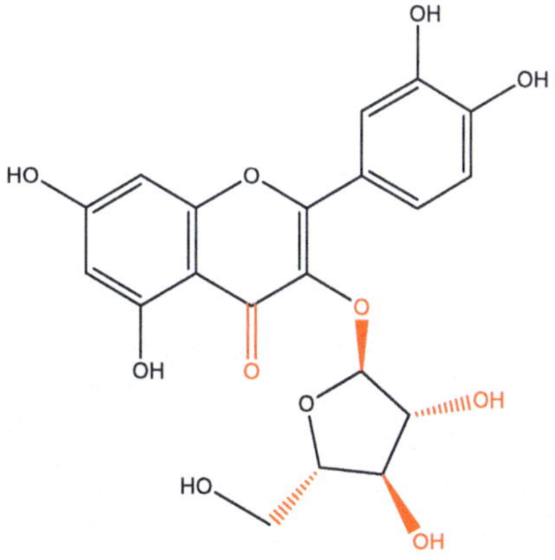

Guavinoside C

Fig. 2 (continued)

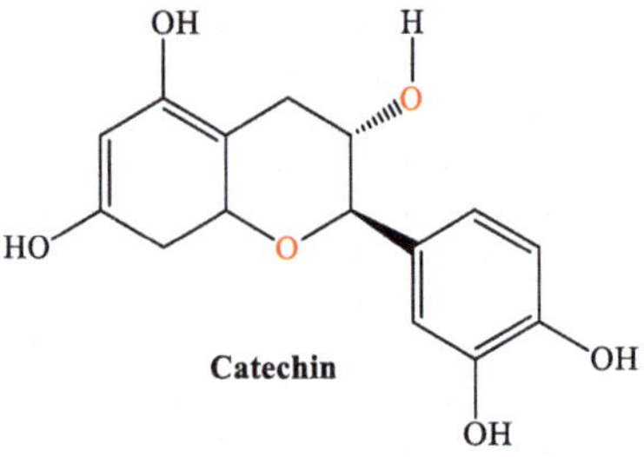

Catechin

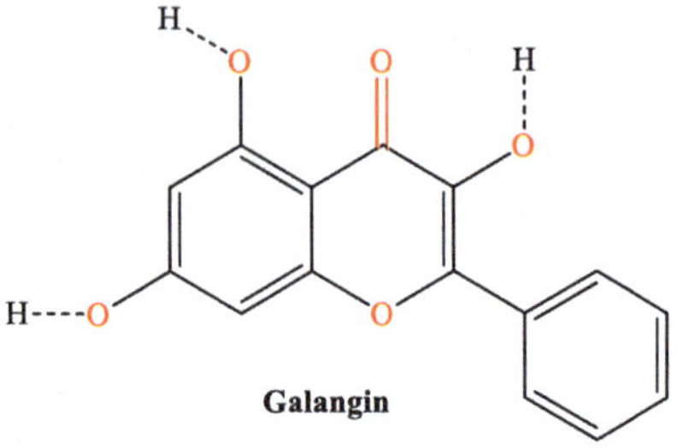

Galangin

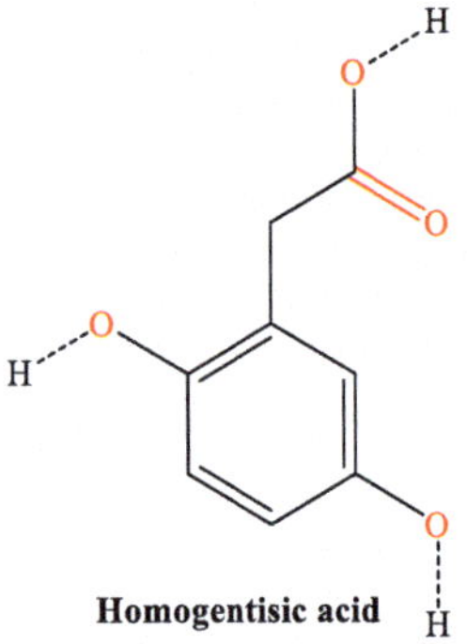

Homogentisic acid

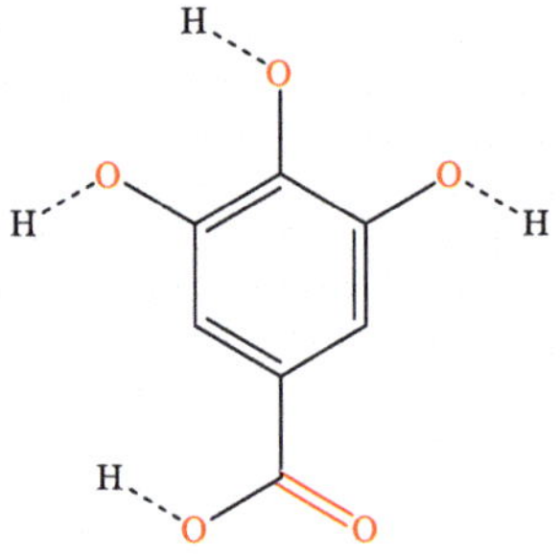

Gallic Acid

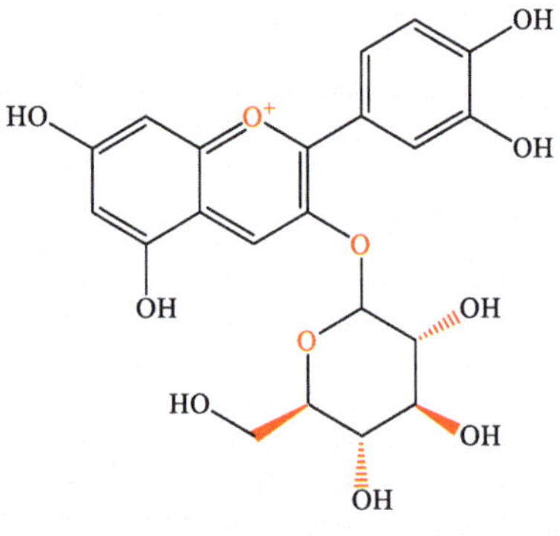

Cyanidin 3 glucoside

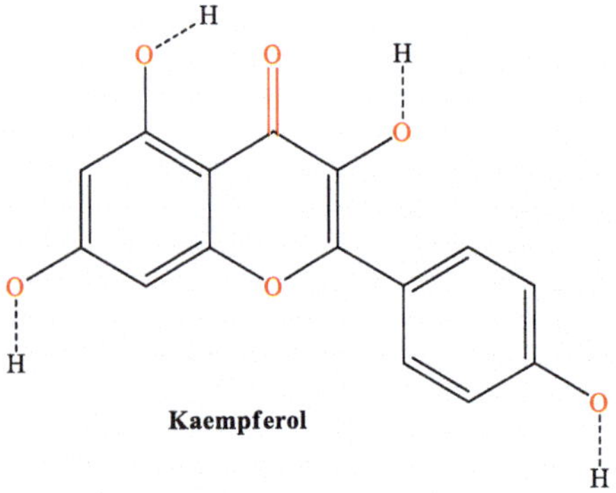

Kaempferol

Fig. 2 (continued)

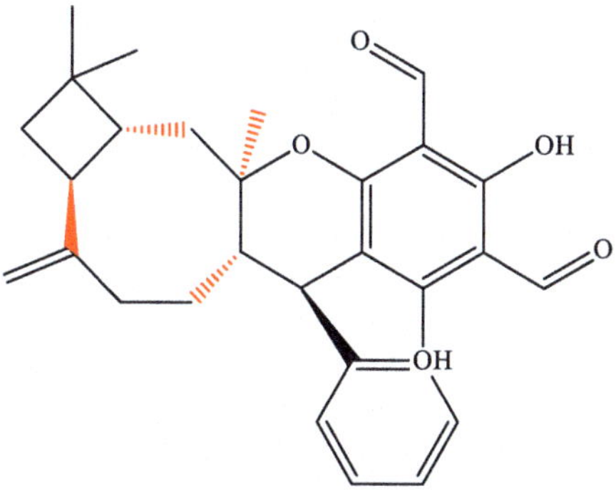

Guajadial

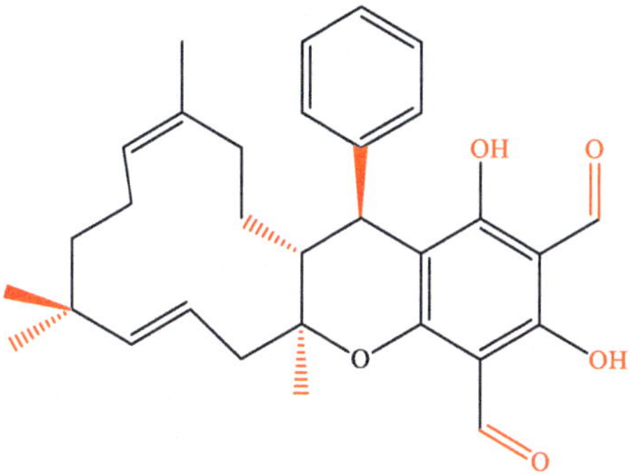

Guajadial B

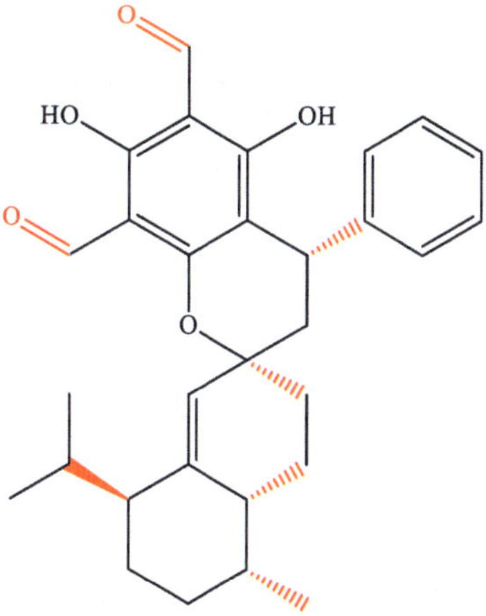

Guajadial D

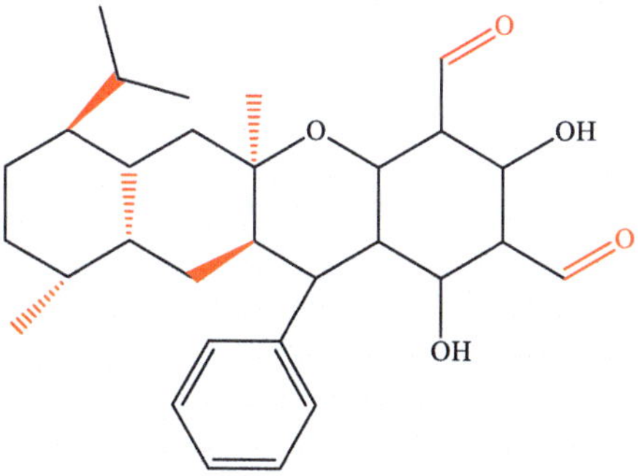

Guajadial F

Fig. 2 (continued)

pentacyclic triterpenoid compound, and it is identified as potential anticancer agent through molecular docking study against HDAC6 and HDAC10 which are cancer receptors [47]. Catechin, galangin, homogentisic acid, gallic acid, kaempferol, and cyanidin 3-glucoside have been identified in fruit samples of guava [30]. According to the literature, catechin [48], galangin [49], homogentisic acid [50], gallic acid [51], kaempferol [52], cyanidin 3-glucoside [53], and gallic acid [51] possess anticancer activity.

In Vitro and In Vivo Anticancer Potential of Entire Fruit and Peel, Pulp, Seed, and Their Isolated Phytocompounds

Cancer cell lines have been widely used for research purpose and proved to be an excellent model for the study of the biological mechanisms engaged in cancer [54]. These models are essential for deciphering molecular mechanisms of key events such as tumor growth, metastasis, drug resistance, and aspects of immune evasion. They are also required for anticancer drug screening and development [55]. In vivo studies provide valuable information concerning the effects of a particular substance or disease progression in a whole, living organism. Animal studies act as a kind of middle ground between in vitro experiments and human trials. Most animal studies use laboratory-bred mice or rats that are nearly genetically identical. As a result, researchers can observe various biological effects within complex organisms. In vitro and in vivo studies present valuable information for all stages of biomedical research. Researchers often use in vitro methods for initial investigations to examine drug interactions and disease processes at the cellular level. In vivo studies expand the data collected from in vitro studies by monitoring biological responses in living organisms (https://www.medicalnewstoday.com/articles/in-vivo-vs-in-vitro#in-vivo). Table 3 briefly summarizes the various in vitro and in vivo studies regarding the entire fruit *P. guajava* and its different parts as peel, pulp, and seed and isolated phytocompounds involved in the anticancer effects.

Anticancer Mechanism of *P. guajava* Fruit

Cancer is a worldwide health problem with high morbidity and mortality and creates both economic and psychological challenges. Therefore, cancer treatment and prevention remain a high priority for the scientific and medical community across the world. Chemotherapy is the main treatment of cancer using chemical substances. Even though they are capable of destroying cancer cells by keeping them from growing and spreading and relieving the cancer symptoms, chemotherapy can destroy or slow down the growth of normal cells. Therefore, oncologists are still exploring for new anticancer drugs with more effective inhibition and less side effects. It is possible that plant-derived chemoprevention agents might target cancer-inducing receptors by regulating the cell cycle, cellular senescence, and apoptosis [71]. Various parts of guava fruit such as peel, pulp, seed, and seed oil contain

Table 3 Different extracts and isolated phytocompounds of guava fruits with in vitro and in vivo anticancer potential

Sl. No.	Plant part used	Extracts/isolated Phytocompound	In vitro cell lines	Result	References
1.	Fruit	Ethanolic extract	Erythroid leukemic cell line (K562), human promyeloid leukemia (HL60), human monocytic leukemia (U937), and human lymphoblastic cell line (Molt4)	Inhibited the proliferation of HL60 cell line with IC50 of 31.8 ± 2.5 μg/mL	[56]
2.	Fruit	Isolated compounds: lycopene and apigenin	LNCaP (prostate cancer) and UACC257 (melanoma)	Study demonstrated the synergistic mode of action of lycopene and apigenin in combination to obtain a high anticancer activity	[57]
3.	Peel and pulp	Aqueous extracts	Anticancer activity was examined by XTT assay	Higher anticancer activity was found in peel extract than pulp extract	[58]
4.	Fruit	Isolated phytocompound: -lycopene	Breast cancer cells (MCF-7)	Lycopene influenced over G_2-M phase of the cell cycle and increased apoptotic cell percentages compared to untreated cells with IC50 value 19.1 μg/mL in dose-dependent manner	[59]
5.	Fruit (peel, flesh, and seed)	Ethanol-water (50:50, v/v) Isolated compounds: Catechin Galangin Homogentisic acid Gallic acid Kaempferol Cyanidin 3-glucoside	A549 (human lung cancer cells), MCF-7 (human breast cancer cells), HepG2 (human hepatoma cells), and HT-29 (human colon cancer cells)	Several bioactive compounds of guava have been identified and quantified. The contents of these compounds in guava peel and seed were found higher than in guava flesh	[30]

(continued)

Table 3 (continued)

Sl. No.	Plant part used	Extracts/isolated Phytocompound	In vitro cell lines	Result	References
6.	Fruit (red guava)	Total extract and smaller molecular weight (<30 kDa) extracts	Cytotoxicity was determined with human mammary epithelial cell line (MCF-10A) and human TNBC cell lines (MDA-MB-231 and MDA-MB-468)	Both the extracts decrease the viability of the TNBC cell lines with more necrotic cell death in MDA-MB-231 cell lines and apoptotic cell death in MDA-MB-468 cell lines	[60]
7.	Fruit	Pulp extract	MCF-7 and MDAMB-435 cell lines	Guava extracts caused MDA-MB-435 and MCF-7 cell count reduction in G0/G1 and G2/M phases and increased apoptosis	[61]
8.	Fruit	Lycopene-rich extract (LEG)	Human breast cancer cells (MCF-7) Fibroblast cells (NIH-3 T3)	Treatment with LEG for 24 and 72 h significantly affected the viability of MCF-7 cells (IC50) = 29.85 but not for NIH-3 T3 cells (IC50 = 911.5 μg/mL). LEG was able to induce cytostatic and cytotoxic effects on breast cancer cells probably via induction of an apoptotic-like pathway	[62]
9.	Entire fruit, peel, pulp, and seed	Acetone-based extracts of entire fruit (crude), peel, pulp, and seed	NB4 (promyelocytic leukemia cells)	Crude acetone extracts were able to induce cell cycle arrest and cell death in NB4 cells. Acetone extracts of peel and pulp were capable of inducing CD11c expression and apoptosis in NB4 cells	[63]
10.	Fruit	Poly-E-caprolactone lipid-core nanocapsules containing fruit extract	MCF-7 (human breast cancer cells)	Lipid-core nanocapsules improved its toxicity against MCF-7 cancer cells, inhibited the production of intracellular ROS and NF-κB in human microglial cells, and did not affect the membrane integrity of erythrocytes, emphasizing its potential to be used as a cancer treatment	[64]

11.	Fruit (seed polysaccharide fraction 3, GSF3)	Seed extract	MCF-7 (human breast cancer cells)	Inhibited cell growth and decreased cell viability. GSF3 is a potent anticancerous polysaccharide that inhibited MCF-7 cell growth through either by direct action or indirectly modulating immune cell *cytokine secretion* profiles	[65]
12.	Fruit (polysaccharides from seed (GSPS)	Seed extract	MCF-7 (human breast cancer cells)	GSPS increase the percentage of M1 macrophages in the tumor environment and had the highest potential for anti-breast cancer via cancer immunotherapy by modulating macrophage secretion	[66]
13.	Fruit (seed oil)	Hexane extract Compound: linoleic acid	K562 (human erythroleukemic cell)	The seed oil inhibited the proliferation of the human erythroleukemic cells in a dose-dependent manner with IC50 value of 155 and 137 μg/mL at 24 and 48 h, respectively. *P. guajava* seed oil, which is abundant with linoleic acid, can promote skin wound healing and inhibit the proliferation of leukemic cells	[67]
14.	Dried seeds	80% aqueous extract and two isolated compounds: *Compound (1)*: 1-O-3,4-dimethoxy-phenylethyl-4-O-3,4-dimethoxy cinnamoyl-6-O-cinnamoyl-β-d-glucopyranose *Compound (2)*: 1-*O*-3,4-dimethoxyphenylethyl-4-*O*-3,4-dimethoxy cinnamoyl-β-D-glucopyranose	P388 (leukemia cell line) Ehrlich ascites carcinoma cells (EAC)	The acetone extract and the two new isolated compounds showed that the extract has moderate inhibition activity (180% and ED_{50} = 14.6) against both the EAC and leukemia P388, while the two new isolated compounds (*1* and *2*) showed high inhibition activity against EAC (220% and 240%, respectively) and against leukemia P388 (ED_{50} = 17.3 and 16.1, respectively) in a dose-dependent manner	[34]

(continued)

Table 3 (continued)

Sl. No.	Plant part used	Extracts/isolated Phytocompound	In vitro cell lines	Result	References
15.	Seed	Crude polysaccharides from guava seed (GSPS)	PC-3 (prostate cancer cell line) MCF-7 cell line	GSPs showed growth inhibition of PC-3 by decreasing the proinflammatory/cytokine secretion ratio and MCF-7 cell lines by increasing the Th2/Th1 cytokine secretion	[68]
16.	Fruit	Fruit oil Isolated compounds: β-Caryophyllene Limonene	HepG2 (hepatocellular carcinoma cell line)	The in vitro cytotoxicity of the oil against HepG2 was tested using the SRB assay, and 49.76 μg/mL IC50 value was noted	[69]
17.	Pulp	Fresh extract	MCF-7, MDA-MB-231, and MDA-MB-453 cell lines	Extract showed decreased cell proliferation in a time-dependent manner with IC50 values (820 ± 1.22 μg/mL, 680 ± 1.34 μg/mL 50, and 600 ± 1.03 μg/mL for 24 h, 48 h, and 72 h, respectively). In conclusion, extract was found to be more potent cytotoxic in MCF-7 estrogen-positive human breast cancer cell line than MDA-MB-453 and MDA-MB-231 cell lines	[44]
18.	Fruit	Self-emulsifying drug delivery system was prepared using purified lycopene of red guava fruit, i.e., nanoLPG	The toxicity and tissue distribution in orally treated mice were examined	No deleterious effects were found on clinical, behavioral, hematological, biochemical, and histopathological parameters of mice treated for 28 days. Orally administered nanoLPG successfully delivered lycopene into the liver, kidney, and prostate of mice	[70]

numerous anticancer bioactive compounds such as catechin, galangin, homogentisic acid, gallic acid, kaempferol, cyanidin 3-glucoside, β-caryophyllene, limonene, etc. [30, 69]. Figure 3 shows an overview of varied mechanism of action contributing to the anticancer properties of *P. guajava* fruit.

Patents Related to Fruit

A nutraceutical is defined as any substance that is food or a part of food that gives medical or health benefits or the prevention and treatment of ailments [72]. Nutraceuticals comprise a broad range of categories such as dietary supplements, functional foods, and herbal products [73]. The bioactive components or phytochemicals in plants, especially fruits, have been correlated with plentiful health benefits [74] and are employed as ingredients in many nutraceutical and pharmaceutical products today.

Guava is one of the therapeutically important fruits rich in lectins, saponins, tannins, phenols, triterpenes, flavonoids, vitamins, dietary fiber, and carotenoids [75]. The fruit pulp is rich in ascorbic acid and carotenoids (lycopene, β-carotene, and β-cryptoxanthin), and the seeds and peel possess glycosides, carotenoids, and phenolic compounds [18]. Because of these nutritive contents, various by-products have been prepared from guava fruit. Of them, some are patented. For example, E.P.C. (Beijing) Plant Pharmaceutical Technology Co., Ltd. has introduced an invention detailing the preparation, isolation, and utilization of guava fruit extract for managing and regulating sugar levels, available in the form of tablets or capsules. This innovation is documented in a China Patent Application Publication (CN101883575B) dated July 17, 2013 (https://g.co/kgs/m9PNgtu). Guangdong Ocean University submitted an application with the title "A Method for Preparing Instant Dry Powder of Immature *Psidium guajava* Fruit" which discusses a process for producing guava fruit powder using vacuum freeze-drying. This application was published in a China Patent Application Publication CN106819979A on June 13, 2017 (https://g.co/kgs/jhinFU8). Guava fruit extract was used to treat metabolic diseases and suggested its administration through oral in the form of suspensions, syrups, emulsions, power, granules, tablets, and capsules. This invention has been patented (US2018015134A1) [76].

Commercial Products Derived from Guava Fruit

Guava has a very special place due to its nutrient content. The fruit contains the highest quantity of ascorbic acid and large levels of essential oils, triterpenes, phenols, flavonoids, saponins, lectins, fiber, and pectin as well as fatty acids. It has significant mineral contents such as calcium, phosphorus, and iron; vitamins like niacin, pantothenic acid, thiamin, riboflavin, and vitamin A; and two major antioxidants, i.e., polyphenolic compounds and carotenoids [39, 77]. In recent years, the

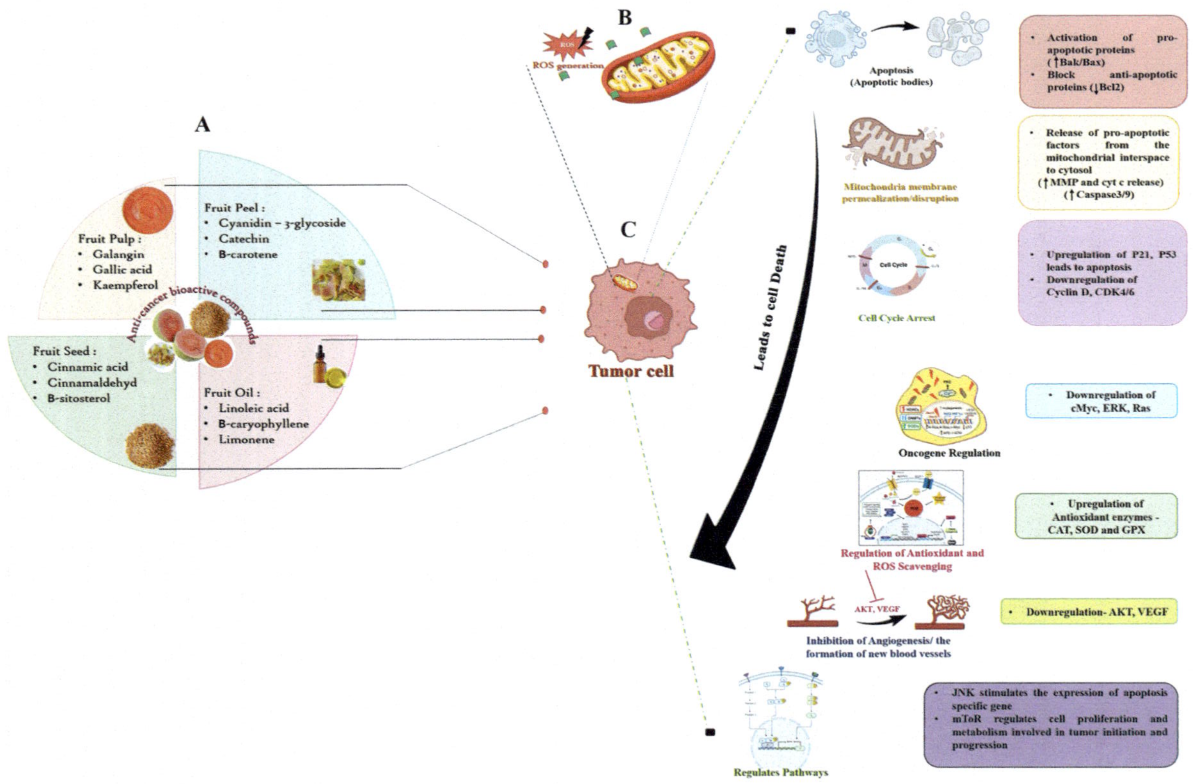

Fig. 3 An overview of the diverse biomolecules and the varied mechanism of action contributing to the anticancer properties of *P. guajava* fruit

Fig. 4 Guava fruit-derived products

global market for functional foods and customer yearning for a better lifestyle is increasing. Because of this, many value-added products are prepared from guava fruit and sold in the market. For example, the fruit has been used to make various items such as jam, jelly, juice, ice cream, dehydrated slice, beverage, flakes, toffee, guava lather sauce, puree of baby food, cheese, yogurt, dahi, whey, etc. [78]. Some of the images of value-added products of guava fruit have shown in Fig. 4.

Future Prospective and Conclusion

This review has highlighted the distribution and varieties grown, health benefits and nutritional composition, and in vitro and in vivo anticancer potential of various phytocompounds of entire fruit of guava and its various parts such as peel, pulp, and seed. In addition, the review discusses the anticancer mechanism of bioactive components of guava fruit and their patents and commercial products that use guava fruit as a raw material. This review could be a good reference for researchers who are enthusiastic to undertake further exploration on drug development using bioactive compounds of guava fruit for alleviation of cancer burden as well as on production of various value-added products from this fruit.

References

1. World Health Organization. Traditional medicine; 2003. Accessed July 20, 2020, from https://web.archive.org/web/20080727053337/http://www.who.int/mediacentre/factsheets/fs134/en/
2. Glade MJ. Food, nutrition, and the prevention of cancer: a global perspective. American institute for cancer research/world cancer research fund, American institute for cancer research, 1997. Nutrition. 1999;15:523–6.
3. Shinde A, Ganu J, Naik P. Effect of free radicals & antioxidants on oxidative stress: a review. J Dental Allied Sci. 2012;1(2):63.
4. Mahmoud M, Mohamed EM, Aboul-Enein AM, Diab AA, Shalaby EA. Anticancer and antioxidant activities of ethanolic extract and semi-purified fractions from guava and mango seeds. Biomass Convers Biorefinery. 2023;1–17.
5. Rudzińska A, Juchaniuk P, Oberda J, Wiśniewska J, Wojdan W, Szklener K, Mańdziuk S. Phytochemicals in cancer treatment and cancer prevention—review on epidemiological data and clinical trials. Nutrients. 2023;15(8):1896.
6. Pereira FMP, Usman M, Mayer NA, Nachtigal JC, Maphanga OR, Willemse S. Advances in guava propagation. Rev Bras Frutic. 2017;39:e-358.
7. Rajan S, Hudedamani U. Genetic resources of guava: importance, uses and prospects. In: Conservation and utilization of horticultural genetic resources; 2019. p. 363–83.
8. Landrum LR. Psidium guajava L.: taxonomy, relatives and possible origin. In: Guava: and possible origin. Botany, production and uses. 1 CABI; 2021.
9. Morton J. Guava. In: Julia F, editor. Fruits of warm climates. Miami, FL: Morton; 1987. p. 356–63. https://hort.purdue.edu/newcrop/morton/guava.html.
10. Mercadante AZ, Steck A, Pfander H. Carotenoids from guava (Psidium guajava L.): isolation and structure elucidation. J Agric Food Chem. 1999;47(1):145–51. https://doi.org/10.1021/jf980405r.
11. Angulo-López JE, Flores-Gallegos AC, Torres-León C, Ramírez-Guzmán KN, Martínez GA, Aguilar CN. Guava (Psidium guajava L.) fruit and valorization of industrialization by-products. PRO. 2021;9(6):1075.
12. Sánchez-Barrantes E, Mora-Newcomer E, Barrantes-Santamaría W. High quality DNA isolation in Psidium guajava L. for genomic studies. Agron Mesoam. 2021;32:638–49. https://doi.org/10.15517/am.v32i2.41606.
13. Daswani PG, Gholkar MS, Birdi TJ. Psidium guajava: a single plant for multiple health problems of rural Indian population. Pharmacogn Rev. 2017;11(22):167.
14. Gutiérrez RMP, Mitchell S, Solis RV. Psidium guajava: a review of its traditional uses, phytochemistry and pharmacology. J Ethnopharmacol. 2008;117(1):1–27.
15. Díaz-de-Cerio E, Verardo V, Gómez-Caravaca AM, Fernández-Gutiérrez A, Segura-Carretero A. Health effects of Psidium guajava L. leaves: an overview of the last decade. Int J Mol Sci. 2017;18(4):897.
16. Ugbogu EA, Emmanuel O, Uche ME, Dike ED, Okoro BC, Ibe C, et al. The ethnobotanical, phytochemistry and pharmacological activities of Psidium guajava L. Arab J Chem. 2022;15(5):103759.
17. Gutierrez-Montiel D, Guerrero-Barrera AL, Chávez-Vela NA, Avelar-Gonzalez FJ, Ornelas-García IG. Psidium guajava L.: from byproduct and use in traditional Mexican medicine to antimicrobial agent. Front Nutr. 2023;10:1108306.
18. Sato R, Dang KM, McPherson BG, Brown AC. Anticancer activity of guava (Psidium guajava) extracts. J Altern Complement Med. 2010;7(1).
19. Correa MG, Couto JS. Anticancer properties of Psidium guajava-a mini-review. Asian Pac J Cancer Prev. 2016;17(9):4199–204.
20. Jamieson S, Wallace CE, Das N, Bhattacharyya P, Bishayee A. Guava (Psidium guajava L.): a glorious plant with cancer preventive and therapeutic potential. Crit Rev Food Sci Nutr. 2022;63(2):192–223.

21. Lok B, Babu D, Tabana Y, Dahham SS, Adam MAA, Barakat K, Sandai D. The anticancer potential of Psidium guajava (Guava) extracts. Life. 2023;13(2):346.
22. Nakasone HY, Paull RE. Tropical fruits. Queensland, Wallirgford: CAB; 1998. p. 93–8.
23. Singh HP, Palaniswami MS. Horticulture in different agro-climatic conditions. New Delhi: Four Decades of Coordinated Research. Published by Westville Publishing House; 2010. p. 55–78.
24. Phadnis NA. Improvement of guava (Psidium guajava L.) by selection in Maharashtra. Indian J Hort. 1970;27:99–105.
25. Subramanyan MD, lyer, C. P. A. Improvement of guava. In: Chadha KL, Parcek OP, editors. Advances in horticulture. 1 New Delhi: Malhotra Publishing House; 1993. p. 143–7.
26. Prabhudesai AP, Biyani DM, Umekar MJ. Psidium guajava: multipurpose medicinal herb. Int J Pharm Sci Rev Res. 2019;59(1):125–32.
27. Rai MK, Jaiswal VS, Jaiswal U. Shoot multiplication and plant regeneration of guava (Psidium guajava L.) from nodal explants of in vitro raised plantlets. J Fruit Ornam Plant Res. 2009;17(1):29–38.
28. Rai MK, Asthana P, Jaiswal VS, Jaiswal U. Biotechnological advances in guava (Psidium guajava L.): recent developments and prospects for further research. Trees. 2010;24:1–12.
29. Jiménez-Escrig A, Rincón M, Pulido R, Saura-Calixto F. Guava fruit (Psidium guajava L.) as a new source of antioxidant dietary fiber. J Agric Food Chem. 2001;49(11):5489–93.
30. Chen Y, Zhou T, Zhang Y, Zou Z, Wang F, Xu D. Evaluation of antioxidant and anticancer activities of guava. Int J Food Nutr Safety. 2015;6(1):1–9.
31. Patel P, Sunkara R, Walker LT, Verghese M. Effect of drying techniques on antioxidant capacity of guava fruit. Food Nutr Sci. 2016;7(07):544.
32. Hartati R, Nadifan HI, Fidrianny I. Crystal guava (Psidium guajava L."Crystal"): evaluation of in vitro antioxidant capacities and phytochemical content. Sci World. 2020;2020.
33. Shahid A, Inam-Ur-Raheem M, Aadil RM, Israr B. Phytochemical screening and in vitro radical scavenging activities of "Gola" guava fruit and leaf extracts. J Food Process Preserv. 2022;46(12):e16989.
34. Lin HC, Lin JY. M1 polarization but anti-LPS-induced inflammation and anti-MCF-7 breast cancer cell growth effects of five selected polysaccharides. Evid Based Complement Alternat Med. 2020;2020(1):9450246.
35. Hardimarta FP, Yuniarti CA, Aini MN. Effectiveness of red guava juice in increasing erythrocyte index for prevention of anemia in adolescents. Adv Health Sci Res. 2018;12:38–41.
36. Balode L, Strazda G, Jurka N, Kopeika U, Kislina A, Beinare M, et al. LSC 2011 Abstract: The role of lipoxin A4 in the chronic obstructive pulmonary disease; 2011.
37. Moreno MA, Zampini IC, Costamagna MS, Sayago JM, Ordóñez RM, Isla MI. Phytochemical composition and antioxidant capacity of Psidium guajava fresh fruits and flour. Food Nutr Sci. 2014;05:725.
38. Parvez GM, Shakib U, Khokon M, Sanzia M. A short review on a nutritional fruit: guava. Open Access: Toxicol Res. 2018;1:1–8.
39. Omayio DG, Abong GO, Okoth MW, Gachuiri CK, Mwang'ombe AW. Current status of guava (Psidium Guajava L.) production, utilization, processing and preservation in Kenya: a review. Curr Agric Res J. 2019;7(3):318–31.
40. Kafle A, Mohapatra SS, Reddy I, Chapagain M. A review on medicinal properties of Psidium guajava. J Med Plants Stud. 2018;6(4):44–7.
41. Shu JiCheng SJ, Chou GuiXin CG, Wang ZhengTao WZ. Triterpenoid constituents in fruits of Psidum guajava; 2009.
42. Ravi K, Divyashree P. Psidium guajava: a review on its potential as an adjunct in treating periodontal disease. Pharmacogn Rev. 2014;8(16):96.
43. Pino JA, Bent L. Odour-active compounds in guava (Psidium guajava L. cv. Red Suprema). J Sci Food Agric. 2013;93(12):3114–20.

44. Polinati RM, Teodoro AJ, Correa MG, Casanova FA, Passos CLA, Silva JL, Fialho E. Effects of lycopene from guava (Psidium guajava L.) derived products on breast cancer cells. Nat Prod Res. 2022;36(5):1405–8.
45. Qin XJ, Yu Q, Yan H, Khan A, Feng MY, Li PP, et al. Meroterpenoids with antitumor activities from guava (Psidium guajava). J Agric Food Chem. 2017;65(24):4993–9.
46. Simas Frauches N, Montenegro J, Amaral T, Abreu JP, Laiber G, Junior J, et al. Antiproliferative activity on human colon adenocarcinoma cells and in vitro antioxidant effect of anthocyanin-rich extracts from peels of species of the Myrtaceae family. Molecules. 2021;26(3):564.
47. Adewole KE, Ishola AA. A computational approach to investigate the HDAC6 and HDAC10 binding propensity of psidium guajava-derived compounds as potential anticancer agents. Curr Drug Discov Technol. 2021;18(3):423–36.
48. Sun H, Yin M, Hao D, Shen Y. Anti-cancer activity of catechin against A549 lung carcinoma cells by induction of cyclin kinase inhibitor P21 and suppression of cyclin E1 and P–AKT. Appl Sci. 2020;10(6):2065.
49. Sulaiman GM. Molecular structure and anti-proliferative effect of galangin in HCT-116 cells: in vitro study. Food Sci Biotechnol. 2016;25:247–52.
50. Jurič A, Turković AH, Karačonji IB, Prđun S, Bubalo D, Durgo K. Cytotoxic activity of strawberry tree (L.) honey, its extract, and homogentisic acid on CAL 27, HepG2, and Caco-2 cell lines. Arch Ind Hyg Toxicol. 2022;73(2):158–68.
51. Devi YP, Uma ADDEPALLY, Narasu ML, Kalyani C. Anticancer activity of gallic acid on cancer cell lines, HCT-15 and MDA MB 231. Int J Res Appl Nat Soc Sci. 2014;2:269–72.
52. Imran M, Salehi B, Sharifi-Rad J, Aslam Gondal T, Saeed F, Imran A, et al. Kaempferol: a key emphasis to its anticancer potential. Molecules. 2019;24(12):2277.
53. Chen PN, Chu SC, Chiou HL, Chiang CL, Yang SF, Hsieh YS. Cyanidin 3-glucoside and peonidin 3-glucoside inhibit tumor cell growth and induce apoptosis in vitro and suppress tumor growth in vivo. Nutr Cancer. 2005;53(2):232–43.
54. Louzada S, Adega F, Chaves R. Defining the sister rat mammary tumor cell lines HH-16 cl. 2/1 and HH-16. cl. 4 as an in vitro cell model for Erbb2. PLoS One. 2012;7(1):e29923.
55. Katt ME, Placone AL, Wong AD, Xu ZS, Searson PC. In vitro tumor models: advantages, disadvantages, variables, and selecting the right platform. Front Bioeng Biotechnol. 2016;4:12.
56. Ampasavate C, Okonogi S, Anuchapreeda S. Cytotoxicity of extracts from fruit plants against leukemic cell lines. Afr J Pharm Pharmacol. 2010;4(1):13–21.
57. Salib JY, Michael HN. Cytotoxic phenylethanol glycosides from Psidium guaijava seeds. Phytochemistry. 2004;65(14):2091–3. https://doi.org/10.1016/j.phytochem.2004.06.009.
58. Bontempo P, Doto A, Miceli M, Mita L, Benedetti R, Nebbioso A, et al. Psidium guajava L. anti-neoplastic effects: induction of apoptosis and cell differentiation. Cell Prolif. 2012;45(1):22–31.
59. El-Ahmady SH, Ashour ML, Wink M. Chemical composition and anti-inflammatory activity of the essential oils of Psidium guajava fruits and leaves. J Essent Oil Res. 2013;25(6):475–81.
60. Lin HC, Lin JY. Immune cell–conditioned media suppress prostate cancer PC-3 cell growth correlating with decreased proinflammatory/anti-inflammatory cytokine ratios in the media using 5 selected crude polysaccharides. Integr Cancer Ther. 2016;15:NP13.
61. Karia PD, Patil LA, Vakani MS, Chauhan GM, Gupta SS, Rathod SP, Patel KV. Chemoprevention of breast cancer by Psidium guajava Linn. Asian J Pharm Pharmacol. 2018;5:58–68.
62. Prommaban A, Utama-ang N, Chaikitwattana A, Uthaipibull C, Srichairatanakool S. Linoleic acid-rich guava seed oil: safety and bioactivity. Phytother Res. 2019;33(10):2749–64.
63. Dos Santos RC, Ombredane AS, Souza JMT, Vasconcelos AG, Plácido A, das GN Amorim A, et al. Lycopene-rich extract from red guava (Psidium guajava L.) displays cytotoxic effect against human breast adenocarcinoma cell line MCF-7 via an apoptotic-like pathway. Food Res Int. 2018;105:184–96.
64. Liu X, Yan X, Bi J, Wu X, Liu J, Zhou M. Identification of phenolic compounds and antioxidant activity of guava dehydrated by different drying methods. Dry Technol. 2020;38(8):987–1000.

65. Correa MG, Couto JS, Trindade BB, Abreu JP, Nakajima VM, Oliveira FL, et al. Antiproliferative effect of guava fruit extracts in MDA-MB-435 and MCF-7 human breast cancer cell lines. An Acad Bras Cienc. 2020;92:e20191500.
66. Vasconcelos AG, Valim MO, Amorim AG, do Amaral CP, de Almeida MP, Borges TK, et al. Cytotoxic activity of poly-ε-caprolactone lipid-core nanocapsules loaded with lycopene-rich extract from red guava (Psidium guajava L.) on breast cancer cells. Food Res Int. 2020;136:109548.
67. Lin HC, Lin JY. GSF3, a polysaccharide from guava (Psidium guajava L.) seeds, inhibits MCF-7 breast cancer cell growth via increasing Bax/Bcl-2 ratio or Fas mRNA expression levels. Int J Biol Macromol. 2020;161:1261–71.
68. Priam F, Marcelin O, Marcus R, Wijkhuisen A, Smith-Ravin EJ. Evaluation of anti-cancer effects of lycopene extracted from pink guava Psidium guajava L. and its combination with apigenin or resveratrol; 2021.
69. Suwanwong Y, Boonpangrak S. Phytochemical contents, antioxidant activity, and anticancer activity of three common guava cultivars in Thailand. Eur J Integr Med. 2021;42:101290.
70. Vasconcelos AG, Barros ALA, Cabral WF, Moreira DC, da Silva IGM, Silva-Carvalho AÉ, et al. Promising self-emulsifying drug delivery system loaded with lycopene from red guava (Psidium guajava L.): in vivo toxicity, biodistribution and cytotoxicity on DU-145 prostate cancer cells. Cancer Nanotechnol. 2021;12:1–29.
71. Pakpahan EL, Alveolie Gabriella BDS, Melisa BDS. In vitro anticancer activity of guava (Psidium guajava L.) against HSC-2 human oral cancer cells. Dental Students Res. 2013;1:16–20.
72. DeFelice SL. The nutraceutical revolution: its impact on food industry R&D. Trends Food Sci Technol. 1995;6(2):59–61.
73. Radhika PR, Singh RBM, Sivakumar T. Nutraceuticals: an area of tremendous scope. Int J Res Ayurveda Pharm. 2011;2:410–5.
74. Lachance MA, Daniel HM, Meyer W, Prasad GS, Gautam SP, Boundy-Mills K. The D1/D2 domain of the large subunit rDNA of the yeast species Clavispora lusitaniae is unusually polymorphic. FEMS Yeast Res. 2003;4:253–8.
75. Pathare P, Nilegaonkar S, Agte V. Factors influencing the nutraceutical activity of guava fruits. Adv Food Sci Eng. 2017;1:107–11.
76. Kim K, Jung H, Kwon KH. Composition for increasing exercise performance ability and enhancing physical strength, containing composite extract (United States Patent No. US20180015134A1); 2018. https://patents.google.com/patent/US20180015134A1/en
77. Bogha TT, Sawate AR, Kshirsagar RB, Bochare SS. Studies on physical, chemical and mineral evaluation of guava (Psidium Guajava L.). Pharma Innov J. 2020;9(3):117–9.
78. Rani R, David J. A review on utility of an astonishing fruit: Psidium guajava (guava). J Sci Technol (JST). 2021;6(1):60–72.
79. Singh MP, Bhansali RR, Kaul RK, Tripathi RS. 22 Integrated pest management in perennial plants. In: Fifty years of arid zone research in India. 375 ; 1998.
80. Mathpal D, Rathore G. An analysis of health benefits of guava. Int J Innov Res Eng Manag. 2022;9(1):239–42.
81. Kandasamy P, Shanmugapriya C. Medicinal and nutritional characteristics of fruits in human health. J Med Plants Stud. 2015;4(4):124–31.
82. Porwal K, Pal S, Dev K, China SP, Kumar Y, Singh C, et al. Guava fruit extract and its triterpene constituents have osteoanabolic effect: stimulation of osteoblast differentiation by activation of mitochondrial respiration via the Wnt/β-catenin signaling. J Nutr Biochem. 2017;44:22–34.
83. Vijaya Anand A, Velayuthaprabhu S, Rengarajan RL, Sampathkumar P, Radhakrishnan R. Bioactive compounds of guava (Psidium guajava L.). In: Bioactive compounds in underutilized fruits and nuts; 2020. p. 503–27.
84. Mishra SK, Ishfaq PM, Tripathi S, Gupta N. Fruits as boosters of the immune system. In: Plants and phytomolecules for immunomodulation: recent trends and advances. Singapore: Springer; 2022. p. 391–411.

85. Li Y, Bai D, Lu Y, Chen J, Yang H, Mu Y, et al. The crude guava polysaccharides ameliorate high-fat diet-induced obesity in mice via reshaping gut microbiota. Int J Biol Macromol. 2022;213:234–46.
86. Blancas-Benitez FJ, González-Aguilar GA, Sáyago-Ayerdi SG. Guava (Psidium guajava). In: Fruit and vegetable phytochemicals: chemistry and human health. 2nd ed. ; 2017. p. 1067–107.
87. Tay CI, Ong ST. Guava leaves as adsorbent for the removal of emerging pollutant: ciprofloxacin from aqueous solution. J Phys Sci. 2019;30(2):137.
88. Norazmir MN, Ayub MY. Beneficial lipid-lowering effects of pink guava puree in high fat diet induced-obese rats. Malays J Nutr. 2010;16(1):171.
89. Tay CI, Ong ST. Guava leaves as adsorbent for the removal of emerging pollutant: ciprofloxacin from aqueous solution. J Phys Sci. 2019;30(2).
90. Norazmir MN, Ayub MY. Beneficial lipid-lowering effects of pink guava puree in high fat diet induced-obese rats. Malays J Nutr. 2010;16(1).

An Overview on Wild Edible Fruits of Servarayan and Kalrayan Hilly Ranges

V. Sivasankari and P. Viswanathan

Abstract

Fruits are a natural gift to humanity; they are tasty, reviving, and full of nutrients, which help people's everyday diets. Wild edible fruit (WEF) plant species have a major role in enhancing human welfare. These plants are rich in nutrients and a source of unique alleles and genes that are crucial for creating enhanced agricultural cultivars that will support long-term food security. It has been shown that these fruits are rich in antioxidants, vitamins, minerals, and other nutrients. Vital minerals like salt, potassium, magnesium, iron, calcium, and phosphorus are abundant in them. Numerous bioactive substances, including flavonoids and anthocyanins, are present in these fruits. Tribal people and woodland inhabitants can easily access WEFs. Most WEFs were eaten both raw and after being cooked or processed. Nevertheless, the majority of WEFs are less well known, underutilized, or both. Indigenous knowledge and WEF usage have been sharply dropping among younger generations in recent years. In the Tamil Nadu district of Salem, the Servarayan and Kalrayan Hills are a component of the southern ranges of the Eastern Ghats System. These hilly ranges are home to the "Malayali" tribes. They are well versed in the uses, availability, and wild edible fruits. In total, 51 wild edible plant species have been found to grow in the mountainous regions of the Servarayan and Kalrayan. They are divided into 45 genera in 29 families. More edible fruit genera are found in families like Anacardiaceae, Boraginaceae, Rutaceae, and Rhamnaceae than in any other family that has been documented. The ethnobotanical knowledge of wild edible fruits that have historically been consumed by tribal and local rural people in the Servarayan and Kalrayan Hills of Tamil Nadu, India, is summarized in this article.

V. Sivasankari (✉) · P. Viswanathan
Department of Botany, Government Arts College (Autonomous),
Coimbatore, Tamil Nadu, India

P. Thangaraj (ed.), *Pomotherapeutic Insights on Wild Edible Fruits*,
https://doi.org/10.1007/978-981-96-0543-9_11

Keywords

Wild edible fruits (WEFs) · Servarayan Hills · Kalrayan Hills · Food security · Medicinal property

Introduction

The sustainable use of natural resources for food, fuel, medicine, and other purposes is essential to human existence, and man has long been intimately involved in the development of these resources. Because they are so reliant on the local flora and other natural resources, tribal people are extremely knowledgeable about them. While trees and forests are essential for their function in providing ecosystem services to agriculture, wild foods are critical for food security and nutrition [1]. Fruits are a gift from nature to humanity; they are tasty, reviving, and full of nutrients, which help people's everyday diets. Edible fruit species that are harvested from their natural environments, such as forests, but are not grown are referred to as wild edible fruits or WEFs. It has been found that these fruits are excellent providers of minerals, vitamins, and antioxidants [2]. Since ancient times, the aboriginal tribal people and local groups living in the forest catchment area have primarily consumed them. They are either conserved for use during the off-season or sold in rural markets. With the introduction of more exotic fruits in recent times, the common public is gradually becoming more aware of the significance of these fruits. They have several benefits beyond their customary usage as food. Vital minerals like salt, potassium, magnesium, iron, calcium, and phosphorus are abundant in them. They are immune to a wide range of illnesses and are frequently used into various "Ayurvedic" formulations in Indian traditional medicine [3]. Like any other widely available fruit, they offer fibres that help avoid constipation.

Because improved varieties are easily accessible and because habitat damage from deforestation and infrastructure development is reducing species variety [4], the reliance on wild edibles is expected to shrink over time [5]. Because of this, new generations are becoming less and less interested in indigenous knowledge and WEF consumption. There is a connection between the decline in plant diversity and the disappearance of traditional knowledge [6]. There is a chance that improved fruit varieties will completely replace wild fruit varieties, upsetting human-forest coexistence and hastening the loss of traditional knowledge. This is due to the growing deterioration of indigenous knowledge on WEFs and increased reliance on improved fruit varieties.

Availability of Wild Edible Fruits, Collection, and Their Consumption

People who live in forests, tribes, and rural areas have an excellent understanding of wild fruit species, including how each kind and location ripens and fruits. These individuals take use of the edible wild fruits to feed themselves when the weather is bad. For example, certain *Ficus* species have a very narrow harvest window, but other species have a much wider one [7]. Certain species, such as *Flacourtia indica* and *Carissa carandas*, ripen more than once a year, with the main fruiting season producing a bountiful harvest. Furthermore, there are considerable differences in the fruiting seasons between sites due to variations in meteorological and edaphic conditions. Fruit is available all year round as a result of the significant ripening overlap between various species, both within and between locations. Because of this, locals may obtain the seasonal minerals, vitamins, and micronutrients they need by eating a variety of wild fruits all year long. The majority of rural women and children harvest wild fruits, which are then either sold at the community market or eaten by the entire family. Generally speaking, kids are more accustomed to wild fruits than adults.

Children and women routinely gather firewood, small forest products, and wild fruits in tribal and rural communities while herding cattle through woods, village thickets, and woodlands. As a result, they are more able to obtain and get familiar with wild fruit resources than others [8]. In the past few decades, forests have retreated from villages and undergone significant degradation, which has left many wild fruits scarce or non-existent. The shifting socio-economic landscape of rural and tribal communities has resulted in women living near forests choosing to work in jobs that pay a regular wage over foraging for wild fruits, which are rarely sold for much money.

The once-ideal habitats for a variety of wild edible fruit species and once close to settlements wastelands, sacred groves, scrub forests, and bush lands have been changed to agricultural fields or other non-forestry uses. As a result, there is now less wild fruit available, less wild food used, and less awareness about using wild foods. Lack of detailed information and a poor grasp of the taxonomy and biology of these taxa limit the use and advancement of underutilized edible fruit species [9]. Therefore, before indigenous species and their traditional knowledge disappear, it is imperative to record the diversity of wild species and their indigenous potential for sustainable management of natural resources [10].

However, little effort has been made to utilize the genetic resources of these wild fruits since the introduction of modern agriculture to India. This is especially important given the serious genetic erosion that is causing the extinction of valuable species due to urbanization, forest degradation, and intensive agriculture.

Servarayan Hilly Ranges

In the Salem district of Tamil Nadu, the Servarayan Hills (sometimes spelled Shevaroy Hills) and the Kalrayan Hills are two of the southern ranges of the Eastern Ghats System. It encompasses 400 km^2. At 1620 m mean sea level, the highest peak is Solaikaradu. The hill range, which spans an area of roughly 400 km^2, is located between longitudes 11°45 and 78°20 E [11]. On this range, Yercaud is a well-liked hill station that is connected to Salem town via road. The Archean charnockites are abundant in the Shevaroys, along with a small amount of granite gneiss [12]. The area's laterite is abundant in hydrated alumina, and it is thought to contain over 15 million tons of bauxite reserves. The deposits range in thickness from 5 to 20 m. The rocks in the area all have a general 30° E strike and dip sharply toward the southeast along the Shevaroy slopes. The soil is red sandy loam. Despite being incredibly rare and poorly shaped, the valleys and shoals are rich in humus [13]. Nearly the whole plateau of the hill range is traversed by the river Vanniyar and its tributaries and branches. These hilly ranges are home to the "Malayali" tribes.

Kalrayan Hilly Ranges

The Kalrayan ranges, in conjunction with the Alavaimalai, Javadi, Pachaimalai, and Shevaroy Hills, divide the Kaveri and Palar river basins from south to north. The hills cover an area of 1095 km^2 and range in height from 2000 to 3000 ft. The hills stretch northeast from the Salem district and cross several districts in Tamil Nadu [14]. The range forms the border between the districts of Kallakurichi and Salem. The Kalrayans are classified into two regions: the Periya ("big") Kalrayans are located in the south, while the Chinna ("little") Kalrayans are located in the north. The Periya Kalrayans are on average 4000 ft. tall, while the Chinna Kalrayans are on average 2700 ft. The temperature ranges from a low of 19 °C to a high of 27 °C. The soil is ideal for plant growth, and the range is often rather smooth. Woods of shrubs and bushes can grow as high as 400 m, whereas deciduous woods are found at elevations above 800 m. Grown on remote plateaus, sholas are a form of high-altitude stunted evergreen forest. Because of "habitat uniqueness, human impacts, and cultural tradition", the forest stand is expanding, but conservation efforts are still required. The people who live in hills and forests are referred to as "Malayali" tribes [15].

Wild Edible Fruits of Servarayan and Kalrayan Ranges

Fruits are an essential supplementary food source for the Servarayan and Kalrayan hill communities. It is anticipated that this chapter would highlight important wild edible fruit plant resources in the research region. As a result, it seeks to educate botanists, agriculturists, conservationists, and naturalists about the fascinating world of fruits and how best to exploit them for human welfare.

In total, 51 wild edible plant species have been found to grow in the mountainous regions of the Servarayan and Kalrayan. They are divided into 45 genera in 29 families. More edible fruit genera are found in families like Anacardiaceae, Boraginaceae, Rutaceae, and Rhamnaceae than in any other family that has been documented. Some of the edible fruit plants found in the present study area have gained potential recognition and are now either cultivated in orchards or intercropped in agroforestry and social forestry.

Cultivated Wild Edible Fruit Plants

Table 1 provides the list of cultivated or popularized wild edible fruits growing wild in the forest ranges of Servarayan and Kalrayan Hills. Few of these plants are cultivated throughout the world. For example, *Mangifera indica*, *Psidium guajava*, and few others are cultivated in some regions only.

Table 1 List of cultivated wild edible fruit-yielding plants

S. No.	Botanical name	Vernacular name	Family	Habit
1.	*Anacardium occidentale* L.	Munthri	Anacardiaceae	Tree
2.	*Artocarpus heterophyllus* Lam.	Pala	Moraceae	Tree
3.	*Canavalia ensiformis* Dc.	Kattuavarai	Fabaceae	Climber
4.	*Azadirachta indica* L.	Vembu	Miliaceae	Tree
5.	*Borassus flabellifer* L.	Panai maram	Arecaceae	Tree
6.	*Carissa carandas* L.	Kila/Kela	Apocynaceae	Small tree
7.	*Coccinia indica* Wight & Arn.	Kovai chedi	Cucurbitaceae	Climber
8.	*Cyphomandra betacea* Cav.	Marathakkali	Solanaceae	Shrub
9.	*Ficus racemosa* L.	Athi	Moraceae	Tree
10.	*Limonia acidissima* Linn.	Vilankay maram	Rutaceae	Tree
11.	*Mangifera indica* L.	Maa	Anacardiaceae	Tree
12.	*Murraya koenigii* (L.) Spreng.	Karuveppilai	Rutaceae	Shrub
13.	*Phyllanthus emblica* L.	Malai nelli	Euphorbiaceae	Small tree
14.	*Pithecellobium dulce* (Roxb.) Benth.	Kona puliyanga	Fabaceae	Tree
15.	*Psidium guajava* L.	Koyya	Myrtaceae	Small tree
16.	*Pyrus communis* L.	Berikai	Rosaceae	Tree
17.	*Quercus incana* Roxb.	Masikai	Fagaceae	Tree
18.	*Semecarpus anacardium* L.f.	Senkottai	Anacardiaceae	Tree
19.	*Solanum nigrum* L.	Manathakkali	Solanaceae	Herb
20.	*Solanum surattense* Burm f.	Kandankathiri	Solanaceae	Herb
21.	*Solanum torvum* Swartz.	Sundakkai	Solanaceae	Shrub
22.	*Terminalia bellirica* Roxb.	Thandi maram	Combretaceae	Tree
23.	*Terminalia chebula* Retz.	Kadukkai maram	Combretaceae	Tree
24..	*Ziziphus mauritiana* Lamk.	Yellandai	Rhamnaceae	Small tree

Table 2 List of wild edible fruit plants of Servarayan and Kalrayan Hills

Sl. No	Botanical name	Vernacular name	Family	Habit
1.	*Alangium salviifolium* Linn.	Alinjil	Alangiaceae	Shrub/small tree
2.	*Artocarpus hirsutus* Lam.	Kattupala	Moraceae	Tree
3.	*Atalantia monophylla* (L) Corr.	Kurunthu, Kattu kolinji	Rutaceae	Shrub
4.	*Azima tetracantha* Lam.	Sangam	Salvadoraceae	Shrub
5.	*Buchanania angustifolia* Roxb	Seeramaram	Anacardiaceae	Tree
6.	*Canthium parviflorum* Lam.	Kaarai palam	Rubiaceae	Shrub
7.	*Capparis zeylanica* L.	Aathandai	Capparaceae	Herb
8.	*Clausena dentata* Willd.	Ana palam	Rutaceae	Shrub
9.	*Cordia myxa* L.	Neruvili	Boraginaceae	Small tree
10.	*Crateva religiosa* G. Forst.	Mavalli maram	Capparidaceae	Tree
11.	*Diospyros ferrea* (Willd.) Bakh.	Irumbuli pazham	Ebenaceae	Tree
12.	*Ehretia microphylla* Lam.	Kurangu vetrilai	Boraginaceae	Shrub
13.	*Flacourtia indica* (Burm.f.) Merr.	Sothaikala	Salicaceae	Shrub
14.	*Gmelina arborea* Roxb.ex Smith	Kumula maram	Lamiaceae	Tree
15.	*Grewia disperma* Rottl.	Uduppai	Tiliaceae	Herb
16.	*Hardwickia binata* Roxb.	Aacha maram	Fabaceae	Tree
17.	*Holoptelea integrifolia* (Roxb.) Planch.	Aali maram/Aaya maram	Ulmaceae	Tree
18.	*Morinda tinctoria* Roxb.	Nuna maram	Rubiaceae	Tree
19.	*Opuntia elatior* Mill.	Sappathaikkalai	Cactaceae	Herb
20	*Passiflora foetida* L.	Kurangu pazham	Passifloraceae	Climber
21.	*Phoenix farinifera* Roxb.	Icham	Palmaceae	Shrub
22.	*Psydrax dicoccos* Gaertn.	Nikkanai maram	Rubiaceae	Tree
23.	*Scutia myrtina* Merr.	Thovatti mul	Rhamnaceae	Shrub
24.	*Securinega leucopyrus* (Willd.) Mull. Arg.	Poolapazhalam	Euphorbiaceae	Shrub
25.	*Tarenna asiatica* (L.) Kuntze ex K.Schumann	Therani chedi	Rubiaceae	Shrub
26.	*Ziziphus glabrata* Heyne.	Karukattankai	Rhamnaceae	Shrub
27.	*Ziziphus oenoplia* Mill.	Surai pazham	Rhamnaceae	Shrub

Wild Edible Fruit Plants

Given below is the list of edible fruit-yielding wild plants that are growing in the forest ranges of the Servarayan and Kalrayan Hills (Table 2). Each species has been given a botanical name with author citation along with its family name, local names, information on when it flowers and fruits, brief but notable morphological characters, and notes on distribution.

Detailed Information on the Wild Edible Fruiting Plants

Field surveys, conversations with locals and subject matter specialists, and the collection of images and information on phenology and local applications of several particular species were utilized to create the book.

Alangium salviifolium Linn

It grows up to 18 m tall and is a deciduous shrub or small to medium-sized prickly tree. It can be found up to 750 m above sea level in riverine and lowland rainforests. In addition to being edible, the fruit has antidotes for poisoning, carminative, expectorant, purgative, and eye problems; fruit juice is also utilized for these purposes. In India, sweet flag (*Acorus calamus* L.) rootstocks and mature *A. salviifolium* fruits combined with honey are used to control pests on agricultural crops (Plate 1a). Astringent, antiulcer, pungent, purgative, laxative, and antiepileptic activities [16], anthelmintic, emetic, anti-protozoa, and hypoglycaemic properties [17], and alleviation of spasms are only a few of its medical uses. According to reports, the tribes in the Maldives utilized it as a contraceptive for cattle and pig rearing as well as a treatment for skin illnesses including scabies and leprosy. In Ayurveda, external haemorrhoids and rheumatism are treated with roots and fruits [18]. This tree bears fruit in March and May, right before the onset of the rainy season, and blossoms from February to April. When it flowers, it sheds all of its leaves, and when it starts to fruit, new leaves emerge.

Artocarpus hirsutus Lam

The tropical evergreen tree *Artocarpus hirsutus* Lam., which is a member of the Moraceae family and genus *Artocarpus*, is well known for its edible wood and fruit. Fruiting and flowering is from December to March. India is the native home of *Artocarpus hirsutus* Lam., especially in South India (Plate 1b). It is found in areas with rocky and sandy soils that are between sea level and 1000 m above sea level, with an average annual rainfall of 1500 mm. Up to the last part of the twentieth century, impoverished peasants consumed the ripe fruit and roasted seeds as a staple diet [19]. Flowering and fruiting are December to March.

Atalantia monophylla (L.) Correa

Atalantia monophylla (L.) Correa is a tiny, evergreen tree with many branches that can reach a height of 6–8 m. Fruit is a berry, which turns orange and aromatic when fully ripe (Plate 1c). The Malayali tribes in the area use the leaves of this plant to cure snakebite [20]. September–December and February–March are the two peaks for flowering and fruiting.

Plate 1 Fruits of (**a**) *Alangium salvifolium* (**b**) *Artocarpus hirsutus* (**c**) *Atalantia microphylla* (**d**) *Azima tetracantha* (**e**) *Bauchanania angustifolia* (**f**) *Canthium parviflorum*

Azima tetracantha Lam

The armed straggling shrub *Azima tetracantha* is common in Tamil Nadu's scrub forests. Blooming is from March to May. Branchlets are quadrangular, pale, pubescent; spines are straight, paired at the axils of the leaves; leaves are greyish-green, ovate to lanceolate, obtuse with a spinulose tip. Flowers are little white or yellowish-white in panicles along the axilla. Berries are white with one seed and are the tiny fruits. The inhabitants in the area consume the ripe berries (Plate 1d). Flowering is from March to May.

Buchanania angustifolia Roxb

Buchanania angustifolia is a medium-sized, deciduous tree that grows to a height of 10 m. Its bark is deeply fissured and has glabrous, robust branches. Tree is small; bark is tough, either dark grey or black. Leaves are petioled, oblong or obvate-oblong, pubescent underneath, highly nerved, and alternating. Flowers are greenish-white in pyramidal panicles at the axillary and terminal ends. Sepals are five, almost at no cost. Petals are 4–5, round or triangular in shape. Stamens are ten near the disc's base, erect, and as long as the petals that are spreading. Carpels are hairy, ovoid or globose, black, and hard as stones. Fruits are globose, obliquely shaped drupes (Plate 1e). The oil that is collected from the kernels is applied to skin conditions and also utilized to eliminate facial spots and blemishes. The gum is used as an analgesic after being combined with goat milk. Powdered kernels are used as an aphrodisiac and to treat fever and burning sensations when mixed with milk [21]. Potent panicles begin to flower in January and continue through March, with ripe fruits appearing in May or June.

Canthium parviflorum Lam

Canthium parviflorum Lam. (syn: *Plectoria parviflora*) is a prickly sub-Savannah shrub that can reach a height of 3 m and spreads its branches widely throughout dry plains and shrub woods in India. Leaves are oval or lanceolate-ovate that are shiny above and pale below. Flowers are sessile or peduncled, thickly umbellate, and white or greenish-white with a strong scent. When ripe, the rectangular, two-chambered fruits turn yellow. Blooming is throughout May through August. When ripe, the globose drupes will turn orange, with wrinkled pyrenes (Plate 1f). Fruiting starts in July. Every component of the plant, including the stems, leaves, roots, and fruits, has medicinal use. One way to treat headaches is with fruits. Traditionally, snake bites are treated with it [22]. High concentrations of beta and total carotenoids, which are crucial for vitamin A action, are present in this plant.

Capparis zeylanica L.

The climbing shrub *Capparis zeylanica* L. has several branches and two curving spines on either side of the leaf bases. It produces fruit and flowers from January to October. Thorns are frequently straight; leaves are elliptic, oblong, or oblanceolate, coriaceous, and hairless. Flowers are axillary, single, and 3–4 cm in diameter. Sepals vary in shape from ovate to oval shaped. Petals are cream or white with a yellow or maroon core. Gynophore length is 2–5 cm. When completely ripe, the ellipsoid, beaked berry has deep red colour and seeds immersed in white pulp (Plate 2a). Indian capers, or *Capparis zeylanica* Linn., are widely distributed throughout India.

Plate 2 Fruits of (**a**) *Capparis zeylanica* (**b**) *Clausena dentata* (**c**) *Cordia myxa* (**d**) *Crateva religiosa* (**e**) *Diospyros ferrea* (**f**) *Ehretia microphylla*

Pickles can also be prepared from fruits. Ripe fruits are consumed twice a fortnight to manage diabetes [24].

Clausena dentata (Willd.) Roem

Clausena dentata is a tiny tree plant that reaches a height of 2–4 m, produces excellent fruit, and has odorous leaves. It grows in the Eastern Ghats' evergreen and semi-evergreen forests. It is mostly located in the Kalrayan and Servarayan ranges. Berries are white, oval-shaped fruits. The fruit is pulpy and tastes good [25]. It ripens in late June after blooming in April (Plate 2b).

Cordia myxa L.

The tree *Cordia myxa* L. is modest to moderate in size, with branches that droop. Leaves are elliptic-oblong, sinuate, orbicular, and somewhat dentate. Flowers are white and fragrant in corymbose, often clustered, glabrous cymes. When young, the fruit is drupe-conical; when ripe, it is ovoid, glossy, yellowish, or pinkish, and it has a viscous, sweetish, nearly transparent pulp around a rocky core; it has one seed (Plate 2c). This herb has great ethnomedical value. There have been reports of anti-diabetic, antiulcer, anti-inflammatory, immune-modulating, and analgesic properties in plant components such as leaves, fruit, bark, and seeds. In many cultures, *Cordia myxa* trees, leaves, and fruit are used in significant ways. The fruit itself is well known for its therapeutic properties. From March to May, flowers bloom. Due to the fruits' anti-inflammatory properties, they are edible and utilized as traditional medicine in Nepal and India [26].

Crateva religiosa G. Forst

Crateva religiosa is a deciduous, attractive tree with many branches that grow to a considerable size. It is used from the time of the Vedas. It usually prefers rich, moist, neutral to acid soil and grows well in either partial shade or full sun. It can be found in forests that are periodically flooded, mainly at elevations of less than 100 m, but it can also be found up to 700 m. There are several medical uses for *C. religiosa* [27]. Season of flowering is from March through April. Season of fruiting is from July through January (Plate 2d).

Diospyros ferrea (Willd.) Bakh

It is a tiny tree or shrub with many branches that have pubescent new shoots. The leaves are coriaceous, glabrous, obovate, oblong or round, and obtuse. Flowers are three-merous, white, and tiny. Male flowers are hairy, in extremely short cymes, and smaller than female blooms. Female flowers are 1–3 in a group, with a corolla and calyx that are wider and more obtusely lobed. The ovary has three cells. Fruiting calyx is cup shaped; berry is globose, golden, or reddish and one-seeded; and endosperm is uniform. It is extensively present in degraded forest areas and tropical dry scrub woodlands. The fruit is a globose berry that ripens to orange (Plate 2e). When mature, the fruit pulp can be eaten. It is considered a meal of starvation in southern India. It can be used for a variety of things, such as valuable timber, edible fruits, and decorative purposes. Fruiting and flowering are from February to May.

Ehretia microphylla **Lam**

Ehretia microphylla are shrubs or tiny trees that grow up to 2 m tall in scrub jungles and dry deciduous woods. It is a massive plant with many branches and hairy, strigose stems. Leaves are fascicled, on short stalks, oblong or oblanceolate, somewhat serrated at a distance, obtuse at the tip, and scabrous above. Flowers are small, axillary, white or cream coloured, unisexual, solitary, or paired. Fruit is drupaceous and globose (Plate 2f). November to December is when it flowers and fruits. In Siddha Materia Medica [28], Kuruvichi poondu is mentioned. It is used to treat toxic diarrhoea in children, chronic dysentery, leprosy, eczema brought on by sexually transmitted infections, and infertility [29].

Flacourtia indica **(Burm.f.) Merr**

Flacourtia indica are small, dioecious trees or shrubs that grow up to 10 m tall. They have rough, yellowish, or orange-brown bark that is frequently covered with simple, branching thorns. Leaves are coarsely crenate and alternating or crowded. Flowers are yellowish that can be solitary or in short racemes with few flowers. The ovary is globular with 4–7 nearly free styles; stigmas are slightly bilobed. When ripe, the fruit is globose to ellipsoid and turns dark purple or scarlet (Plate 3a). Many seeds are imbedded in luscious pulp. As bushes or little trees, it treats spleen problems and jaundice in Ayurvedic, Siddha, Tibetan, and traditional medicine. Fruits have a lovely flavour and a sweet taste. They can be used to make jams and jellies because they have a high pectin content and a moderate acidity. Fruits are palatable and aid in digestion and those with jaundice and enlarged spleen [30]. The plant is grown as hedge plants occasionally. Fruiting and blooming are December–March/May–August.

Gmelina arborea **Roxb.ex Smith**

The Lamiaceae family includes the quickly growing deciduous tree *Gmelina arborea*, which is a major plantation species in many tropical regions of the world. The species is found in tropical and subtropical parts of Southeast Asia in naturally

Plate 3 Fruits of (**a**) *Flacourtia indica* (**b**) *Gmelina arborea* (**c**) *Grewia disperma* (**d**) *Hardwickia binata* (**e**) *Holoptelea integrifolia* (**f**) *Morinda tinctoria*

occurring semi-deciduous forests. The fruits have a fragrant, sweet pulp and are oblong drupes (Plate 3b). In addition to being edible, fruits are used to treat colitis, ulcers, anaemia, leprosy, and constipation [31]. They are also believed to promote hair development. According to Indian medical systems, the tree is also a significant medicinal plant. Flowers occur in panicles from February to April.

Grewia disperma Rottl

Grewia disperma are stellate-tomentose shrubs or trees. Leaf blades are similar in size, three-nerved, lanceolate to ovate-lanceolate, serrulate, acuminate, and delicately tomentose beneath. Peduncles are 1–4 and thin. Stamens are 15–25; flowers are white or yellowish, dioecious or polygamous, 3–5 mm in diameter, pilose, strongly bilobed, and covered with long, deciduous hairs. Ecology is quite frequent in scrub jungles and open, dry deciduous woods. Seeds are used to propagate the species. Fruits are frequently consumed. Fruits are considered to be heart healthy [32]. They are given in cases of dysentery and diarrhoea. The tasty, acidic fruits are frequently used to make sherbets (Plate 3c).

Hardwickia binata Roxb

Native to India, *Hardwickia binata* Roxb. reaches heights of 25–30 m. When medium sized, it can be an up to 15-m-tall deciduous tree. Bark is dark grey and rough and contains deep cracks; the trunk has a girth of 1.8–3 m and a clean, cylindrical bole that extends up to 12–15 m as well as drooping branches [33]. It is found in the Eastern and Western Ghats of dry deciduous forests and is native to Peninsular India. The fruit resembles a strap-shaped, samaroid pod that narrows at both ends (Plate 3d). Fruiting and flowering are from June to April.

Holoptelea integrifolia (Roxb.) Planch

The large deciduous tree *Holoptelea integrifolia* can grow to a height of 25 m. The smooth, whitish-grey, 6–8-mm-thick bark has pubescent branchlets. The leaves are acuminate, ovate or elliptic-ovate, solitary, alternating, and stipulate. An awful smell is released when the leaves are crushed and the bark is cut. The fruits are indehiscent, single-seeded, light brown, obliquely elliptic, or orbicular, with wings and a stem (Plate 3e). They are 2.5–3.5 cm in length and 1.5–2.5 cm in width. February and March are when flowers and fruits are observed. The seeds are kidney shaped, tiny, and pale. The plant is a multipurpose medicinal herb that is used to treat common medical conditions in a number of traditional medical systems. It has long been used to cure and prevent a number of skin aliments including eczema, leukoderma, ringworm, and chronic wound and also to cure malaria, rickets, scabies, rheumatism, inflammation, and intestinal cancer. Unripe fruits are also eaten [34].

Morinda tinctoria Roxb

Morinda tinctoria are tiny trees with corky, grey bark. Leaves are oval, obovate, or elliptic with pubescence on both surfaces. Flowers are white and fragrant in dense, ovoid, peduncled heads. Fruits are with succulent, expanded hypanthia that are whitish-green in colour, forming a pseudocarp that has lines defining its limits (Plate 3f). This tree is very common along roadsides, in wastelands, in dry deciduous forests, and close to villages. In India, this is wild and occasionally cultivated. The mature fruits can be eaten. Pickled green fruits are consumed with curries. Ayurveda has employed the fruits, flowers, leaves, and heartwood of *M. tinctoria* to cure a variety of illnesses [35]. Fruit juice is utilized as an antidote for headache, high blood pressure, diarrhoea, and diabetes in tropical Pacific islands [36].

Opuntia elatior Mill

Large, fleshy shrubs with large, flat, jointed limbs are called *Opuntia elatior*. Areoles feature many barbed bristles and 5–6 pale yellow spines. Flowers are funnel shaped, pale yellow or greenish-yellow, with a scarlet core. When ripe, pyriform, truncate fruits show with black seeds inside, and white, spongy tissues inside are deep red or reddish purple in colour. This is abundant along sandy riverbanks, sand dunes, and dry waste areas. Originally brought to India as a hedge plant, it is now found wild in a number of regions. Fruits are suggested as an expectorant and treatment for syphilis, ulcers, tumours, asthma, gonorrhoea, whooping cough, and diarrhoea [37]. Prickly pears, *Opuntia elatior*, are used as a highly nutritious meal.

Passiflora foetida L.

They are pubescent, slender climbers with tendrils that release a foul odour when they get injured. Three lobes are found on each palmate leaf; glandular hairs ciliate the margins. Flowers are greenish white in colour, with a corona that processes white and purple. The epicalyx is made up of pinnatifid bracteoles that are divided into segments that are gland-tipped pectinate. When ripe, the fruit is globose, shiny, and orange-yellow, with a persistent epicalyx surrounding it (Plate 4a). It is abundant in waste areas, scrub jungles, thickets, and hedges, generated from seeds, and not found in agricultural settings. While unripe fruits contain cyanogenic glucoside and are dangerous, ripe fruits are edible. They have emetic properties, and their decoction is used to treat biliousness and asthma [38].

Plate 4 Fruits of (**a**) *Passiflora foetida* (**b**) *Phoenix farinifera* (**c**) *Psydrax dicoccus* (**d**) *Scutia myrtina* (**e**) *Securinega leucopyrus* (**f**) *Tarenna asiatica* (**g**) *Zizyphus glabrata* (**h**) *Zizyphus oenoplia*

Phoenix farinifera Roxb

It is a tiny wild date palm and a palm tree, solitary or in clusters, up to 6 m tall, 30 cm diameter. Leaves are up to 3 m long. Seed is oval with rounded apices, pinkish-brown when fresh, glossy chestnut-brown when drying, and 8–12 × 6 mm; endosperm is homogenous; fruit is ovoid, 11–15 × 5–8 mm, maturing from green to red to purple-black, and somewhat juicy and tasty (Plate 4b). It is locally gregarious amid forests of dry scrub. The fruits are heavily devoured by children and tribes due to their little but extremely sweet content. It is regarded as a food for starvation.

Psydrax dicoccos Gaertn

Psydrax dicoccos is an evergreen shrub or tree growing to a height of 15 m and is locally employed as quality timber. This plant is categorized as "vulnerable" according to IUCN Red List of Threatened Species (2013) [39]. These are broad-leaved woods with sparse foliage found at low to intermediate heights, ranging in elevation from 100 to 600 m moist evergreen woodlands on the lowlands. Blooming season is from December to February, and fruiting is from April to May (Plate 4c).

Scutia myrtina Merr

Scutia myrtina is an evergreen shrub or small tree. It has prickly branches that can be upright, scandent, or straggling. South India has a large supply of it. Its height is between 1 and 5 m and cultivated as a hedge. The size, shape, and denticulation of the leaves in this species vary greatly. When mature, astringent fruit is consumed. Sometimes locals pick the edible fruit from the wild for their own purposes (Plate 4d).

Securinega leucopyrus (Willd) Muell

It has angular branchlets on shrubs that terminate in spines. Leaves are distichous, obovate, with basal cuneate and emarginate apex, membranous, and measure up to 2.5 × 1.5 cm. Male flowers are in axillary clusters; tepals are 0.7 mm, concave, and obtuse; there are five stamens, filaments are 1.5 mm and pistillode 1 mm; blooms are greenish white. Bracts are many and minute; pedicels are 2 mm. Fruit is 5 mm in diameter, globose, and in a white capsule. It is common only in India's scrub jungles. The Indian tribes have been using it for a very long time. Flowering is from June to September. Fruiting is from November to April (Plate 4e). The plant also attracts beetles and butterflies. It is an effective deworming medication.

Tarenna asiatica (L) Kuntze ex Schumann

It belongs to a broad genus of shrubs and small trees with elliptic or oblong-lanceolate, coriaceous leaves; black, multi-seeded berries; and greyish, brown bark that grows in plains and hilly areas. The species is utilized for its timber and therapeutic properties. *Tarenna asiatica* is a large evergreen shrub or small tree with greyish brown bark, elliptic or oblong, lanceolate, coriaceous leaves, fragrant white flowers in cymes, and black, multi-seeded fruit that grows in plains and hilly areas. The species is utilized for its timber and therapeutic properties. It is applied to skin conditions and boils to treat suppuration [40]. Fruits that have ripened are edible (Plate 4f). Crows consume these fruits. Fruiting and flowering are from February to September.

Ziziphus glabrata Heyne

Ziziphus glabrata are dry deciduous trees up to 10 m tall and have tomentose branchlets and dark brown bark with an orange-red blaze. Leaves simple, alternate with filiform, caducous stipule; petiole is stout and pubescent, 3–5 mm long; and lamina measuring 1.5–10.3 × 1.5–5 cm, elliptic, elliptic-oblong, elliptic-orbicular, or ovate. The leaf base is obtuse, round, or asymmetrical; apex is obtuse and retuse; and margin is crenulate, with callous tips tipping the crenations. Leaves glabrous, shining above, tomentose below, coriaceous, and is prominent and three-ribbed from the base. Flowers are bisexual greenish-yellow in axillary, peduncled cymes measuring 7 mm in diameter; bracteoles are subulate; pedicels are up to 2 m; calyx tube is 1.5–2 mm. Sepals 5, pubescent, triangular, keeled within, deflexed, and valvate. Petals 5, 1.5-mm long and obovate, and disc is ten lobed and entire. Stamens 5, opposite, enclosed in the petals, and typically longer than them. Filaments are flattened, and two-celled, single-ovule ovary is inserted in the disc. There are two connate styles to the centre and two stigmas. Fruit is drupe, globose, obovoid, up to 6 mm in diameter, has 1–2 cells, and is yellow when ripe; seeds are 1 or 2 (Plate 4g). Fruiting and flowering are from March to December.

Ziziphus oenoplia Mill

Ziziphus oenoplia Mill. is a prickly shrub that can reach a height of 1.5 m. It spreads and occasionally climbs. The leaves are acute, oblique, ovate-lanceolate, simple, and alternating. Green blooms in subsessile axillary cymes are present. When ripe, the fruit is a globose drupe with a solitary seed within that is glossy and black (Plate 4h). An important shrub known by its English name, jackal jujube, is widely found in the torrid regions of tropical Asia and northern Australia. This traditional herbal remedy is intended to treat diarrhoea and relieve pain in the abdomen [41].

Conclusion

Wild edible plants serve as a cheap and easily available source of micronutrients and additional food, making them an essential part of the diets of rural and tribal communities. Tribal people can gather wild edible fruits from the forests and sell them in the nearby marketplaces, which gives them another source of money. Wild fruits constitute a major source of high-quality nutrition for the tribal people. Since the agricultural season is shorter in the study region, the wild plants are essential for their food security. The abundance of wild fruits is well known among forest residents, making it a viable choice for households or those without land to gather and harvest them. However, wild fruit plants are ignored, and more significantly, there is scant research and documentation on the indigenous knowledge that goes along with using them. It is anticipated that this analysis will highlight important wild edible fruit resources that are less well known, generating interest in both in situ and ex situ conservation. Its goal is to get botanists, horticulturists, nutritionists, and naturalists interested in the delicious world of fruits and how they might benefit human happiness and sustainability.

References

1. Sunderland T, Powell B, Ickowitz A, Foli S, Pinedo-Vasquez M, Nasi R, Padoch C. Food security and nutrition: The role of forests. Discussion Paper. CIFOR; Bogor: Indonesia; 2013;8.
2. Chalise JP, Acharya K, Gurung N, Bhusal RP, Gurung R, Skalko-Basnet N. Antioxidant activity and polyphenol content in edible wild fruits from Nepal. Int J Food Sci Nutr. 2010;61(4):425–32.
3. Maroyi A. The gathering and consumption of wild edible plants in Nhema communal area, Midlands Province. Zimbabwe. Ecology of Food and Nutrition. 2011;50:6.
4. Chauhan SH, Yadav S, Takahashi T, Luczaj L, D'Cruz L, Okada K. Consumption patterns of wild edibles by the Vasavas: a case study from Gujarat. India. J Ethnobiol Ethnomed. 2018;14:57.
5. Joshi N, Kehlenbeck K, Maass BL. Traditional, neglected vegetables of Nepal: their sustainable utilization for meeting human needs. In: Conference on International Agricultural Research for Development. Tropentag. 2007;1–10.
6. Uprety Y, Poudel RC, Shrestha KK, Rajbhandary S, Tiwari NN, Shrestha UB. Diversity of use and local knowledge of wild edible plant resources in Nepal. J Ethnobiol Ethnomed. 2012;8:16.
7. Bagra G, De LC, Singh VB. Collection and evaluation of some indigenous lesser known fruits of Arunachal Pradesh. Horticulture for sustainable income and environmental protection. 2006;1:224–31.
8. Bhatia H, Sharma YP, Manhas RK, Kumar K. Traditionally used wild edible plants of district Udhampur, J&K. India. J Ethnobiol Ethnomed. 2018;14:73.
9. Khasbagan S. Indigenous knowledge for plant species diversity: A case study of wild plants' folk names used by the Mongolians in Ejina desert area, Inner Mongolia, P R China. J Ethnobiol Ethnomed. 2008;4:11.
10. Jasmine B, Singh Y, Onial M, Mathur VB. Traditional knowledge systems in India for biodiversity conservation. Indian Journal of Traditional Knowledge. 2016;15(2):304–12.
11. Sivaraj N, Krishnamurthy KV. Flowering phenology in the vegetation of Shervaroys. South India. Vegetation. 1989;79:85–8.
12. Krishnan MS. Geology of India and Burma. Higginbothams, Madras. 1956.

13. Subramanyam K, Thothathri K, Henry AN. A collection of ferns from Shevaroy Hills, Salem District, Chennai state. Bulletin of the Botanical Survey of India. 1960;2:323–27.
14. Kannan M, Senthil Kumar T, Rao MV. Ethnobotanical survey on wild edible plants of kalrayan hills, Salem District, Tamil Nadu, India. Global Journal of Research on Medicinal Plants & Indigenous Medicine. 2015;4(12):236–46.
15. Sakthivel R, Manivel M, Raj NJ, Pugalanthi V, Raju DK. Role of remote sensing in geomorphic mapping: A case study from Kalrayan hills, parts of Eastern Ghats. Tamilnadu. Indian Journal of Geomorphology. 2006;11:103–12.
16. Balakrishnan N, Kumar S, Balasubramaniam A, Sangameswaran B, Chaurey M. Antiepileptic activity of *Alangium salvifolium* leaf extracts. Herbal Tech Industry. 2010;31(6):20–3.
17. Sreekanth P, Sudhakara K, Gouse Basha G, Murali K, Kumar SA. Anti ulcer effect of *Alangium salvifolium* ethanolic leaf extract on gastric lesion induced by ethanol in rats. Asian Journal of Pharmaceutical and Clinical Research. 2011;4(2):112–4.
18. Jain A, Katewa SS, Chaudhary BL, Galav P. Folk herbal medicines used in birth control and sexual diseases by tribals of southern Rajasthan. India. J Ethnopharmacol. 2004;90(1):171–77.
19. Mathew SP, Mohandas A, Shareef SM, Nair GM. Biocultural diversity of the endemic 'Wild Jack Tree' on the malabar coast of South India. Ethnobotany Research and Applications. 2006;4:025–40.
20. Premalatha S, Karthi A. Phytochemical screening and antimicrobial activity of leaf of *Atalantia monophylla* (L.) Corr. (Rutaceae). International Journal of Pharmacy and Biological Sciences. 2017;7(4):179–84.
21. Patil RN, Rothe SP. *Buchanania lanzan*: An enormous medicinal value. International Journal of Advance Research and Innovative Ideas in Education. 2017;3(1):395–96.
22. Parinitha M, Srinivasa BH, Shivanna MB. Medicinal plant wealth of local communities in some villages in Shimoga District of Karnataka. India. J Ethnopharmacol. 2005;98:307–12.
23. Rajyalakshmi P. *Canthium parviflorum*: an underexploited carotene rich fruit. Indian Journal of Natural Products and Resources. 2003;2:70.
24. Amit L, Kumar CA, Vikas G, Parveen B, Renu B. Phytochemistry and pharmacological activities of *Capparis zeylanica*: An overview. Int J Res Ayurveda Pharm. 2010;1(2):384–9.
25. Musa A, Aminah NS, Davies-Bolorunduro OF, Kristanti AN, Suhaili, Islami AI, Wai TS, Naing TTSP. Antimicrobial activities of the extracts and secondary metabolites from *Clausena genus* –A Review. Open. Chemistry. 2022;20:627–50.
26. Yahia EM. Postharvest biology and technology of tropical and subtropical fruits. Volume 1. Fundamental issues. Woodhead Publishing Limited, Oxford, England. 2011;215–232.
27. Mohanapriya R, Venkata Ratina Kumar T, Sunil Kumar KN. Standardization of *Crataeva religiosa* G. Forst. leaf by pharmacognostical studies. Journal of Research in Siddha Medicine. 2021;4:67–77.
28. Murugesa Mudaliyar KS, Gunapadam. Part-I. Chennai: Tamil Nadu Siddha Medical Board; 1956;274.
29. Kannusamy PC. Pathartha Guna Vilakkam. Chennai: B Rathina Nayakar & Sons; 1939;265–66.
30. Satyanarayana V, Kurupadanam GL, Srimanaraya G. A butyrolactone lignan disaccharide from *Flacourtia ramontchi*. Phytochemistry. 1991;30:1026–29.
31. Rohit K, Vaibhav P, Manodeep C, Kamath JV. Phytochemical and pharmacological profile of *Gmelina arborea*: An overview. Int Res. J Pharm. 2012;3(2):61–4.
32. CSIR. Wealth of India: raw materials. New Delhi: Council of Scientific and Industrial Research; 1998;1–10.
33. Shingade SP, Kakde RB. A Review on "Anjan" *Hardwickia binata* Roxb.: Its Phytochemical studies, Traditional Uses and Pharmacological Activities. Pharmacognosy Reviews. 2021;15(29):65–8.
34. Martin FW, Campbell CW, Rubertbe RM. Perennial edible fruits of the tropics: An inventory. USDA – Agricultural Research Service. 1987.
35. Kirtikar KR, Basu BD. Indian Medicinal Plants (2nd edn). Vol. II, M/S Bishen Singh Mahendra Pal Singh. 1935;1294–1295.

36. Siddiqui BS, Sattar FA, Ahmad F, Begum S. Isolation and structure determination of two new constituents from the fruits of *Morinda citrifolia* Linn. Nat Prod Res. 2008;22:1128–36.
37. Patel KN, Ishnava KB. Evaluation of Nutritional and Medicinal Properties of *Opuntia elatior* Mill. Phytochemicals – Source of Antioxidants and Role in Disease Prevention. 2018;121–138.
38. Patil AS, Paikrao HM, Patil SR. Passiflora foetida Linn: A complete morphological and phytopharmacological review. Int J Pharma Bio Sci. 2013;4:285–96.
39. IUCN Red List of Threatened Species. http://www.iucnredlist.org/
40. Amutha D, Shanthi S, Mariappan V. Antiflammatory effect of *Tarenna asiatica* (L) in carrageenan induced lung inflammation International. J Pharm Pharm Sci. 2012;4(5):344–7.
41. Afsana Nahrin M, Junaid SS, Afrose AB, Akter Y, Alam MS, Sharmin T, Sakia Ferdousy SM, Hosen Z. *Zizyphus oenoplia* Mill.: A Systematic Review on Ethnopharmacology, Phytochemistry and Pharmacology of an Important Traditional Medicinal Plant. Mini Rev Med Chem. 2022;22(4):640–60.

A Review on Nutritional and Metabolic Analysis of Wild Edible Fruits of Western Ghats of Tamil Nadu, India

Vetri Velavan Sundararajan, Ashanath Fowmitha Nazeer, and Parimelazhagan Thangaraj

Abstract

India, which is struggling with chronic malnutrition, especially in vulnerable populations like children and rural people, is delving towards underutilized resources like edible wild fruits, which are abundant in a number of ecosystems like the Western Ghats. Rich in phytochemicals like phenolics, flavonoids, and other antioxidants, these fruits provide vital nutrients and bioactive compounds that are important for a balanced diet, supporting general health and well-being. Indigenous communities are aware of their therapeutic uses in treating a range of illnesses and extending life. Research suggests that consumption of wild fruits may assist with preventing and treating diseases associated with malnutrition, such as neurological disorders, cardiovascular diseases, and illnesses causing inflammation. They also appear promising to controlling weight, enhancing the immune system, and hindering cancer. In conclusion, areas such as the Western Ghats, with their rich biodiversity of wild edible fruits, provide an invaluable resource for addressing nutritional deficiencies and promoting holistic well-being. To fully reap their benefits and protect supporting ecosystems, it is essential to make use of scientific research, traditional knowledge, and community involvement. For wild edible fruits to remain accessible for future generations, an interdisciplinary approach is crucial.

Keywords

Malnutrition · Wild fruits · Phytocompounds · Tribal communities · Western Ghats

V. V. Sundararajan · A. F. Nazeer · P. Thangaraj (✉)
Bioprospecting Laboratory, Department of Botany, Bharathiar University,
Coimbatore, Tamil Nadu, India

P. Thangaraj (ed.), *Pomotherapeutic Insights on Wild Edible Fruits*,
https://doi.org/10.1007/978-981-96-0543-9_12

Introduction

India, the world's second-most populous country with population nearly 1.3 billion, has achieved sustainable food grain production. Still, chronic malnutrition remains a serious issue, impacting over 195 million people worldwide, 47 million of whom are children. Improving dietary variety and nutrient consumption is critical in addressing this enduring problem. Regarding the complex relationship between environment and human health, focus has recently switched to the wide range of edible wild fruits found in many ecosystems. These nutrient-dense fruits offer a viable way to satisfy the population's daily dietary requirements, enhancing nutritional security in the process. Even though there are more than 3000 species of edible plants that are known to exist, just around 30 crops provide more than 90% of the calories consumed worldwide, and only about 120 crops are significant economically on a national level [1]. Numerous edible plant species, especially wild edible fruits, are still unknown in many of the nation's environments. These fruits could provide a sustainable approach to agriculture and diversify human diets.

One of India's most important biodiversity hotspots, the Western Ghats are home to a wide variety of wild edible fruits that could potentially alleviate nutritional insecurity among tribal and rural populations. These fruits' overall worth is increased by the numerous nutraceutical qualities they possess. Ensuring that everyone has access to enough food that is healthy, inexpensive, and produced sustainably is a major concern in the modern world. The preservation and responsible use of biodiversity become worldwide imperatives, vital to the upkeep of a healthy planet and the provision of fundamental benefits to all. Global food and nutritional security are based on the wide genetic variety seen in food crops, which creates a vital link in the food chain. Investigating underutilized fruits is a viable way to address urgent social problems like poverty and insecure health and nutrition. For tribal and other forest-dwelling populations, who depend on these wild fruits and transmit traditional knowledge about their valued qualities from one generation to the next, these neglected plants serve as a major source of nourishment. The ethnobotanical lens reveals these fruits' diverse functions, which include nutritional supplements and treatment for a range of ailments. The scientific study of the interactions between humans and plants, known as ethnobotany, provides important new information about the customs surrounding the usage of wild edible fruits [2].

Even while tropical fruit biodiversity is well known in India, there is currently a dearth of thorough research regarding the nutritional content, antioxidant potential, and other health benefits of many native and underutilized fruits. Despite the fact that certain fruits are known to provide health benefits, there is still a large amount of information lacking about these benefits. Fruit crops that are not fully utilized are vital sources of micronutrients like vitamins and minerals as well as macronutrients like proteins, lipids, and carbohydrates. They also include bioactive non-nutrients that support a healthy diet in general. In addition to being delicious, these fruits are valued for possible medicinal benefits; they are a natural pharmacy full of essential nutrients, bioactive chemicals, and folk wisdom. It becomes critical to comprehend the complex interactions between phytochemicals, cultural contexts, and contemporary scientific research that together form our knowledge of these remarkable gifts from nature as we go deeper into investigating the therapeutic potential of wild edible fruits [3].

Edible wild fruits have an abundance of bioactive substances known as phytochemicals, which are closely related to their medicinal properties. These substances include flavonoids, alkaloids, terpenes, polyphenols, and a variety of antioxidants. Each phytochemical contributes to the unique tastes, hues, and most importantly health-promoting qualities of the fruits. These fruits are rich in nutrients because, in addition to phytochemicals, they are stores of vital nutrients such as vitamins, minerals, dietary fibre, and macronutrients. This makes them essential parts of a balanced diet in addition to making them delicious [4]. Phytochemicals and minerals typically work in concert to increase the bioavailability of phytochemicals and the health advantages that follow. The potential for therapeutic benefits present in wild edible fruits is deeply ingrained in cultural customs and traditional knowledge that is passed down through the centuries. Not only have indigenous tribes worldwide relied on these fruits for nutrition, but they have also acknowledged their therapeutic qualities. The symbiotic interaction between nature and civilization is the foundation of traditional healing systems, which often use wild fruits as treatments for a wide range of diseases. It is evident how important wild edible fruits' antioxidant qualities are in shielding the body from oxidative stress.

Antioxidants are essential for the elimination of free radicals, which are unstable chemicals that can damage cells and are linked to a number of chronic illnesses. These fruits serve as a natural barrier against oxidative damage by enhancing the body's natural defence mechanisms. Many edible wild fruits have shown encouraging effects on cardiovascular health. Antioxidants, dietary fibre, and other bioactive substances work together to reduce blood pressure, reduce inflammation, and improve lipid profiles. These effects work together to lower the risk of cardiovascular illnesses [5].

Certain wild edible fruits may have neuroprotective properties, according to recent studies. These fruits contain bioactive substances called flavonoids and polyphenols, which have been related to cognitive health and may lower the risk of neurodegenerative diseases. By consuming these fruits, one may be able to prevent age-related cognitive decline and preserve brain function. Chronic inflammation has been linked to a number of illnesses, including cancer and arthritis. Edible wild fruits are highly acclaimed for their ability to reduce inflammation and provide a natural solution for inflammatory diseases. Quercetin and other compounds are found in many fruits and have been shown in preclinical research to have anti-inflammatory properties. The antioxidant qualities of wild edible fruits are closely related to their possible anticancer effects. These fruits may help prevent DNA damage and slow the growth of cancer cells by scavenging free radicals and reducing oxidative stress. Many bioactive substances that can be found in wild edible fruits, like resveratrol and ellagic acid, have been studied for their ability to cause cancer cells to undergo apoptosis. These substances have the power to cause programmed cell death, which stops malignant cells from growing and proliferating uncontrollably.

Edible wild fruits are rich in vitamins, particularly vitamin C, which is well known for strengthening the immune system. By including these fruits in one's diet, one can boost their immune system and improve their body's resistance to illnesses and infections. The possible significance of a few edible wild fruits in managing weight has been studied [6]. These fruits include compounds that may influence metabolic functions, such as appetite control and fat metabolism, which may be beneficial for people trying to keep a healthy weight. It is imperative to establish a bridge between traditional wisdom from ancient times and modern scientific study in order to properly utilize the medicinal properties of wild edible fruits. A more thorough knowledge of the medicinal qualities connected to these fruits may result from cooperative efforts that balance rigorous scientific methods with local knowledge. The growing popularity of wild edible fruits highlights the need to focus on conservation and sustainable collecting methods. The abundance and diversity of these fruits can be negatively impacted by overharvesting, habitat degradation, and the effects of climate change. Protecting biodiversity and cultural heritage requires giving top priority to the use of strategies meant to preserve and responsibly use these resources.

Distribution of the Western Ghats in Tamil Nadu

Thousands of plant species can be found in the Western Ghats, an area known for its high endemism or the fact that a large percentage of its animal and plant species are unique to this area and cannot be found anywhere else on Earth. The Western Ghats, which span a range of forest types such as semi-evergreen, tropical evergreen, moist deciduous, and montane grasslands, have had a significant influence on the cultural and historical evolution of the areas they cross. There are three primary components that make up Tamil Nadu's Western Ghats. The Nilgiri Hills, which are part of the Nilgiri Biosphere Reserve, are situated in Tamil Nadu's northern region [7]. This region, which includes the districts of Nilgiris, Coimbatore, and Erode and portions of Salem, is well known for its mist-clad peaks, shola-grassland ecosystems, and famous hill stations like Ooty and Coonoor. With a height of 2637 m, Doddabetta is the highest peak in Tamil Nadu. The Anaimalai Hills, which are located in the centre of Tamil Nadu's Western Ghats, are well known for its biodiversity, tea plantations, and Anaimudi, South India's highest peak at 2695 m. These hills, which are located in portions of Idukki district, greatly enhance the natural diversity of the area.

The Cardamom Hills, also called the Palani Hills, are a unique combination of tropical evergreen forests, grasslands, and various ecosystems that are spread over the southern region of Tamil Nadu's Western Ghats. These hills add to the state's environmental diversity, spanning areas in districts like Theni and Dindigul and portions of Virudhunagar and Tirunelveli. Tamil Nadu's Western Ghats' distribution is extremely important ecologically, since it supports the state's abundant biodiversity and environmental integrity. The Western Ghats, which include the Tamil Nadu region, are known worldwide as hotspots for biodiversity. They have a variety of ecosystems, from high-altitude grasslands to moist deciduous forests. Numerous species of plants and animals that are unique to this area are supported by this variety. The Western Ghats' high endemism level is mostly due to the distinctive biological niches that their varied topography creates. The Nilgiri Hills, which are renowned for their shola-grassland ecosystems, are especially noteworthy. At higher elevations, sholas, distinctive, stunted evergreen forests, play a vital function in controlling water flow, halting soil erosion, and offering a home to endemic plants and animals. Tamil Nadu's Western Ghats are not only a remarkable natural feature, but they also constitute an important ecological resource that faces conservation problems and calls for coordinated efforts to maintain this rare hotspot of biodiversity [8].

Wild Fruits in the Western Ghats

Nestled along Tamil Nadu's western border, the Western Ghats reveal a diverse range of habitats, lush forests, and unusual flora that combine together to create an extensive array of biodiversity. The wild edible fruits of the Western Ghats are a lesser-known gem hidden away in this lush tropical paradise. With an emphasis on their diversity, nutritional worth, cultural significance, and sustainable harvesting methods, this article seeks to explore the fascinating world of these fruits. A plethora of wild edible fruits with unique aromas, textures, and nutritional profiles may be found in Tamil Nadu's Western Ghats. For those who are ready to explore, the area offers a culinary journey, with flavours ranging from the tart notes of native fruits to the delicious sweetness of wild berries [9]. The Western Ghats are known for their high endemism, and this uniqueness extends to their edible fruits, many of which are unique to the area and contribute even more exclusivity to the local culinary scene. The dietary diversity of the indigenous communities is greatly enhanced by the presence of wild edible fruits from the Western Ghats. These fruits offer a healthy substitute for fruits that are sold in stores because they are frequently loaded with vitamins, minerals, antioxidants, and dietary fibre. Numerous wild edible fruits are prized for their therapeutic qualities in addition to their culinary appeal. These fruits are commonly used in traditional healing techniques to address a variety of health issues, demonstrating the interdependent relationship between nature and well-being. The ingestion of edible wild fruits is an integral part of the cuisines of the native people who live in the Western Ghats. Customary recipes, which have been handed down over the years, highlight the many ways in which these fruits are used in regional cuisines. Gathering and consuming wild fruits are a common part of indigenous festivals and cultural rituals, which serve to celebrate nature's bounty and promote environmental care and a sense of community. Harvesting wild edible fruits sustainably is essential to protecting the Western Ghats' fragile ecosystems. Ethical foraging principles promote ethical behaviours to guarantee the continued availability of these riches. They are based on traditional knowledge and ecological awareness [10].

Community-organized initiatives are essential to the preservation of wild edible fruits because they involve local people in environmentally friendly collecting techniques. This strategy not only protects biodiversity but also gives locals the ability to take care of their natural heritage. There are about 344 officially recognized species of wild edible fruits in the Western Ghats (Raj). The majority of these species are members of the plant families Anacardiaceae, Apocynaceae, Moraceae, Myrtaceae, Sapindaceae, and Sapotaceae. Varied parts of the Western Ghats have varied levels of variety within each family. Sadly, many of the unidentified species may become extinct before they are even named. There are still many undiscovered species out there waiting to be identified. These trees bear fruit for 2–6 months of the year every year.

Wild edible fruits are becoming more and more important because of their nutraceutical qualities and high nutritional content. They are important parts of sustainable agriculture since they are resistant to both biotic and abiotic stressors and climate change. These fruits also serve as possible gene donors in efforts to improve crops under different stress conditions. Comparable to commercially available fruits, the majority of wild edible fruits are reported to be abundant in dietary nutrients, such as proteins, carbs, vitamins, and minerals. They play a critical function in helping tribal cultures vary their diets and ward off deficiency illnesses. Both raw and processed varieties of these fruits are consumed, helping to meet daily calorie requirements. Wild fruits that are edible are important for human nutrition because they contain minerals, dietary fibre, vitamins C, A, and E, thiamine, niacin, pyridoxine, and folacin, as well as carbs and proteins [11]. They also have a lot of medicinal potential. Interestingly, in addition to their immediate qualities, like protein and carbs, they also contain comparatively large amounts of other vital minerals, such calcium, manganese, iron, copper, and others. With a balanced combination of macro- and micronutrients that are vital for good health, fruits stand out as a key source of nutrition in human diets. Fruits' high calorie content and sweetness are attributed to their abundance of fast-absorbing sugars. The micronutrient content and dietary fibres of fruits are principally responsible for their nutritional value. They provide a variety of minerals, such as magnesium, phosphorus, manganese, iron, calcium, and potassium. Furthermore, fruits are a good source of provitamin A in the form of β-carotenoids, vitamin C, vitamin E, and B group vitamins. Antioxidants, such as carotenoids, anthocyanins, and phenolic compounds, as well as vitamins C and E, can alter the way that carcinogens are metabolically activated and function as anti-ageing agents [12].

Based on scientific research, eating fruit on a daily basis may lower your risk of heart disease and other malignancies. Because fruits contain a high concentration of the precursors of bicarbonate ions, which act as buffers against acidity, they also help to maintain the body's natural pH. As excellent sources of macro- and micronutrients, wild fruits have been used historically to supplement the diets of a wide range of human societies [13]. Recent years have seen an increase in interest in researching and analysing the nutritional qualities of new crops, particularly wild fruits. Fruit consumption is encouraged generally because new research constantly shows how healthy and nutrient dense they are. According to recent studies, a wide variety of native fruits may be able to satisfy rural communities' nutritional demands.

Tribal Settlements of the Western Ghats

The residences of indigenous communities in Tamil Nadu are tucked away among the verdant vegetation and foggy surroundings of the Western Ghats, where they have managed to preserve their rich cultural legacy and peaceful coexistence with the natural world. These tribal communities, which are frequently tucked away in the hills and forests, offer a distinct perspective on the variety of Indian culture. The tribal communities in the Western Ghats have a long history and represent a generation-long bond with the land. These native groups have coexisted peacefully with the Western Ghats, depending on its resources for both cultural activities and food [14]. There are numerous tribal communities in the Western Ghats, each with its own language, traditions, and customs. Prominent tribes in the area that contribute to the rich cultural diversity of the Western Ghats are the Irulas, Paniyas, Kurumbas, Kanikkars, Badagas, etc., and many cultural customs of the tribal people residing in the Western Ghats are indicative of their strong bond with the natural world. An essential component of their culture is their traditional dances, rituals, music, and artwork, which frequently centre around religious festivals, hunting seasons, and agricultural cycles. The Western Ghats are home to several indigenous populations, many of which speak languages other than those of the general public. One essential component of their cultural identity that has been carried down through the ages as an intangible heritage is the preservation of oral traditions, myths, and folklore. The primary source of income for tribes in the Western Ghats is agriculture.

Shifting cultivation is a method used by the tribes, who cultivate tiny areas of land for a few years and then let them return to their original state of vegetation. Their historic notion of the harmony between forest conservation and agriculture is in line with this sustainable practice. Tribal tribes rely heavily on the woods of the Western Ghats for their means of subsistence. They gather non-timber forest products, including as honey, traditional crafts, bamboo, and medicinal plants. Their way of life depends on sustainable harvesting methods, which guarantee the availability of these resources for future generations [15]. Nestled hidden in the varied scenery of Tamil Nadu's Western Ghats, the relationship between tribal people and edible wild fruits becomes an engrossing story of mutual reliance. Over the course of generations, these aboriginal communities have developed a profound understanding of the intricate ecosystems that surround them in addition to relying on the abundance of natural resources for their livelihood. In this investigation, we embarked on a journey through the lush landscapes and traditional practices of the Western Ghats, revealing the intricate connections between foraged fruits and the indigenous tribes residing in this area.

Generation after generation, the tribal groups living in the Western Ghats have been passing on a comprehensive knowledge of the local flora. The identification, gathering techniques, and culinary applications of wild edible fruits are all part of this traditional ecological knowledge, which forms the basis for their peaceful coexistence with the environment. For tribal societies, wild edible fruits are more than just food; they play a significant role in their culinary customs. These fruits add unique flavours and textures to the many tribal cuisines of the Western Ghats, whether they are eaten fresh, dried, or processed into jams and pickles. Apart from their culinary value, wild fruits are essential for maintaining the dietary health of indigenous populations. Rich in nutritional fibre, vitamins, minerals, and antioxidants, these fruits add variety and balance to a diet. Wild edible fruits add nutritional diversity to the tribal diet in areas where access to other food sources may be limited [16].

Tribal societies have included wild edible fruits into their spiritual and cultural traditions. These fruits are frequently offered to gods or served as a communal feast during festivals that commemorate the harvest season or certain ecological phenomena. These customs highlight how important wild fruits are to the tribe's spiritual and cultural beliefs. Numerous fruits have symbolic implications in tribal customs and folklore. The jackfruit, for example, is frequently regarded as a symbol of fertility and wealth, while other fruits could be connected to folktales that impart ecological knowledge and cultural values. The indigenous populations residing in the Western Ghats have developed a profound comprehension of the therapeutic qualities present in naturally occurring edible fruits. Many fruits are thought to have medicinal properties that can help with anything from respiratory disorders to digestive troubles. These communities' healthcare procedures heavily rely on this ancient healing understanding. Additionally, studies examining the ethnopharmacological characteristics of wild fruits that can be eaten have revealed the existence of bioactive substances that may have therapeutic use in the Western Ghats [17]. These investigations demonstrate the fruits' potential for therapeutic use and provide a link between conventional wisdom and contemporary scientific validation.

Distribution of Tribal Communities in the Western Ghats of Tamil Nadu

There are numerous tribal communities living in Tamil Nadu's Western Ghats, each with its own unique cultural character. Among them are prominent groups such as the Irulas, Paniyas, Kurumbas, Kanikkars, Badagas, and many others who together add to the region's rich cultural fabric. These tribes enhance the Western Ghats' cultural tapestry by adding distinctive components to the social structure. Tribal groups in the Western Ghats have deeply ingrained customs that form their socio-economic basis. Pastoralism, shifting farming, and the gathering of non-timber forest products are the main sources of livelihood sustenance. Recognizing the resilience displayed by these communities in the face of modern difficulties requires a thorough comprehension of these socio-economic structures. Nestled in the Western Ghats and recognized as a UNESCO World Heritage Site, the Nilgiri Biosphere Reserve in Tamil Nadu, Kerala, and Karnataka provides essential habitat for a variety of tribal people, including the Paniyas, Kanikkars, Irulas, Kurumbas, and Badagas. Analysing their existence within the reserve provides important context for understanding the mutually beneficial link that exists between tribal traditions and the preservation of vital ecosystems.

Wild Edible Fruits Used by the Irula Tribes of the Western Ghats

The Irula tribes have been an essential component of the Western Ghats for generations, developing a distinct cultural identity and extensive history in the process. Nestled in the shade of verdant hills and dense forests, the Irula people have developed a close relationship with the many natural resources in the area, especially the wide variety of edible wild fruits. They are a small tribal people in southeast India that belong to the Dravidian linguistic group. The Indian government has recognized them as a scheduled tribe (ST). They were originally hunter-gatherers who relied on forest items for their daily sustenance and medicinal requirements. Historically, they were seminomadic and dispersed throughout the Nilgiri Biosphere Reserve. Known as the ultimate jungle people, the Irula people are a treasure trove of information about plants and animals. Poochamarathur, Baralikadu, Nellimarathur, Sethumadai, Shengalur, Nellithurai, Situkunni, Surki, Kodiyur, Neeradi, Veerakal, Anaipallam, Chinnalacombai, Thondai, and Mannar are among the 15 villages inhabited by the Irula tribes. Their primary source of income is the collection of non-timber forest products (NTFPs), as they are dependent on forest resources for their existence. The forest provides for many of their everyday needs, and they often trade forest products in the marketplace. Their food security is also influenced by agriculture, and a number of the villages such as Poochamarathur and Neeradi being the most populous are dedicated to the collecting of honey. The old people in the village are quite knowledgeable about forest plants, and they use a large range of them for food and income.

The Irula tribes in the Western Ghats, which experience diverse seasons, have adapted to the special presents of wild edible fruits that these seasons provide. Gaining an understanding of this seasonal diversity is essential to understanding the dynamic relationship that exists between the Irula tribes and the variety of fruits that are available all year round. Their extensive traditional knowledge includes how to recognize, gather, and prepare tasty wild fruits. The Irula tribes' diverse diet is greatly influenced by these fruits, which guarantee a balanced intake of vital nutrients such vitamins, minerals, and antioxidants. Many of the edible wild fruits they eat are said to have therapeutic qualities, which support folk remedies for conditions ranging from respiratory disorders to stomach troubles. This information is frequently passed down orally through customs. The sustainable collecting methods used by the Irula tribes guarantee the life of wild edible fruits. Conventional practices, such as selective harvesting, leaving plant portions whole, and adhering to seasonal cycles, support the preservation of the species and the habitats [18]. The list of wild edible fruits utilized by Irula tribes is listed in Table 1.

Table 1 Wild edible fruits utilized by the Irula tribes of the Western Ghats

SL. No	Plant name	Vernacular name	Family	Phytocompounds	Medicinal properties	References
1.	*Alangium salviifolium* (L.f.) Wangerin	Alanji	Cornaceae	Cephaeline, deoxytubulosine, alangine, campesterol, myristic acid, emetine, anabasine, ecosyne, phytol	Antineoplastic, antieczematic, antischistosomal, antihelmintic, antibacterial, antiglaucomic, antiasthmatic	[24]
2.	*Artocarpus hirsutus* Lam.	Wild jack	Moraceae	Artocarpin, tannic acid, ellagic acid, gallic acid, beta-sitosterol, myristic acid	Antineoplastic, antiulcerative, anti-inflammatory, antiparasitic, antileukaemic	[25]
3.	*Clausena heptaphylla* Wight & Arn.	Karanpha	Rutaceae	Myrcene, estragole, beta-pinene, carbazole	Antieczematic, antineurotic, antiuremic, antipyretic, antiparasitic	[26]
4.	*Cochlospermum religiosum* (L.) Alston	Silk-cotton tree	Bixaceae	Safrole, naringetol, aldobiouronic acid, neochlorogenic acids, hydrocinnamic acid	Antieczematic, antiulcerative, antiischaemic, antithrombotic, antineoplastic, antiviral, antihypercholesterolemic	[27]
5.	*Elaeocarpus tuberculatus* Roxb.	Rudraksham	Malampinnai	Kaempferol, ethyl gallate, gallic acid, n-hexadecanoic acid	Antihaemorrhagic, antineoplastic, anti-inflammatory, antiallergic, antifibrinolytic, antioxidant, antidiabetic	[28]

Wild Edible Fruits Used by the Paniya Tribes of the Western Ghats

The Paniya tribes have a close and nuanced relationship with the abundant natural resources that surround them, especially the wild, edible fruits that thrive in this biodiverse region. These edible wild fruits are much more than just a basic food source; they are an integral part of the Paniya tribes' culture. Many fruits have symbolic meanings and are included in a variety of ceremonies, festivals, and everyday practices. Harvesting these fruits has become more than just a utilitarian necessity; it has become an ingrained cultural custom that strengthens their ties to the environment and their ancestry. The Paniya community's festivities and rituals heavily rely on wild, edible fruits. When some fruits become abundantly available, the fruit-harvesting season becomes a time for joyful celebration and group cohesion. The act of offering these fruits to the gods expresses gratitude for the land's generosity. Stories, which have been passed down through the years, are extremely important in forming the tribe's cultural legacy and highlighting the value of wild edible fruits above and beyond their nutritional value.

Edible wild fruits have cultural and nutritional value beyond measure, and they have a direct impact on the socio-economic dynamics of the Paniya tribes. Many Paniya people engage in the collection and trade of non-timber forest products, such as edible fruits found in the wild. This approach contributes to the local economy in addition to ensuring their survival. Overabundance of wild fruits often turns into a kind of money or commodity for trade, which can be used as goods for barter. Paniyas may trade these fruits in neighbourhood markets, promoting business dealings both inside and outside of their community. The Paniya tribes hold a wealth of traditional knowledge regarding the seasonality, harvesting techniques, and long-term planning related to foraging for wild fruits. Based on this age-old knowledge, the Paniyas skilfully manage the varying fruit seasons. By harvesting fruits at their peak, they maximize their flavour and nutritional content. The Paniyas intentionally leave parts of the plant intact in order to promote natural regeneration through the use of selective harvesting techniques. The biodiversity of the Western Ghats is preserved in part by such sustainable methods [19]. The list of wild edible fruits utilized by Paniya tribes is listed in Table 2.

Wild Edible Fruits Used by the Kurumba Tribes of the Western Ghats

The Kurumbas are an indigenous group living in the Western Ghats who rely heavily on the wide variety of wild fruits that they eat. This reliance affects their food habits, customs, and way of life in general. Wild edible fruits play an important role in the diet of the Kurumba tribe because of a variety of factors, including its nutritional worth, cultural importance, and vital function in preserving the community's well-being. The Kurumbas rely heavily on wild edible fruits for their diet because they provide essential elements that other food sources might not provide. These

Table 2 Wild edible fruits used by the Paniya tribes of the Western Ghats

Sl. No	Plant name	Vernacular name	Family name	Phytocompounds	Medicinal properties	References
1.	*Acronychia pedunculata* Miq.	Mutta nari	Rutaceae	Carvacrol, myrcene, p-cymene	Antieczematic, antipyretic, anti-inflammatory, antihypercholesterolemic, antipsoriatic, antiulcerative, antituberculosic, antiosteoporotic, antiparkinsonian, antineoplastic, antiarthritic	[29]
2.	*Bridelia retusa* (L.) A.Juss.	Malai-vengai	Phyllanthaceae	Ellagic acid, gallic acid, beta-sitosterol	Antineoplastic, antiviral, antipyretic, antinociceptive, antiviral, anti-inflammatory, antirickettsial, antipsoriatic	[30]
3.	*Phyllanthus emblica* L.	Nelli	Phyllanthaceae	Geraniin, isocorilagin, procyanidin, proanthocyanidin, leucodelphidin, mucic acid 6-methyl ester 3-*O*-gallate	Antihaemorrhagic, antineoplastic, antifibrinolytic, antiviral, antileukaemic, antiosteoporotic, antiulcerative, anti-inflammatory, antipyretic, antidiabetic, antithyroid, antituberculosic	[31]
4.	*Melastoma malabathricum* L.	Indian rhododendron	Melastomataceae	Myristic acid, lauric acid, stigmasterol, cyanidin dihexoside	Antieczematic, antiulcerative, antiuremic, antipyretic, anti-inflammatory, antihyperlipoproteinemic, antipsoriatic, antiparkinsonian, anticataract, antileprosy, antituberculosic, antithyroid, antirickettsial	[32]
5.	*Nicandra physalodes* (L.) Gaertn.	Apple of Peru	Solanaceae	beta-Sitosterol, withanicandrin, solanesol	Antihypercholesterolemic, antieczematic, antipsoriatic, antiosteoporotic, antiparkinsonian, antineoplastic, antileukemic	[33]
6.	*Solanum pseudocapsicum* L.	Jerusalem cherry	Solanaceae	Solacapine, sinomenine, 9,12-octadecadienoic acid (*Z*, *Z*)-, methyl ester, cedran-diol	Antiosteoporotic, antineoplastic, antipsoriatic, antihypercholesterolemic	[34]
7.	*Solanum indicum* L.	Brihati	Solanaceae	Indioside F, diosgenin, indioside A, protodioscin	Anti-inflammatory, antieczematic, antihypercholesterolemic, antileukaemic, antipsoriatic	[35]

8.	*Carica papaya* L.	Pappali	Caricaceae	β-Cryptoxanthin, thiamine, myristic acid, heptadecanoic acid, carpaine	Antieczematic, antiviral, antiulcerative, antiuremic, antipyretic, antipsoriatic, antiparkinsonian, anticataract, antileprosy, antihelmintic	[36]
9.	*Momordica charantia* L.	Pagakkai	Cucurbitaceae	Taiwacin A, myrtenol, 1-penten-3-O-L, diosgenin	Anti-inflammatory, antineurotic, antirickettsial, antieczematic, antihypercholesterolemic, antileukaemic, antipsoriatic antihypercholesterolemic, antiulcerative	[37]
10.	*Passiflora edulis* Sims	Passion fruit	Passifloraceae	β-Myrcene, D-limonene, 2-tridecanone, diacetone alcohol, stearic acid, vomifoliol	Antiulcerative, anti-inflammatory, antiuremic, antipsoriatic, antiviral, antipyretic, antifibrinolytic, anticataract, antihelmintic, antituberculosic, antileprosy, antithyroid, antihyperlipoproteinemic, antineoplastic	[38]
11.	*Sida cordata* (Burm.f.) Borss. Waalk.	Kurunthotti	Malvaceae	Phenethylamine, malvalic acid	Antiparkinsonian, antifibrinolytic, antihelmintic, antileprosy, antithyroid, antirickettsial, antiulcerative, anticataract, antituberculosic, antidiabetic, antihypertriglyceridemic, antineoplastic	[39]

fruits contribute to a complete nutritional profile because they are rich in vital components such vitamins, minerals, antioxidants, and dietary fibre. Eating a wide array of wild fruits ensures that the Kurumbas get a range of nutrients that are essential to maintaining their general health. These fruits provide a comprehensive source of nutrition for the tribe since they are rich in fibre, potassium, vitamin C, vitamin A, and other phytochemicals.

The Kurumba tribe's cultural tradition is deeply rooted in the consumption of wild, edible fruits. Many fruits are deliberately included into their ceremonies, rituals, and daily practices and have symbolic connotations. Gathering wild fruits is more than just a means of obtaining food; it is a cultural custom that strengthens the tribe's ties to the environment and its ancestors. Wild fruit rituals are frequently connected to harvest seasons, acting as thanksgiving for the land's bountiful produce and building a sense of community among participants. Like many other aboriginal communities, the Kurumbas actively engage in sustainable harvesting practices. They use environmentally friendly harvesting methods and adhere to traditional understanding regarding the seasonal availability of various wild fruits. The Kurumbas' commitment to maintaining the delicate balance between human needs and environmental conservation is further demonstrated by their commitment to sustainable practices. Eating wild fruits has cultural and nutritional value, but for the Kurumba tribe, it also has economic value. Overabundance fruits are regularly exchanged or sold in the surrounding marketplaces, generating revenue for the neighbourhood. The financial exchanges associated with gathering wild fruits contribute to the local economy and may have an impact on more general aspects of the Kurumbas' subsistence [20]. The list of wild edible fruits utilized by Kurumbas tribes is listed in Table 3.

Wild Edible Fruits Used by the Kanikkar Tribes of the Western Ghats

The Kanikkar tribe has a strong bond with the forests and emphasizes their ancestral rights to these lands. They are located in the southern tribal zone along the southeastern slopes of the Western Ghats. The word "Kanikkar" itself refers to their inherited ownership of the land. These native people have unique physical qualities; they are frequently small in stature, and some of them have Negroid traits. The Kanikkars were once engaged in migratory cultivation, but this activity has mostly been abandoned over time. The community has moved from relying on migrant farming for their livelihoods to other sources of income. Many Kanikkar tribal people are generally knowledgeable about medicinal plants; they use them for first aid treatments, for common illnesses like colds, fevers, headaches, and coughs, as well as for minor injuries like poisonous bites. They gather honey, bee wax, and other small forest products during the harvest season. The Kanikkars also grow economic crops like cashew, areca nut, and pepper as well as edible plants like tapioca, bananas, and millets. Kanyakumari district villages such as Mothiramalai, Thottamalai, Chappangupparai, Koovaikadu of Thiruvattar, Kothaiyar, Alamparai, Koruvakuzhi, Puravilai, Valayamthanki, Vannachipparai, and Thovalai are primarily inhabited by the Kaani people, another

Table 3 Wild edible fruits used by the Kurumba tribes of the Western Ghats

Sl. No	Plant name	Vernacular name	Family	Phytocompounds	Medicinal properties	References
1.	*Garcinia gummi-gutta* (L.) N.Robson	Kodaikaapulli	Clusiaceae	Citric acid, d-tartaric acid, hydroxycitric acid, lactone, caprylic acid, linoleic acid	Antidiabetic, antifibrinolytic, antipsoriatic, antiparkinsonian, antithyroid, antirickettsial, anticataract, antiulcerative, antiosteoporotic, antituberculosic, anti-inflammatory	[40]
2.	*Citrullus colocynthis* (L.) Schrad.	Kumatti	Cucurbitaceae	Myristic acid, dotriacontane, cucurbitacin	Antieczematic, antiviral, antihypercholesterolemic, antiulcerative, antipyretic, antipsoriatic, antiparkinsonian, anticataract, antituberculosic, antithyroid, antidiabetic	[41]
3.	*Cucumis melo* L.	Mulambalam	Cucurbitaceae	Benzyl alcohol, acetoin, 2,3-butanediol, lauric acid	Antiviral, antidiabetic, antiparkinsonian, antipyretic, anti-inflammatory, antituberculosic, anticataract, antithyroid, antiulcerative, antiosteoporotic, antirickettsial, antileprosy, antileukaemic, antiglaucomic	[42]
4.	*Juglans regia* L.	Akarottu	Juglandaceae	1,2,3-Trimethylbenzene, syringic acid, octanal	Antipyretic, antipsoriatic, antiuremic, antiallergic, antituberculosic, antirickettsial, antithyroid, antiulcerative, antidiarrhoeal, anticataract	[43]
5.	*Lantana camara* L.	Unnichedi	Verbenaceae	Hexadecane, beta-bisabolene, eugenol	Anticataract, antipsoriatic, antituberculosic, antiasthmatic anti-inflammatory, antiulcerative, antibacterial, antithyroid	[44]

(continued)

Table 3 (continued)

Sl. No	Plant name	Vernacular name	Family	Phytocompounds	Medicinal properties	References
6.	*Helicteres isora* L.	Itampuri	Malvaceae	Helisorin, diosgenin, flavone, 5,8 dihydroxy-7,4-dimethoxy flavone, hibifolin	Anti-inflammatory, antieczematic, antileukaemic, antipsoriatic, antiosteoporotic, antipyretic antihypercholesterolemic, antidiabetic, antiulcerative, antiallergic, antirickettsial	[45]
7.	*Glycosmis pentaphylla* (Retz.) DC.	Amutam	Rutaceae	Phlobaphene, mupamine, 1-methyl-9(10H)-acridinone, glycopentalone	Antihypoxic, antinociceptive, antineoplastic, antieczematic, antidiarrhoeal, antileukaemic, antirickettsial, antiparkinsonian, antituberculosic, antileprosy	[46]
8.	*Naringi crenulata* (Roxb.) Nicolson	Magavilvam	Rutaceae	Estragole, 1,3,4,5-tetrahydroxycyclohexanecarboxylic acid	Antiviral, antieczematic, antihelmintic, antipsoriatic, antihypercholesterolemic, antiulcerative, antiglaucomic, antidiabetic	[47]
9.	*Physalis peruviana* L	Manathakkali keerai	Solanaceae	Riboflavin, retinol, methyl-2-methylbutyrate, isophorone, hydrocinnamic alcohol	Antiviral, antipsoriatic, antiulcerative, antieczematic, antileukaemic, antihelmintic, antiosteoporotic, antituberculosic, antidiabetic, antirickettsial	[48]
10.	*Solanum viarum* Dunal	Tropical soda apple	Solanaceae	Solasodine, solamargine, solasonine, farnesol	Anti-inflammatory, anticarcinogenic, antileukaemic, antineoplastic, antinociceptive, antihypercholesterolemic, antieczematic	[49]

indigenous group in the area. There are between 50 and 100 Kaani households living in each of the forest regions of Pechiparai Hills, Perunchani Hills, Arrukani Hills, Kothaiyaru, Keeriparai, and Pathukani Hills [21]. These groups make use of the abundant biodiversity found in the Kanyakumari area Wildlife Forest, where the Kaani tribes' primary healthcare practices heavily rely on medicinal plants. These medicinal plants are used to cure a wide range of illnesses, such as skin conditions, hepatitis, diabetes, piles, ulcers, snake and bug bites, piles, wounds, and respiratory conditions like asthma. The tribes use leaves, fruits, seeds, roots, bark, whole plants, flowers, stems, rhizomes, bulbs, and gum as their most often consumed plant parts. Although many people view traditional herbal knowledge as being kept hidden, modern exchanges of ideas and conversations have encouraged the sharing of this information. The two Kaani groups' use value indices differ, according to a quantitative examination of survey data. The retention of traditional herbal knowledge within families for both religious and financial reasons, as well as for economic purposes, is closely linked to cultural behaviours. The exchange of this knowledge illustrates how the relationship between customary practices and the larger framework of community health and well-being is changing [22]. The list of wild edible fruits utilized by Kanikkar tribes is listed in Table 4.

Wild Edible Fruits Used by the Badaga Tribes of the Western Ghats

The wild edible fruits that grow in the surrounding forests of the Badaga tribe have a deep and intricate relationship with the tribe. Their dietary habits, economic survival, and cultural heritage all play a significant role in this relationship. The Badaga people have a long-standing custom of consuming edible wild fruits that has been handed down through the generations. These fruits have symbolic meaning and are frequently used in customs related to food, celebrations, and rituals, which enhance the community's cultural fabric. The Badaga people rely heavily on wild edible fruits for their diet because they are a rich source of vital vitamins, minerals, and other nutrients. They enhance the community's agricultural produce, adding variety to their diet and promoting general health and wellbeing. In terms of economy, the Badagas rely on wild edible fruits as an additional source of income. Local markets can be used to sell extra fruit, giving residents another source of income and bolstering their community's economic stability. Furthermore, the Badagas' relationship with edible wild fruits demonstrates their profound awareness of and reverence for the environment. Sustainable harvesting methods protect these fruits and the ecosystems they live in, encouraging community environmental stewardship and biodiversity. In summary, there are many facets to the relationship between the Badaga tribe and wild edible fruits, including cultural, nutritional, economic, and environmental aspects. Their cultural identity and bond with the land they call home are strengthened in addition to meeting their basic needs [23]. The list of wild edible fruits utilized by the Badaga tribes is listed in Table 5.

Table 4 Wild edible fruits used by the Kanikkar tribes of the Western Ghats

Sl. No	Plant name	Vernacular name	Family	Phytocompounds	Medicinal properties	References
1.	*Goniothalamus wightii* Hook.f. & Thomson	Goniothalamus	Annonaceae	beta-Bisabolene, carvacrol, myrcene, viridiflorol, sabinene	Antieczematic, antineoplastic, anti-inflammatory, antiulcerative, antinociceptive, antihypercholesterolemic, antipsoriatic	[50]
2.	*Psychotria adenophylla* Wall.	Samiruka	Rubiaceae	Betulin, bauerenyl acetate, ursolic acid	Antineoplastic, antieczematic, antileukaemic, antiviral, antipsoriatic, antihypercholesterolemic, antihelmintic, antiosteoporotic, antiparasitic, anti-inflammatory	[51]
3.	*Aegle marmelos* (L.) Corrêa	Vilvam, bael	Rutaceae	Umbelliferone, marmin, isoimperatorin	Antihelmintic, antiuremic, antineoplastic, antiparasitic, antiosteoporotic, antifibrinolytic, antiparasitic, antiulcerative	[52]
4.	*Citrus limon* (L.) Osbeck	Elumichai, lemon	Rutaceae	Myrcene, linalool, naringin	Antieczematic, antiulcerative, antituberculosic, antihypercholesterolemic, antithyroid, anticataract, antibacterial	[53]
5.	*Rubia cordifolia* L.	Manjishtha, Indian madder	Rubiaceae	1,4-Dihydroxy-2-carboethoxy anthraquinone, rubiatriol	Antiuremic, antieczematic, antihypercholesterolemic, antituberculosic, antiulcerative, antineoplastic	[54]
6.	*Sapindus mukorossi* Gaertn.	Indian soapberry	Sapindaceae	Stigmasterol, saponin, stearic acid	Antihypercholesterolemic, antipsoriatic, antiosteoporotic, antiviral, antileukaemic	[55]
7.	*Semecarpus anacardium* L.f.	Marking nut tree	Anacardiaceae	Anacarduflavanone, catechol	Antihelmintic, antiuremic, anti-inflammatory, antituberculosic	[56]
8.	*Solanum virginianum* L.	Indian nightshade	Solanaceae	Solasonine, scopolin	Antieczematic, antineoplastic, antiprotozoal, antileukaemic, antirickettsial, antidiabetic	[57]
9.	*Syzygium cumini* (L.) Skeels	Jamun, Malabar plum	Myrtaceae	Ellagic acid, quercetin, maslinic acid, gallic acid	Antineoplastic, antiviral, antipyretic, antinociceptive, antiviral, anti-inflammatory, antirickettsial	[58]

Table 5 Wild edible fruits used by the Badaga tribes of the Western Ghats

S. No	Plant name	Vernacular name	Family	Phytocompound	Medicinal properties	References
1.	*Annona reticulata* L.	Ramachita	Annonaceae	Michelalbine, carvacrol, 1,4-cineole, perillyl alcohol, limonene	Antiparkinsonian, antiischaemic, antineoplastic, antinociceptive, antidepressant, antiepileptic, antiulcerative, antihelmintic, antipyretic, Anti-inflammatory, antihypercholesterolemic	[59]
2.	*Berberis tinctoria* Lesch.	Oosikala	Berberidaceae	Berbamine, palmatine, berberine, jatrorrhizine	Antineurotic, antituberculosic, antileukaemic, antiparkinsonian, antipyretic	[60]
3.	*Physalis minima* L.	Balloon cherry	Moraceae	Hyperoside, withaminimin	Antihypercholesterolemic, anti-inflammatory, antidiabetic, antiulcerative, antileukaemic, antituberculosic, antithyroid, anticataract, antirickettsial	[61]
4.	*Ficus carica* L.	Simaiyatti	Moraceae	Cadalene, angelicin, methoxsalen, catechin	Antieczematic, anti-inflammatory, antipyretic, antiulcerative, antipsoriatic, antileprosy, antihelmintic, antithyroid, antituberculosic, anticataract antihypercholesterolemic, antirickettsial	[62]
5.	*Ficus exasperata* Vahl	Maramthinni atthi	Moraceae	Bergaptol, carotol, oxalic acid, malic acid, fumaric acid	Antihelmintic, anti-inflammatory, antihypercholesterolemic, antipsoriatic, antiparkinsonian, antipyretic, antimyopathy, antituberculosic, antirickettsial, antiulcerative	[63]
6.	*Morus alba* L.	Kambli chedi	Moraceae	Cholesterol, campesterol, beta-sitosterol, catechin, syringic acid	Antihypercholesterolemic, antieczematic, antipsoriatic, antiosteoporotic, antiparkinsonian, antineoplastic, antileukaemic	[64]

(continued)

Table 5 (continued)

S. No	Plant name	Vernacular name	Family	Phytocompound	Medicinal properties	References
7.	*Persicaria chinensis* (L.) H.Gross	Creeping smartweed	Polygonaceae	Kaempferol, 3-O-methylellagic acid	Antihaemorrhagic, anti-inflammatory, antihypercholesterolemic, antiulcerative, antileukaemic, antiosteoporotic, antituberculosic, antiallergic, antithyroid	[65]
8.	*Rubus ellipticus* Sm.	Indian raspberry	Rosaceae	Octacosanoic acid, rubitic acid, ursolic acid, gallic acid, ferulic acid	Anti-inflammatory, antiulcerative, antileukaemic, antifungal, antipsoriatic, antiarthritic, antidiabetic, antiosteoporotic, antileukaemic	[66]
9.	*Solanum sisymbriifolium* Lam.	Sticky nightshade	Solanaceae	Choline, nuatigenin, isonuatigenin, solacaproine	Anti-inflammatory, antifungal, antieczematic, antipsoriatic, antineoplastic, antiosteoporotic, antiparkinsonian, antiulcerative, anticataract, antiglaucomic	[67]

Conclusion and Future Prospects

A fruit-rich diet is necessary for human nutrition. As a result, the growing number of people who are consuming fruits with medicinal and nutritional benefits is driving up their demand. Micronutrients, various phytochemical substances, and dietary supplements could be obtained from these underutilized fruits. Many natural ingredients in them include antibacterial, anti-inflammatory, antidiabetic, and antioxidant properties. Additionally, underutilized fruits may be a good source of fibre, vitamins, fat and carbohydrates. These fruits may contain more nutritious contents than their native and exotic counterparts. The utilization of these fruits presents a challenge or obstacle because a large quantity is needed if the industry is to use them. The other issue associated with using these crops as dietary supplements is that toxicity profiling of the underutilized edible fruits requires a lot of research. Strengthening botanical knowledge, germplasm collection, improvement, production, and processing technologies, expanding the supply of improved planting materials, and encouraging on-farm cultivation of wild edible fruit-based agroforestry systems are essential future strategies for domestication and more comprehensive cultivation of underutilized edible fruit species.

In summary, the wild edible fruits of Tamil Nadu's Western Ghats are not only a culinary treat but also a microcosm of the complex interrelationships between sustainability, culture, and the environment. The goal of this review is to shed light on the many aspects of these fruits, including their nutritional worth, therapeutic qualities, cultural significance, and difficulties. Conservationists and fruit enthusiasts as well should take note of and respect the abundance of wild edible fruits found in the Western Ghats. It is important to make sure that future generations continue to enjoy the flavours of these undiscovered gems while maintaining the ecological integrity of this UNESCO World Heritage Site by integrating traditional knowledge, scientific research, and community involvement and embracing our responsibility to take care of the vibrant and resilient ecosystems of the Western Ghats and celebrate the bounty of nature as we set out on this culinary and ecological journey. Empowering these communities is critical to the future of edible wild fruits relationships with tribal communities.

References

1. Kaur S, Roy A. A review on the nutritional aspects of wild edible plants. Curr Tradit Med. 2021;7(4):552–63.
2. Sreekumar VB, Sreejith KA, Hareesh VS, Sanil MS. An overview of wild edible fruits of Western Ghats, India. Genet Resour Crop Evol. 2020;67:1659–93.
3. Karuppusamy S, Muthuraja G, Rajasekaran KM. Antioxidant activity of selected lesser known edible fruits from Western Ghats of India. Indian J Nat Prod Resources. 2011;2:174–8.
4. Deshmukh BS, Waghmode A. Role of wild edible fruits as a food resource: traditional knowledge. Int J Pharm Life Sci. 2011;2(7).
5. Valvi SR, Rathod VS, Yesane DP. Screening of three wild edible fruits for their antioxidant potential. Curr Bot. 2011;2(1):48–52.

6. Saad B. A review of the anti-obesity effects of wild edible plants in the mediterranean diet and their active compounds: from traditional uses to action mechanisms and therapeutic targets. Int J Mol Sci. 2023;24(16):12641.
7. Vignesh A, Amal TC, Sivalingam R, Selvakumar S, Vasanth K. Influence of ecological factors on the phytochemical composition and bioactivity of Berberis tinctoria Lesch. wild edible fruits. Biochem Syst Ecol. 2024;112:104771.
8. Komala HP, Prasad AD, Deepashree CL, Gopal S. Bioactivities of selected edible fruits of Western Ghats region of Karnataka. J Adv Sci Res. 2022;13(04):23–30.
9. Jose R, Praveena S, Anilkumar C, Jothish PS. Edibility and nutrient analysis of Diospyros crumenata Thw. fruit: an additional fruit resource from the Western Ghats, India. Genet Resour Crop Evol. 2024;1–6.
10. Zepelin KA. Foraging culture: ethics, practice, and identity among contemporary wild food foragers in the Southwest United States (Doctoral dissertation, University of Colorado at Boulder).
11. Mahapatra AK, Mishra S, Basak UC, Panda PC. Nutrient analysis of some selected wild edible fruits of deciduous forests of India: an explorative study towards non conventional bio-nutrition. Adv J Food Sci Technol. 2012;4(1):15–21.
12. Prakash D, Upadhyay G, Gupta C, Pushpangadan P, Singh KK. Antioxidant and free radical scavenging activities of some promising wild edible fruits. Int Food Res J. 2012;19(3).
13. Sardeshpande M, Shackleton C. Wild edible fruits: a systematic review of an under-researched multifunctional NTFP (non-timber forest product). Forests. 2019;10(6):467.
14. Basavarajaiah DM, Narasimhamurthy B, Bharathi M, Naik J. Tribal livelihood status in Western Ghats. Fores Res. 2020;9:234.
15. Roy N, Debnath A, Nautiyal S. Climate change, shifting cultivation and livelihood vulnerabilities in India.
16. Devi VM, Kalirajan K, Prasad PN. Existence of vitamin C fruit sources in the Kani tribal settlements of south TamilNadu. J Phytology. 2010;2(4).
17. Silambarasan R, Sureshkumar J, Krupa J, Amalraj S, Ayyanar M. Traditional herbal medicines practiced by the ethnic people in Sathyamangalam forests of Western Ghats, India. Eur J Integr Med. 2017;16:61–72.
18. Aruna R, Nithyapriya J, Ramachandran VS, Gopakumar K, Ramaswamy RS. A study on the sustainable utilization of edible plants by Irular tribes of Anaikatty, Western Ghats, India. Res J Pharmacogn Phytochem. 2015;7(2):95–100.
19. Ramachandran VS, Udhayavani C. Knowledge and uses of wild edible plants by Paniyas and Kurumbas of Western Nilgiris, Tamil Nadu. Indian J Nat Prod Resour. 2013;4(4):412–8.
20. Alex A, Vidyasagaran K, Geethakutty PS. Gender dimensions in NTFP collection among the indigenous communities of Western Ghats of India. Int J Usuf Mngt. 2016;17(1):24–32.
21. Rani SL, Devi VK, Soris PT, Maruthupandian A, Mohan VR. Ethnomedicinal plants used by Kanikkars of agasthiarmalai biosphere reserve, Western Ghats. J Ecobiotechnol. 2011;3(7).
22. Sukumaran S, Sujin RM, Geetha VS, Jeeva S. Ethnobotanical study of medicinal plants used by the Kani tribes of Pechiparai Hills, Western Ghats, India. Acta Ecologica Sinica. 2021;41(5):365–76.
23. Sathyavathi R, Janardhanan K. Wild edible fruits used by Badagas of Nilgiri District, Western Ghats, Tamilnadu, India. J Med Plants Res. 2014;8(2):128–32.
24. Hongal AM, Shettar AK, Hoskeri JH, Vedamurthy AB. GCMS-based phytochemical profiling and in vitro pharmacological activities of plant Alangium salviifolium (Lf) Wang. Futur J Pharm Sci. 2024;10(1):1–9.
25. Mali SN, Yadav S, Pandey A, de Oliveira MS, Jawarkar RD. A decade review of analysis of essential oils from genus artocarpus: its phytochemistry and pharmacology. Pharm Res-Natl Prod. 2024;100016.
26. Peerakam N, Phoowiang P, Chansakaow S, Thongpoon C, Hadpech S. Chemical profiling revealed a dominant compound trans-anethole and biological evaluation of an edible plant Clausena harmandiana containing essential oil. Nat Prod. 2022;16(2):118–27.

27. Altemimi A, Lakhssassi N, Baharlouei A, Watson DG, Lightfoot DA. Phytochemicals: extraction, isolation, and identification of bioactive compounds from plant extracts. Plan Theory. 2017;6(4):42.
28. Balamurugan V, Sridhivya M, Dharani R, Selvakumar S, Vasanth K. Phytochemical screening, antioxidant, antidiabetic and anticancer activities of Elaeocarpus variabilis fruit. Turk J Agric Food Sci Technol. 2022;10(8):1352–62.
29. Panyasawat P, Wisetsai A, Lekphrom R, Senawong T, Schevenels FT. Acroquinolones A and B, two polyphenolic isoprenylated acetophenone-quinolone hybrids with anti-proliferative activities from Acronychia pedunculata (L.) Miq. Nat Prod Res. 2022;36(11):2743–52.
30. Patil PS, Jadhav VD. Bioactive potential of Bridelia retusa (L.) A. Juss. plant: a wild edible plant. J Drug Res Ayu Sci. 2022;7(1):55–61.
31. Ruhela G, Dhama P, Shanker K, Ding X, Sharma A. Exploration of metabolomics of Phyllanthus emblica: from Ayurveda food to modern prospectives in quality control. Food Chem Adv. 2023;3:100330.
32. Aslam MS, Ahmad MS, Ahmad MA. An up-to-date review on phytochemical constituents and pharmacological activities of Melastoma malabathricum. Int J Pharm Sci Res. 2017;8:76–91.
33. Wang LiQing WL, Wang Ying WY, Gao SuYu GS, Zhu LiHan ZL, Wang Fei WF, Li Hua LH, Chen LiXia CL. Phenolic amides with anti-Parkinson's disease (PD) effects from Nicandra physaloides. J Funct Foods. 2017;31:229–36.
34. Mahadev R, Ramakrishnaiah H, Krishna V, Kumar NN. Chemical composition of the essential oil from the fruits of Solanum erianthum D. Don. J Essent Oil-Bear Plants. 2012;15(3):387–91.
35. El-Aasr M, Miyashita H, Ikeda T, Lee JH, Yoshimitsu H, Nohara T, Murakami K. A new spirostanol glycoside from fruits of Solanum indicum L. Chem Pharm Bull. 2009;57(7):747–8.
36. Rivera-Pastrana DM, Yahia EM, González-Aguilar GA, Hermosillo S. Identification of phenolic and carotenoid compounds in Carica papaya L.'Maradol'Using liquid chromatography—mass spectrometry. Acta Hortic. 2010;877(877):1197–204.
37. Lin KW, Yang SC, Lin CN. Antioxidant constituents from the stems and fruits of Momordica charantia. Food Chem. 2011;127(2):609–14.
38. Barbosa Santos T, De Araujo FP, Neto AF, De Freitas ST, de Souza AJ, de Oliveira Vilar SB, Brito Araújo AJ, Lima MS. Phytochemical compounds and antioxidant activity of the pulp of two brazilian passion fruit species: Passiflora cincinnata Mast. and Passiflora edulis Sims. Int J Fruit Sci. 2021;21(1):255–69.
39. Gulnaz AR, Savitha G. Phytochemical evaluation of leaf and stem extracts of siddha medicinal plant: sida cordata. J Evol Med Dent Sci. 2013;2(15):2514–22.
40. Krishnakumar NM, Shiney G. Chemical composition and biological activities of seed butter extracted from Garcinia gummi-gutta (L.) Roxb. Acta Pharm Sci. 62(2).
41. Gurudeeban S, Ramanathan T, Satyavani K. Characterization of volatile compounds from bitter apple (Citrullus colocynthis) using GC-MS. Int J Chem Anal Sci. 2011;2(8):108–10.
42. Dagli MM, Yeldan PA, Sevindik O, Guclu G, Kelebek H, Selli S. Investigation of aroma compounds of queen Anne's pocket melon (Cucumis melo L. ssp dudaim) Juice. J Raw Mater Process Foods. 2022;3(2):74–82.
43. Croitoru A, Ficai D, Craciun L, Ficai A, Andronescu E. Evaluation and exploitation of bioactive compounds of walnut, Juglans regia. Curr Pharm Des. 2019;25(2):119–31.
44. Prakash VS, Radhakrishnan N, Vasantha-Srinivasan P, Veeramani C, El Newehy AS, Alsaif MA, Al-Numair KS. In silico analysis of selected nutrition rich fruit of Bunch berry (Lantana camara) constituents as human acetylcholinesterase (hAchE), carbonic anhydrase II (hCA-II) and carboxylesterase 1 (hCES-1) inhibitory agents. Saudi J Biol Sci. 2023;30(12):103847.
45. Kumar N, Singh AK. Plant profile, phytochemistry and pharmacology of Avartani (Helicteres isora Linn.): a review. Asian Pac J Trop Biomed. 2014;4:S22–6.
46. Sreejith PS, Asha VV. Glycopentalone, a novel compound from Glycosmis pentaphylla (Retz.) Correa with potent anti-hepatocellular carcinoma activity. J Ethnopharmacol. 2015;172:38–43.

47. Singh N, Meena MK, Patni V. Phytochemical profiling and GC-MS analysis of bioactive constituents of callus of Naringi crenulata (Roxb.) Nicolson. Int J Pharm Sci Rev Res. 2014;24(1):29–34.
48. Yilmaztekin M. Analysis of volatile components of cape gooseberry (Physalis peruviana L.) grown in Turkey by HS-SPME and GC-MS. Sci World J. 2014;2014.
49. Confortin TC, Todero I, Luft L, Teixeira AL, Mazutti MA, Zabot GL, Tres MV. Valorization of Solanum viarum dunal by extracting bioactive compounds from roots and fruits using ultrasound and supercritical CO 2. Braz J Chem Eng. 2020;(36):1689–702.
50. Vino P, Santhosh C, Maruthupandian A, Viji M, Vadivu CC, Murugesh S, Murugan P. Chemopreventive effects of natural bioactive compounds from the genus goniothalamus: a review and perspectives. In: Phytochem pharmacoogical aspects of ethnomedicinal plants; 2021. p. 269–87.
51. de Carvalho Jr AR, De Carvalho MG, Braz-Filho R, Vieira IJ. Psychotria genus: chemical constituents, biological activities, and synthetic studies. Stud Nat Prod Chem. 2016;48:231–61.
52. Venthodika A, Chhikara N, Mann S, Garg MK, Sofi SA, Panghal A. Bioactive compounds of Aegle marmelos L., medicinal values and its food applications: a critical review. Phytother Res. 2021;35(4):1887–907.
53. Saini RK, Ranjit A, Sharma K, Prasad P, Shang X, Gowda KG, Keum YS. Bioactive compounds of citrus fruits: a review of composition and health benefits of carotenoids, flavonoids, limonoids, and terpenes. Antioxidants. 2022;11(2):239.
54. Ali A, Aslam M, Chaudhary SS. A review a review on pharmacognostic and therapeutic uses of Rubia cordifolia. J Drug Deliv Therap. 2020;10(6):195–202.
55. Ling Y, Zhang Q, Zhong W, Chen M, Gong H, He S, Liang R, Lv J, Song L. Rapid identification and analysis of the major chemical constituents from the fruits of Sapindus mukorossi by HPLC-ESI-QTOF-MS/MS. Nat Prod Res. 2020;34(15):2144–50.
56. Nikam YP. Potent ethanomedicinal plant Semecarpus anacardium Linn: a review. Syst Rev Pharm. 2022;13(3).
57. Saraswathi K, Bharkavi R, Khusro A, Sivaraj C, Arumugam P, Alghamdi S, Dablool AS, Almehmadi M, Bannunah AM, Sahibzada MU. Assessment on in vitro medicinal properties and chemical composition analysis of Solanum virginianum dried fruits. Arab J Chem. 2021;14(12):103442.
58. Madani B, Mirshekari A, Yahia EM, Golding JB, Hajivand S, Dastjerdy AM. Jamun (Syzygium cumini L. Skeels): a promising fruit for the future. Hortic Rev. 2021;48:275–306.
59. Soleman DM, Eldahshan OA, Ibrahim MH, Ogaly HA, Galal HM, Batiha GE, Elkousy RH. GC/MS analysis, cytotoxicity, and antiviral activities of Annona glabra hexane extract supported by in silico study. Molecules. 2023;28(4):1628.
60. Gaurav ZS, Parveen B, Ibrahim M, Sharma I, Sharma S, Sharma AK, Parveen R, Ahmad S. TLC-MS bioautography-based identification of free-radical scavenging, α-amylase, and α-glucosidase inhibitor compounds of antidiabetic tablet BGR-34. ACS Omega. 2020;5(46):29688–97.
61. Wu J, Zhang T, Yu M, Jia H, Zhang H, Xu Q, Gu Y, Zou Z. Anti-inflammatory Withanolides from Physalis minima. ACS Omega. 2020;5(21):12148–53.89.
62. Wojdyło A, Nowicka P, Carbonell-Barrachina ÁA, Hernández F. Phenolic compounds, antioxidant and antidiabetic activity of different cultivars of Ficus carica L. fruits. J Funct Foods. 2016;(25):421–32.
63. Mouho DG, Oliveira AP, Kodjo CG, Valentão P, Gil-Izquierdo A, Andrade PB, Ouattara ZA, Bekro YA, Ferreres F. Chemical findings and in vitro biological studies to uphold the use of Ficus exasperata Vahl leaf and stem bark. Food Chem Toxicol. 2018;112:134–44.
64. Bang IS, Park HY, Yuh CS, Kim AJ, Yu CY, Ghimire B, Lee HS, Park JG, Choung MG, Lim JD. Antioxidant activities and phenolic compounds composition of extracts from mulberry (Morus alba L.) fruit. Korean J Med Crop Sci. 2007;15(2):120–7.

65. Seimandi G, Álvarez N, Stegmayer MI, Fernández L, Ruiz V, Favaro MA, Derita M. An update on phytochemicals and pharmacological activities of the genus Persicaria and Polygonum. Molecules. 2021;26(19):5956.
66. Kewlani P, Tiwari D, Rawat S, Bhatt ID. Pharmacological and phytochemical potential of Rubus ellipticus: a wild edible with multiple health benefits. J Pharm Pharmacol. 2023;75(2):143–61.
67. Popova I, Sell B, Pillai SS, Kuhl J, Dandurand LM. High-performance liquid chromatography–mass spectrometry analysis of glycoalkaloids from underexploited solanum species and their acetylcholinesterase inhibition activity. Plan Theory. 2022;11(3):269.

Improving the Microbiological, Sensory, and Chemical Constituents of Stored Crystallized "Jenipapo" (*Genipa americana* L.), a Native Brazilian Fruit

Raquel P. Campos, Marana Soares, Karuppusamy Arunachalam, Suni Liu, Luciana Miyagusku, and Priscila Aiko Hiane

Abstract

The preparation of processed products from native fruits involves providing nutritional products and functions with greater period of marketing and value addition to contribute to environmental preservation. This study aimed to evaluate the production of crystallized candy of jenipapo with two types of cuts, subjected to two drying temperatures, and evaluate the physicochemical, microbiological, and sensory appearance attributes, after 4 months of storage. The fruits of jenipapo (*Genipa americana* L.) were collected in the municipality of Miranda-MS. The samples were selected, washed, peeled, and diced and sliced. Impregnation was performed using sucrose and glucose (3:1), gradually increasing the soluble solids of the syrup concentrations (40, 50, 60, and 70 °Brix). After boiling for 3 min, the samples were stored for 8–12 h between each impregnation. Type dryer was used cabin at 45–60 °C. Processing reduced producers of titratable acidity, vitamin C, phenols, and tannins present in fresh fruits, although the use of cutting into slices has maintained these levels greater than cubed. Treatments with drying temperatures 45 and 60 °C did not affect the quality and storability. The candy jenipapo crystallized has great sensorial acceptance in the taste, texture, and appearance attributes. The candy had adequate sanitary hygienic conditions, after the storage period.

R. P. Campos · M. Soares · S. Liu · L. Miyagusku · P. A. Hiane
Faculty of Pharmaceutical Sciences, Food and Nutrition (FACFAN), Federal University of Mato Grosso do Sul Foundation (UFMS), Mato Grosso do Sul, Brazil

K. Arunachalam (✉)
Center for Studies in Stem Cells, Cellular Therapy and Toxicological Genetics (CeTroGen), Faculty of Medicine (FAMED), Federal University of Mato Grosso do Sul (UFMS), Mato Grosso do Sul, Brazil

P. Thangaraj (ed.), *Pomotherapeutic Insights on Wild Edible Fruits*,
https://doi.org/10.1007/978-981-96-0543-9_13

Keywords

Genipa americana L. · Total phenols · Total tannins · Sensory analysis

Introduction

The Cerrado region is home to a wide variety of fruit species that provide significant agricultural potential, and these fruit species have been traditionally employed by the local population for both consumption and the sustainable development of communities [1]. The Cerrado fruits possess unique flavors characterized by elevated concentrations of carbohydrates, fibers, vitamins, and minerals and a multitude of health advantages, encompassing antioxidant, antibacterial, anticarcinogenic, anti-degenerative, and aging retardant attributes of bioactive substances [2–4]. The food sector has the potential to utilize these fruits in order to produce lucrative commodities such as juices, jellies, ice cream, fruit pastes, and substitutes for conventional oils, capitalizing on their unique flavors and nutritional value. The evaluation of consumption and the development of novel products rely on essential methodologies, such as comprehending the chemical characteristics and nutritional composition of these fruits [5–7].

The jenipapo (*Genipa americana L.*), a member of the Rubiaceae family, is regarded as indigenous to Brazil while not exclusively found in the country. It is distributed in several regions of Brazil, including the Amazon, Caatinga, Cerrado, Atlantic Forest, and Pantanal biomes. The tree exhibits a maximum height of 30 m, characterized by its smooth bark and glossy foliage. The blossoms of this plant exhibit a cream hue against a yellowish backdrop, while the mature fruits display a brownish-gray coloration. The fruit is produced between October and February. The pulp of the fruit is pliable and succulent, making it ideal for producing juices, liqueurs, ice creams, cakes, and confections [7, 8]. Jenipapo's stem bark and fruits have historically been utilized for their therapeutic qualities, exhibit advantageous technological attributes, and possess a high nutritious composition, rendering it well suited for the manufacturing of food items [9, 10].

The fruits were collected at maturity, dehydrated in a 60 °C oven, pulverized, and transformed into jenipapo pulp flour with the following conditions: pH 3.85 ± 0.830, water activity (Aw) 0.401, moisture 7.86 g/100 g, ash 3.38 g/100 g, lipids 1.12 g/100 g, proteins 1.12 g/100 g, carbohydrates 84.82 g/100 g, soluble solids (SS) 64.66 °Brix, and total energy value (TEV) 360.85 g/100 g. Also the mineral composition was founded as potassium, magnesium, calcium, iron, phosphorus, sulfur, and manganese [11]. The primary constituents of the lipid fraction found in jenipapo pulp are essential fatty acids, specifically linoleic acid (35.54%) and linolenic acid (25.66%). The desired ratio of n-6 to n-3 is regarded advantageous as it enhances the bioavailability, metabolism, and integration of these fatty acids into membrane phospholipids [12].

Monoterpenoids isolated from jenipapo fruits have been identified as iridoids, including genipin, geniposidic acid, geniposide [13], genamesides A–D [14], genipacetal, genipamide, and genipaol [15]. The ethanolic extract of *G. americana* fruits included steroids that inhibited the growth of choriocarcinoma placenta cells, indicating antiproliferative effects [16].

Fruit crystallization is a time-honored and significant method for preserving food. It facilitates the preservation of excess produce and allows for fruit consumption during periods when the season is not in season. Crystallization is a method used in the sweet processing industry to convert fruits into delectable and durable goods, such as jams and preserves [17, 18]. The crystallization process entails the continuous and slow infusion of sugars (syrup) into the fruit until it reaches a concentration that inhibits its decay, usually beyond 65 °Brix, thereby prolonging its shelf life. Osmotic dehydration of jenipapo using sucrose solutions has been found to enhance acceptance in comparison to fresh fruit [19]. Additionally, it has been discovered that this method can avoid nutrient loss throughout the drying process, as seen in pumpkins, while also preserving carotenoids more effectively [20].

The World Health Organization (WHO) and the Brazilian Food Guide recommend 400 g of fruits and vegetables per day, which 90% of the Brazilian population does not meet. The south area has the greatest average per capita food consumption of fruit-based sweets, at 5.1 g/day; the elderly consume 2.8 g/day, followed by adults and adolescents (2.4 and 1.6 g/day, respectively). There is a positive relationship between their consumption and income levels, ranging from 1.4 g/day in the lowest income class to 3.1 g/day in the highest income class [21].

This study aimed to assess the preparation of crystallized jenipapo sweets using two different drying temperatures and two different types of cuts. The objective was to evaluate the physical, chemical, sensorial, and microbiological characteristics of the sweets after being stored under ambient conditions, taking into account the consumption potential of fruit-based sweets and the potential for utilizing native fruits in products with an extended shelf life.

Materials and Methods

The fruits were obtained from the Bananal Indigenous village located in the Taunay district of Miranda-MS municipality. The fruits underwent a washing process in flowing water at the UFMS Food Laboratory, followed by natural drying on filter paper. Subsequently, the fruits were peeled, and the seeds were extracted. The pulps underwent a process of slicing, with each fruit yielding 8 slices and 16 cubes.

The completely randomized design utilized plots with a sample size of 100 g, with six replications per treatment. The syrup was prepared using a sucrose to glucose (powder) ratio of 3:1, with an initial concentration of 40 °Brix. Subsequently, the syrup was introduced to the pulps in a mass-to-mass ratio of 2:1.

The syrup samples were positioned on heated plates and subjected to boiling for a duration of 3 min. Subsequently, they were allowed to rest for a period of 2 h, with intermittent stirring to facilitate impregnation. Following this, the methods were replicated using syrup concentrations of 40, 50, 60, and 70 °Brix/plot. The final impregnation involved the addition of sucrose only, as glucose was found to be challenging to dissolve.

Subsequently, the syrup was disposed of, and the pulps were promptly immersed (10 s) in hot water (65 °C), with any surplus water being eliminated. The specimens were organized into nylon grids, positioned on trays, and exposed to a cabin-type drier with air circulation for the purpose of drying. The drying process was continued until the samples attained a final humidity level of 10%, employing a mass loss calculation. This operation was performed at regular intervals of 30 min. The samples were stored at a temperature of 60 ± 2 °C for a duration of 5 h in cube form, while those in slices were kept for an additional 40 min. The cubes and slices were subjected to a temperature of 45 °C for a duration of 8 h and 9.5 h, respectively.

Once the product had attained ambient temperature, it was coated with a layer of crystal sugar and subsequently enclosed within polypropylene bags, with each procedure being repeated six times. The preserves were stored at a temperature of 27 ± 3 °C for a duration of 4 months.

Two replications and triplicate chemical analysis were conducted on fresh fruits and newly processed sweets to assess the efficacy of the crystallization process. The analysis was performed after 2 months and 4 months of ambient storage. The soluble solids, pH, titratable acidity, and ascorbic acid levels were assessed using the procedures established by the Adolfo Lutz Institute in Brazil [22].

The aqueous extracts were acquired following the methodology proposed by Roesler et al. [23] for the purpose of conducting the following determinations: (a) total phenols using the Folin-Ciocalteu method [24], with the results expressed in milligrams of gallic acid equivalents (EAG) per 100 g of fresh sample, and (b) tannins using the Folin-Denis method [25], with the results expressed in milligrams of tannic acid equivalents (EAT) per 100 g of fresh sample. The objective of this study was to compare the impact of treatments administered after a storage period of 4 months.

The sensory, microbiological, and water activity quality of the kept sweets was assessed after a storage period of 4 months. The water activity (aw) of the jenipapo sweets was measured using a Decagon hygrometer. The sample was collected from both the surface and interior of the jenipapo sweets, and the measurement was conducted at a temperature of 20 °C.

The sensory analysis assessed the characteristics of aroma, flavor, texture, and overall appearance using a nine-point hedonic scale. This scale ranges from "I really liked" to "I really disliked," with the term "I didn't even like it" falling in between. According to Brasil [25], the samples were assigned three-digit codes and subsequently randomized. These samples were then presented to a group of 50 judges, who were asked to evaluate their level of preference or aversion toward each sample.

Eaton [26] conducted an evaluation of sanitary hygienic conditions in samples of fresh jenipapo pulp and crystallized sweets. The assessment involved the enumeration of coliforms at a temperature of 45 °C and the identification of *Salmonella* spp. A study on *Salmonella* was conducted using a 1% buffered peptone saline solution for pre-enrichment and three different selective enrichment media: cystine selenite broth, tetrathionate broth, and Rappaport-Vassiliadis broth. The bacteria were then differentially selectively plated on three different media: Hektoen agar,

Salmonella-Shigella agar, and Rambach agar. The plates were incubated at a temperature of 36 ± 1 °C for 24 h. The data underwent analysis of variance, and the means were compared using the Tuckey test ($p \geq 0.05$) with the assistance of the BioEstat 5.0 statistical software [27].

Result and Discussion

The fruits had an average weight of 100 g, and the firm pulp utilized to make the sweet accounted for 33.92 ± 3.41% of the total fruits. When crystallized genipap sweets were cut into cubes, it exhibited a significantly higher yield of around 190%, irrespective of the drying temperature employed. The slices at 45 °C had the lowest yield of 156.47%, which can be attributed to the extended drying duration employed at this temperature.

The moisture content of the fresh fruit was found to be excessive at 79.90%. However, the moisture content was reduced through the application of the crystallization process, resulting in moisture values ranging from 6% to 8.4% in the freshly processed samples. These samples exhibited an increase in moisture over a period of 4 months when stored at a temperature of 27 ± 3 °C, as seen in Table 1. This observation could potentially be attributed to the sealing mechanism employed in the packaging, specifically polypropylene bags. Andrade et al. [28] reported a comparable moisture content of 81.53% in fresh jenipapo. However, in the fruit that underwent osmotic dehydration, the humidity increased to 20.89%, surpassing the moisture content observed in crystallized genipap sweets. This can be attributed to the enhanced impregnation of sucrose and sugar crystals in the coating during the osmotic dehydration process.

The solubility of solids exhibited a range of 74.7–80.9 °Brix, irrespective of the applied treatments, as indicated in Table 1. Overall, a decrease in these levels was noted following the storage duration, which aligns with an elevation in moisture content. Typically, the levels of soluble solids in gabiroba jelly rose over a period of 180 days of storage [20].

The fresh fruit had elevated average titratable acidity values, measuring 0.94 g malic acid. The processing of 100 g^{-1} resulted in a significant decrease, with dry cubes achieving values of 0.26 and 0.31 at 45 °C and 60 °C, respectively, and 0.35 g acid. malic. The slices exhibited a uniform weight of 100 g^{-1}, irrespective of the temperatures employed (Table 1).

Following a storage period of 4 months, the sliced samples exhibited elevated acidity levels, with recorded values of 0.42 and 0.39 g acid. malic. The sample was subjected to drying temperatures of 45 °C and 60 °C, resulting in a mass of 100 g^{-1}. The quantity of samples cut into cubes was found to be 0.34 g acid. malic. The value of 100 g^{-1} remains unaffected by variations in drying temperature [20], and it was also noted an elevation in the titratable acidity level of gabiroba jelly over a 180-day storage period.

Sousa et al. [29] observed comparable attributes in the banana variety Prata (*Musa sapientum* L.), including a pH value nearing 4.0, total soluble solids ranging from 70 to 75 °Brix, water activity ranging from 0.70 to 0.80, and content humidity

Table 1 Physical and chemical characteristics of crystallized jenipapo fruits and sweets, with two types of cut and subjected to two drying temperatures

	Treatments			
Characteristics	Cubes 45 °C	Cubes 60 °C	Slices 45 °C	Slices 60 °C
Performance (%)	190.92 ± 5.50	189.39 ± 6.03	156.47 ± 4.58	174.00 ± 6.31
Moisture (%) Fresh fruit 79.90%				
Post processed	6.00 ± 0.00	6.60 ± 0.00	6.00 ± 0.00	8.40 ± 0.00
2 months	10.05 ± 0.00	9.00 ± 0.00	10.25 ± 0.00	10.25 ± 0.00
4 months	12.50 ± 0.40	12.40 ± 1.25	11.83 ± 0.46	12.43 ± 0.32
Soluble solids (°Brix) Fresh fruit 23.90				
Post processed	79.0 ± 0.00	77.0 ± 0.00	80.9 ± 0.00	79.9 ± 0.00
2 months	77.9 ± 0.00	74.7 ± 0.00	78.1 ± 0.00	76.6 ± 0.00
4 months	77.5 ± 0.00	77.6 ± 0.00	77.2 ± 0.00	75.9 ± 0.00
Titratable acidity (g acid. malic. 100 g^{-1} sample) Fresh fruit 0.94				
Post processed	0.26 ± 0.00	0.31 ± 0.00	0.35 ± 0.00	0.35 ± 0.00
2 months	0.25 ± 0.00	0.27 ± 0.00	0.31 ± 0.00	0.30 ± 0.00
4 months	0.34 ± 0.00	0.34 ± 0.00	0.42 ± 0.00	0.39 ± 0.00
Vitamin C (mg acid. ascorbic. 100 g^{-1} sample) Fresh fruit 34.40				
Post processed	1.23 ± 0.00	1.49 ± 0.00	1.68 ± 0.00	0.92 ± 0.00
2 months	0.92 ± 0.00	0.92 ± 0.00	0.92 ± 0.00	0.93 ± 0.00
4 months	0.76 ± 0.00	0.61 ± 0.00	0.74 ± 0.00	0.72 ± 0.00
Water activity				
After 4 months	0.78 ± 0.00[a]	0.79 ± 0.00	0.76 ± 0.00	0.78 ± 0.00

[a]Not significant ($p \leq 0.05\%$). The data underwent analysis of variance, and the means were compared using the Tuckey test ($p \geq 0.05$)

ranging from 15% to 25%. These values align with the established range for fruits with moderate humidity and stability when stored at room temperature for a duration of up to 4 months.

The fresh fruits had an average vitamin C content of 34.40 ± 9.7 g of ascorbic acid. The pulp content of 100 g^{-1} was significantly decreased during the processing stage, indicating that the cutting and drying temperatures did not have any impact on the results. After 4 months of storage, the cubes treated at 60 °C had the lowest vitamin C content, with a value of 0.61 g of acid. ascorbic.100 g^{-1} sample. The other treatments had varying amounts of vitamin C, ranging from 0.72 to 0.76 g of acid. ascorbic.100 g^{-1} pulp (Table 1).

In terms of the chemical features of the fresh fruits, appreciable values of soluble solids, titratable acidity, and vitamin C were detected in relation to those reported by Figueiredo et al. (1986) in jenipapo picked at the stage near to ripening in the region of Maranguape-CE, with titratable acidity of 0.98 g citric acid.100 g^{-1}, total soluble solids of 20 °Brix, traces of vitamin C, and pH 4.0. The study conducted by Silva et al. [30] focused on ripe fruits that were stored for 7 days at room temperature and 14 days in a refrigerator. The fruits exhibited titratable acidity levels ranging from 0.22 to 0.14 g of acid. citrus 100 g^{-1}, the pulp sample consists with soluble solids ranging from 15 °Brix and a vitamin C content ranging from 1.0 to 0.35 mg of acid ascorbic a quantity of pulp weighing 100 grams per unit.

Genetic diversity, variations in latitude, and differences in soil and meteorological conditions at the production site contribute to the numerous variations observed in the physical and chemical composition of fresh vegetable products. Hansen et al. [31] observed significant diversity in the chemical properties of native jenipapo of the Reconceive region of Bahia that the TA content ranged from 1.40% to 1.42%, while the SS content ranged from 17.18 to 16.8 °Brix. Additionally, the acid ascorbic content ranged from 2.76 to 2.65 mg.100 g^{-1}. These findings suggest that jenipapo can be economically utilized for both fresh consumption and industrial purposes.

Following a storage period of 4 months, it was observed that the levels of total phenolic compounds and total tannins were found to be higher in sliced slices of crystallized jenipapo sweets in comparison to those cut into cubes. The increased contact surface area of the cubes resulted in a higher rate of chemical loss to the syrups, as depicted in Fig. 1.

The amounts of phenols and total tannins, which are chemicals with antioxidant activity and are considered significant for consumer health, were not affected by the drying temperature [32]. The mean initial level of fresh fruits, which was 229.23 ± 22.34 mg EAG.100 g^{-1}, exhibited a decrease of 88% following processing in sweet slices and 82.4% and 82.9% in cubes, which were dried at temperatures of 45 °C and 60 °C, respectively. Donado-pestana et al. [24] reported a comparable decrease in the levels of total phenolic compounds across four biofortified sweet potato cultivars following various heat treatments, including steam, boiling, roasting, and flouring.

A high concentration of tannins can be regarded as a factor that hinders nutrition. The processing of breadfruit, cashew nuts, and pumpkin seeds, including cooking, fermenting, germinating, and toasting, resulted in a decrease in tannin levels ranging from 26.1% to 61.7% [33].

The types of cuts and drying temperature used to obtain crystallized jenipapo sweets did not significantly affect the water activity of the products after 4 months of storage, with values between 0.76 and 0.79 (Table 1).

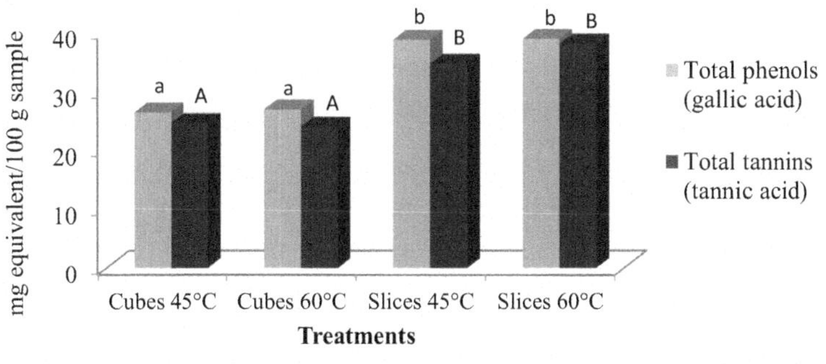

Fig. 1 Content of total phenolic compounds and total tannins in crystallized jenipapo sweets, with two types of cut and subjected to two drying temperatures, after 4 months of storage. Statistically significant differences ($p < 0.05$) among the various treatments. The data underwent analysis of variance, and the means were compared using the Tuckey test

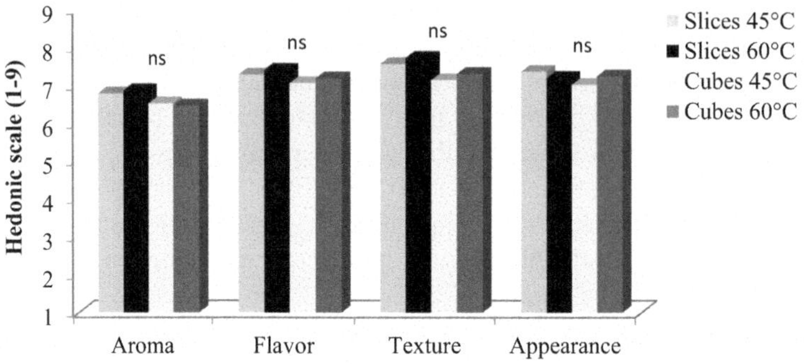

Fig. 2 Comparison of sensory attributes in crystallized jenipapo sweets, with two types of cut and subjected to two drying temperatures, after 4 months of storage. (score 6 = I liked it slightly; 7 = I liked it moderately; and 8 = I liked it a lot) ($n = 50$). Statistically significant differences ($p < 0.05$) among the various treatments. ns: not significant. The data underwent analysis of variance, and the means were compared using the Tuckey test ($p \geq 0.05$)

The sensory analysis of crystallized jenipapo sweets, following a storage period of 4 months, revealed no statistically significant differences ($p < 0.05$) among the various treatments that were administered.

The product received widespread approval in terms of its flavor, texture, and look, with ratings ranging from 7 to 8 on the hedonic scale, which represent "I liked it moderately" and "I liked it a lot"). The aroma attribute presented scores above 6, which is equivalent to "I liked it slightly." Similarly, great acceptance was found by Andrade et al. [28] in osmotically dehydrated jenipapo at 60 °Brix, with sucrose and demerara sugar solution at 60 °C, among other treatments (Fig. 2).

During the assessment of fresh pulp and crystallized jenipapo sweet samples following storage, it was observed that the values fell below the quantification limit of the coliform enumeration method at a temperature of 45 °C. Additionally, the search for *Salmonella* sp. was also conducted. The outcome was nonexistent in all assessed samples. The obtained results agree with the microbiological criteria set out by RDC 12, ANVISA-MS for coliforms at a temperature of 45 °C (NMP/g) of 5×10^2 for fresh fruit and 1×10^2 for processed fruit. Furthermore, the absence of *Salmonella* spp. (in 25 g) is also seen [34]. The demonstration of sufficient hygienic and sanitary conditions during the fruit and product collection process was established [35]. Assis et al. [36] observed that cashew jelly remained microbiologically stable for 4 months when stored at room temperature. Similarly, Andrade et al. [28] showed that osmotically dehydrated jenipapo, held for 90 days at a temperature of 28 °C ± 3 °C, did not contain *Salmonella* spp. or thermotolerant coliforms.

Conclusion

The crystallization of jenipapo sweets exhibits a significant decrease in the concentrations of titratable acidity, vitamin C, phenols, and total tannins found in the fresh fruits. Nevertheless, the utilization of sliced cuts retains these elevated levels in comparison to cubes. The quality and storability of this product are unaffected by the utilization of drying temperatures ranging from 45 to 60 °C. After being stored for 4 months, crystallized jenipapo sweets are highly regarded for its sensory attributes, including flavor, texture, and appearance. It is also considered a microbiologically safe product.

References

1. Moreira-Araújo RS, Barros NV, Porto RG, Brandão AD, Lima AD, Fett R. Bioactive compounds and antioxidant activity three fruit species from the Brazilian Cerrado. Rev Bras Frutic. 2019;3(41):e-011.
2. Oliveira-Alves SC, Pereira RS, Pereira AB, Ferreira A, Mecha E, Silva AB, Serra AT, Bronze MR. Identification of functional compounds in baru (Dipteryx alata Vog.) nuts: nutritional value, volatile and phenolic composition, antioxidant activity and antiproliferative effect. Food Res Int. 2020;131:109026.
3. Borges PR, Edelenbos M, Larsen E, Hernandes T, Nunes EE, Boas EV, Pires CR. The bioactive constituents and antioxidant activities of ten selected Brazilian Cerrado fruits. Food Chem X. 2022;14:100268.
4. Silva JF, Januário JG, Guedes CT, Friedrichsen JD, da Silva LA, de Freitas MM, Felipe MP, Silva PG, da Silva Scapim MR, Madrona GS. Cerrado fruits, study on their technological and medicinal applications: a review. Seven Editora; 2023.
5. Costa ED, Sousa PH, Siqueira AC, Figueiredo EA, Gouveia ST, Figueiredo RW, Maia CS, Gomes DS. Fruit pastes with organic honey texturized with gellan gum: bioacessibility of antioxidant activity and sensory analysis fruit pastes with gellan and organic honey. Food Sci Technol. 2019;39:667–76.
6. Almeida AB, Silva AK, Lodete AR, Egea MB, Lima MC, Silva FG. Assessment of chemical and bioactive properties of native fruits from the Brazilian Cerrado. Nutr Food Sci. 2019;49(3):381–92.
7. Resende NS, de Souza VR, Carvalho EE, de Jesus Junqueira JR, Boas EV. Stability and antioxidant activity of bioactive compounds in Cerrado fruit juices during storage. Res Soc Dev. 2022;11(8):e38211831043.
8. Filardi FL, Barros FD, Baumgratz JF, Bicudo CE, Cavalcanti TB, Coelho MA, Costa AF, Costa DP, Goldenberg R, Labiak PH, Lanna JM. Brazilian Flora 2020: innovation and collaboration to meet Target 1 of the Global Strategy for Plant Conservation (GSPC). Rodriguésia. 2018;69:1513–27.
9. Gomes M. Genipa in Flora do Brasil 2020. Jardim Botânico do Rio de Janeiro; 2020. https://floradobrasil2020.jbrj.gov.br/FB14045.
10. de Sousa Júnior DL, da Silva Benjamim ÍM, Teotônio LE, Gonçalves FJ, Salviano CM, Leandro RC, Lopes MJ, de Aquino PE, Leite NF. Efeito antimicrobiano e modulador do extrato hidroalcoólico de *Genipa americana* (Jenipapo). Saúde (Santa Maria); 2019.
11. Cardoso DR, Pinto LI, de Lima MA, Soares IF, da Silva Rocha FP, da Silva RA, Viana VG. Potencial tecnológico e composição de Farinha de Jenipapo (Genipa americana L.) obtida por secagem em convecção. Braz J Dev. 2020;6(6):33448–67.
12. Cardoso DR, Lima MA, Rocha FP, Silva RA, Viana VG. Composição centesimal, físico-química e mineral da farinha de jenipapo (Genipa americana L.). Hig Aliment. 2019;3321–5.
13. Costa PA, Ballus CA, Teixeira Filho J, Godoy HT. Fatty acids profile of pulp and nuts of Brazilian fruits. Food Sci Technol. 2011;31:950–4.
14. Djerassi C, Gray JD, Kincl FA. Naturally occurring oxygen heterocyclics. IX. 1 Isolation and characterization of genipin2. J Org Chem. 1960;25(12):2174–7.
15. Ono M, Ueno M, Masuoka C, Ikeda T, Nohara T. Iridoid glucosides from the fruit of Genipa americana. Chem Pharm Bullet. 2005;53(10):1342–4.
16. Ono M, Ishimatsu N, Masuoka C, Yoshimitsu H, Tsuchihashi R, Okawa M, Kinjo J, Ikeda T, Nohara T. Three new monoterpenoids from the fruit of Genipa americana. Chem Pharm Bullet. 2007;55(4):632–4.

17. Conceição AO, Rossi MH, de Oliveira FF, Takser L, Lafond J. *Genipa americana* (Rubiaceae) fruit extract affects mitogen-activated protein kinase cell pathways in human trophoblast-derived BeWo cells: Implications for placental development. J Med Food. 2011;14(5):483–94.
18. Silva JN, Freitas BM. Análise sensorial de cristalizado de abacaxi Sensory analysis of candied pineapple. Braz J Dev. 2021;7(9):86815–23.
19. Barros DR, da Silva GA, Rosas AL, de Carvalho AP, da Silva Laranjeira RK, Machado AP, Dias JL, de Araujo SB, Junior EA, Campelo PH. Produção e avaliação da qualidade de doce Cristalizado de cupuaçu (*Theobroma grandiflorum*) tipo "jujuba" sem adição de gelatina. Avanços Em Ciência E Tecnologia De Alimentos. 2021;3(1):148–61.
20. Andrade SA, Metri JC, Barros Neto BD, Guerra NB. Desidratação osmótica do jenipapo (Genipa americana L.). Food Sci Technol. 2003;23:276–81.
21. Nascimento PD. Avaliação da retenção de carotenóides de abóbora, mandioca e batata doce. São José do Rio Preto: 67f. Dissertação (Mestrado em Engenharia e Ciência de Alimentos) - Instituto de Biociências, Letras e Ciências Exatas, Universidade Estadual Paulista; 2006.
22. IBGE. Pesquisa de orçamentos familiares 2008–2009 : análise do consumo alimentar pessoal no Brasil/IBGE, Coordenação de Trabalho e Rendimento. Rio de Janeiro: IBGE; 2011. p. 150.
23. BRASIL. Ministério da Saúde. Agência de Vigilância Sanitária (ANVISA). Instituto Adolfo Lutz. Métodos Físico-Químicos para Análise de Alimentos. Brasil: Ministério da Saúde; 2005. p. 819–77.
24. Roesler R, Malta LG, Carrasco LC, Holanda RB, Sousa CA, Pastore GM. Atividade antioxidante de frutas do cerrado. Food Sci Technol. 2007;27:53–60.
25. Swain T, Hillis WE. The phenolic constituents of Prunus domestica. I.—The quantitative analysis of phenolic constituents. J Sci Food Agric. 1959;10(1):63–8.
26. BRASIL. Agência Nacional de Vigilância Sanitária. Métodos físico-químicos para análises de alimentos. Brasília: Ministério da Saúde; 2005. 1018p.
27. Eaton WD. Microbial and nutrient activity in soils from three different subtropical forest habitats in Belize, Central America, before and during the transition from dry to wet season. Appl Soil Ecol. 2001;16:219–27.
28. Ayres M, Ayres Júnior M, Ayres DL, Santos AD. Aplicações estatísticas nas áreas das ciências bio-médicas. Belém, PA: Instituto Mamirauá; 2007. 324p.
29. Freitas JB, Cândido TL, Silva MR. Geleia de gabiroba: avaliação da aceitabilidade e características físicas e químicas. Pesquisa Agropecuária Tropical. 2008;1:87–94.
30. PHM S, Nassu RT, MSM SF, Maia GA, Figueiredo RW, Souza Neto MA. Obtenção de banana desidratada osmoticamente seguida de secagem em estufa. Comunicado Técnico; 2005. 108, Embrapa, Fortaleza, CE, Dezembro, 2005.
31. Silva MR, Lacerda DBCL, Santos GG, Martins DMDO. Caracterização química de frutos nativos do cerrado. Ciência Rural. 2008;38:1790–3.
32. Hansen DDS, Silva SA, Fonseca AAO, Hansen OADS, França NO. Chemical characterization of native jenipapo fruits from the Recôncavo baiano region aiming in natura fruit consumption and industrialization. Revista Brasileira de Fruticultura. 2008;30:964–9.
33. Donado-Pestana CM, Salgado JM, de Oliveira RA, dos Santos PR, Jablonski A. Stability of carotenoids, total phenolics and in vitro antioxidant capacity in the thermal processing of orange-fleshed sweet potato (Ipomoea batatas Lam.) cultivars grown in Brazil. Plant Foods Human Nutr. 2012;67:262–70.
34. Fagbemi TN, Oshodi AA, Ipinmoroti KO. Processing effects on some antinutritional factors and in vitro multienzyme protein digestibility (IVPD) of three tropical seeds: nreadnut (Artocarpus altilis), cashewnut (Anacardium occidentale) and fluted pumpkin (Telfairia occidentalis). Pak J Nutr. 2005;4(4):250–6.
35. Brasil. Regulamento técnico sobre padrões microbiológicos para alimentos. ANVISA. RDC n°12; 2001.
36. Assis MMM, Maia GA, Figueiredo EAT, Figueiredo RW, Monteiro JCS. Processamento e estabilidade de geléia de caju. Revista Ciência Agronômica. 2007;38(1):46–51.

Biomedical Applications and Health Benefits of Wild Noni and Its Fruits

P. Ponmurugan, A. Anitha, and N. Mathivanan

Abstract

Morinda citrifolia L., a small- to medium-sized tree with a pantropical distribution that is 3–10 m high, is commonly known by the name "noni". Pacific Islanders and people in Southern and Southeast Asia have long used noni fruit and leaves as food. In Java and Thailand, noni leaves were consumed both raw and cooked. The fruit has been utilized more recently to make dietary supplements. It is acknowledged that noni (*Morinda citrifolia* L.) is a sustainable crop with potential market uses. Noni has been utilized for generations by traditional healers in Hawaii and Polynesia to treat and prevent a wide range of ailments. According to reports, there is a growing interest in using noni plants as a novel source of bioactive chemicals. Bioactive substances with antioxidant, anti-mutagenic, anti-tumour, anti-inflammatory, anti-allergic, antiviral, anti-fungal, antimicrobial, and anti-carcinogenic qualities have been demonstrated to exist in noni extracts. Through an explanation of the carried out in vitro and in vivo experiments, this review delves deeply into the industrial applications and pharmacological activity of *M. citrifolia* fruit, seed, leaf, and root, as well as their extracted phytochemical components.

Keywords

Morinda citrifolia · Noni · Pharmacological activity · Health benefits

P. Ponmurugan (✉) · A. Anitha
Department of Botany, Bharathiar University, Coimbatore, Tamil Nadu, India

N. Mathivanan
Department of Centre for Advanced Study in Botany, University of Madras, Chennai, Tamil Nadu, India

P. Thangaraj (ed.), *Pomotherapeutic Insights on Wild Edible Fruits*,
https://doi.org/10.1007/978-981-96-0543-9_14

Introduction

Noni is a Rubiaceae family commercial plant, frequently referred to as Indian mulberry in India and formally as *Morinda citrifolia* L. In Ayurveda, the Indian mulberry is referred as ashyuka, akshi, and atchy and in siddha as nunaa or togaru [1]. Noni is a Polynesian native that is farmed as a sustainable crop in tropical and subtropical regions [2]. Its most common name, noni, was bestowed to it by Polynesian immigrants who arrived on the Hawaiian island some 2000 years ago [3]. One of the reasons that contributed to the spread of knowledge about the potential applications of *M. citrifolia* was the adaptability of using different plant structures for different purposes [4]. The goal of this chapter is to compile the most recent information on *M. citrifolia* and discuss its traditional use, phytochemical properties, ethanobotanical data, as well as pharmacological activities with respect to products derived from noni fruit, in view of various components of *M. citrifolia* having a wide variety of medicinal benefits.

Botanical Description

Morinda citrifolia L is an evergreen shrub with more than 15-cm stem length and a mature height of 3–10 m. The distinguishing features of the plant allow it to be distinguished easily. Sapwood is soft and golden brown. The plant's bark is normally smooth and somewhat rough, ranging in colour from grey to light brown. It has 1.5–2-cm-long leaves that are four-angled, light green, pinnately veined, glossy, and linked by stiff petioles. The stipes are about 12 mm in length, connate or distinct, and either whole or three to four lobed at the top. In sizes ranging from 20 to 45 cm in length and 7 to 25 cm widespread, leaf blades are glabrous and membranous, elliptic-elliptic, and elliptic-ovate. White tuberous flowers with 75–90 ovoid to globose heads, 5 lobes of white corolla, a greenish white tube calculating 7–9-mm long roughly, and oblong deltate lobes of almost 7 mm in stretch [5]. The stamens in five numbers are barely visible, besides the elegance is around 15 mm lengthy.

A yellowish white ovoid bumpy body fruit, about 12 cm in size, has many, fused ovaries that detach from white flowers during ripening. Unripe fruit is dark green, while ripe fruit releases a pungent butyric acid that smells like rotting. When the fruit is mature, the pulp becomes juicy and bitter, and the colour changes to dull yellowish white and gelatinous; a number of firm three-sided reddish brown pits remain also existing [6]. The fruit is covered in dark brown pits, which contain four seeds. On the end of each seed an air sac varies in length from 4 to 9 mm, allowing it to travel into water and pollinate [7, 8]. The plant is similar to citrus and coffee in that it has a broad taproot, as well as long lateral roots.

Phytochemical Characteristics

Morinda citrifolia nutrition is excellent, and over 200 phytochemical substances with bioactive properties have already been discovered and isolated from various portions of the plant [9–11]. A class of biologically active substances in *M. citrifolia*, which include acids, alcohols, phenols, saccharides, anthraquinones, carotenoids, esters, triterpenes, flavonoids, glycosides, lactones, iridoids, ketones, lignans, nucleosides, sulfonates, and additional negligible bioactive compounds has been reported [12–14]. Ali et al. [7] discovered 51 volatile chemicals in the juice extract of the noni fruit, including linalool oxide, acetaldehyde, and methanethiol which belong to the class of terpenes, aldehydes, ketones, and sulphur compounds. According to the authors, the leaves contained tannins, phenols, steroids, alkaloids, and saponins; the roots contained phenols, coumarins, cardiac glycosides, and steroids [15]; the seed contained sucrose, flavonoids, steroids, phenols, carbohydrates, xanthoproteic acids, amino acids, ketose sugars, and reducing sugars; the flower contained iridoids, asperulosidic acid, and flavonoids [16]; and the fruit showed the existence of alkaloids, anthraquinones, carbohydrates, emodin, coumarin, flavonoids, phenols, glycosides, amino acids, steroids, saponin, quinone, tannins, and terpenoids including di- and triterpenoids [17].

According to the nutritional value of the noni plant, the mature fruit includes around 90% water and 10% dry matter. The dry matter is mostly made up of soluble solids, with 5% being glucose and fructose (under reducing sugars) and 1.3% being sucrose (under non-reducing sugar). According to Motshakeri and Ghazali [18], proteins and dietary fibre are the residue that belongs to soluble solids. Nearly carbohydrates of 71%, fibre of 36%, and protein of 5.2% are existing in the 100-g sample of fruit powder [19].

Ethnobotanical Evidences

Noni fruit, leaves, blossoms, stems, barks, and roots have been utilized to treat a variety of illnesses [20]. *M. citrifolia* has been stated to partake in a wide range of healing uses in conditions such as colic, seizures, cough, type 2 diabetes, painful urine output, fever, cirrhosis of the liver, constipation, discharge from the vagina during pregnancy, malarial fever, and nausea and in promoting menstrual bleeding [21]. Smallpox, oedema, enlarged spleen, arthritis, asthma, and other joint and bone issues, cancer, colds, cataracts, digestive difficulties, depression, and stomach ulcers are also treated with it. High pressure in blood, infections, renal diseases, migraine type of headache, premenstrual syndrome, stroke, discomfort, and drowsiness are some of the other applications [22].

The leaves have traditionally been used to relieve symptoms of rheumatoid arthritis and joint swelling, as well as stomach aches, headaches, gastroenteritis, and burns, sores, wounds, and inflammation caused by filariasis, a parasitic condition. The bark was historically used in labour preparedness. The juice has been used to treat arthritis, hypertension, diabetes, muscle pain, menstrual problems, headaches, coronary artery disease, AIDS, cancer, stomach ulcers, sprains, anxiety, senility, digestive issues, atherosclerosis, circulation problems, and addiction of drugs [4]. Noni is sprayed on the skin on occasion to moisturize the skin and reduce the signs of ageing. Noni stem is used to manufacture an ointment for those who have cutaneous leishmaniasis. Seed and oil are used as an insecticide on the scalp, which has been shown to be effective in treating arthritis. Noni has been cultivated for over a thousand years in Polynesia, where it is used as a colouring agent, medicinal product, and foodstuff. The root has been used to make dyes [23] and flowers are used to treat insect bites [24].

Pharmacological Activity of Noni

In order to clarify scientific evidence, the therapeutic indications of *M. citrifolia* shall be explained in detail in the relevant literature, with a description of the components used, the extraction solvent, and the phytochemical component responsible for each action.

Antimicrobial Activity

In vitro studies of *Escherichia coli*, *Candida albicans*, and *Staphylococcus aureus* have been conducted to evaluate the antibacterial activity of *M. citrifolia* fruit that belongs to Tahitian methanol extract. *Candida albicans* remained the most sensitive organism, because of its iridoid content in Tahitian *M. citrifolia*, while *S. aureus* showed the least sensitivity [25]. Noni leaf extracts inhibited the development of *E. coli*, *S. aureus*, *C. albicans*, and *Aspergillus niger* with a reliance on dose order [26]. A number of studies in antibacterial activity of Indian *M. citrifolia* methanol, ethyl acetate, and hexane extract have been conducted against a variety of species belonging to *Staphylococcus aureus*, *Bacillus subtilis*, *Lactococcus lactis*, and other organisms, and the efficacy has been reported. Among the extracts studied, methanol extract remained the utmost effective, followed by ethyl acetate, which was moderately active, and hexane extract, which was unsuccessful contrary to all tested bacteria [27–31].

Methanolic extract of dehydrated noni fruit repressed *Trichophyton mentagrophytes* by 79.3%, while 50% activity was shown over *Penicillium* sp., *Fusarium* sp., and *Rhizopus* sp. [32, 33]. In line with this, antifungal properties of hexane, ethyl acetate, and methanol extract of Indian *M. citrifolia* fruit compared to various fungal cultures, namely, *Aspergillus flavus*, *Aspergillus fumigates*, *Aspergillus niger*, *Candida albicans*, *Fusarium* sp., *Mucor* sp., *Penicillium* sp., *Rhizopus* sp., and *Trichophyton mentagrophytes*, were assessed in vitro by Srinivasahan and Durairaj [28]. The methanolic and ethyl acetate extracts inhibited *T. mentagrophytes* by 79.3% and 62.06%, respectively, while the methanol extract inhibited *Fusarium* sp., *Penicillium* sp., and *Rhizopus* sp. by nearly 50%, and none of the extracts were reported to inhibit *Aspergillus* sp. or *Candida albicans* [31]. Noni roots inhibited the cytopathic impression of HIV-infected MT-4 cells without reducing cell proliferation or endothelial exclusion [34].

Anti-inflammatory Activity

Morinda citrifolia has been revealed to have anti-inflammatory effects equally in in vitro and in vivo models, connecting a wide range of pathological events allied with inflammation. The previous findings from Fletcher et al. [35] and Palu et al. [36] have revealed anti-inflammatory activity of topical application of *M. citrifolia* seed oil in infected persons. It has been observed that *M. citrifolia* juice reduces paw oedema straightly by impeding COX1 and COX2 and lowering nitric oxide generation in the J774.G8 cell line in reliance of dose. Dussossoy et al. [37] and Serafini et al. [38] discovered that water-soluble extracts of *M. citrifolia* leaves were efficient in reducing inflammation, dramatically decreased leukocyte migration, and might be administered for alleviating discomfort and swelling triggered by oxidation. At a dosage of 3 g/kg, *M. citrifolia* root chloroform fraction displayed anti-inflammatory effects that dramatically reduced histamine levels and increased paw oedema. Damnacanthal administered alone reduced oedema at dosages ranging from 10 to 100 mg/kg. A compound derived from fruit extract has shown strong anti-inflammatory action in animal model (mouse) of induced inflammation [39].

Antioxidant Activity

Natural antioxidants are abundant in *M. citrifolia* L. Ali et al. [40] found that various portions of *M. citrifolia*, such as leaf juice and fruit juice with or deprived of seeds, can modify cell-mediated immunity and activity of antioxidant enzymes in vitro. According to Arunachalam [41] noni blossoms have a greater potential towards scavenging the DPPH radical than green tea. A 500 mg/mL quantity of noni blossom extract in water scavenges radicals with a capacity of 88.11 ± 0.01% (mean ± standard deviation), which is greater than green tea's activity of 76.60 ± 0.05% ($P < 0.0001$). Malaysian *M. citrifolia* seedless fruit extracted with methanol scavenges free radicals (RSA) of 55.60% and an overall phenolic concentration of 4.318 mg GAE/1 g, correspondingly [42]. The antioxidant capabilities of extracts of ethanol and ethyl acetate of *M. citrifolia* root, fruit, and leaf were assessed using thiobarbituric acid (TBA) and ferric thiocyanate (FTC) techniques, resulting in the root extract outperforming the fruit and leaf extracts [43]. Matsuda et al. [44] discovered anti-photo-ageing efficacy in seed extract doses extending from 0.5 mg/mL to 1.0 mg/mL.

Anticancer Activity

In general, natural constituents of *M. citrifolia* have been considered to be a natural anticancer remedy in which sulphate polysaccharide metastases are prevented by slowing down the glycosaminoglycan interaction with certain proteins [45] or by elevating the apoptosis in cell lines of colorectal cancer in human [46]. In prevalent medicine, the usage of *M. citrifolia* is widespread as a nutrition supplement in treating several kinds of cancer, and it has been shown to be effective [47]. Treatment with NDAM, an anthraquinone found in *M. citrifolia* stems, reduced the tumour growth in vivo while increasing the total amount of T helper, cytotoxic T, and NK cells. Furthermore, nordamnacanthal inhibits the survival of cancerous MCF-7 breast cells and MDA-MB231 in vivo [48]. According to Lim et al. [49], *M. citrifolia* fruit or leaf ethanolic extract overwhelms the inflammation by suppressing the COX2 (cyclooxygenase 2) gene, a significant indicator of inflammation, which influences the manifestation of tumour suppressor genetic factor and the altitudes of intense fruit.

In B16-F10 melanoma cells (LAN5), *M. citrifolia* fruit ethanolic extract was observed to have anti-proliferative activities [50]. Either noni juice or cisplatin alone, otherwise in combination, was capable of induction of apoptosis on HeLa and SiHa cells via the mitochondrial route in the management of cervical cancer [51]. Damnacanthal can reduce cancer in colorectal tumour genesis by causing cell growth arrest and inducing caspase activity [46]. Alizarin inhibits angiogenesis by preventing blood flow to malignant tumours [52]. Lv et al. [53] discovered that a chemical component, limonene, derived from *M. citrifolia* roots, has durable inhibitory properties on the development of human colon and lung cancer cells by encouraging the thymus gland in the direction of generating more T lymphocytes, which kill the carcinoma cells.

Antidiabetic Activity

Numerous studies are being undertaken in response to the growing number of diabetes patients who are adopting integrated or functional treatment. The phosphorylation of the FOXO1 transcription factor increased glucose metabolism. As a natural antidiabetic drug, *M. citrifolia* is exhibiting encouraging results in vivo. For a period of 10 days, female Wistar rats induced with diabetes by steroid (type 2) were administrated by 1.8 and 3.6 mL/kg of *M. citrifolia* commercial fruit juice orally. Advanced oral doses (3.6 mL/kg, twice in a day) considerably reduced blood glucose levels. *Morinda citrifolia* fruit juice outperformed rosiglitazone, nevertheless inducing liver damages in rats [54]. Fermented *M. citrifolia* juice supplemented at 1.5 μL/g body weight at the rate of twice daily intended for 12 weeks increased insulin and glucose tolerance along with levels of fasting blood glucose in male mice. Insulin-regulated genes of gluconeogenic category such as phosphoenolpyruvate C kinase and glucose-6-phosphatase were likewise suppressed by more than 80% [55]. Streptozotocin induced diabetic rats were also controlled by fermented juice of

M. citrifolia belonging in South America. 360.0 mg/dL of fasting blood sugar in untreated diabetic rats was observed when compared with laboratory rats administered with 2 mL/kg of *M. citrifolia* at the rate of twice a day for 20 days, ensuring glucose levels in blood of 125–150 mg/dL, correspondingly. Saponins in addition to flavonoids found in fruits of *M. citrifolia*, such as rutin, could serve as a secretagogue, accumulating insulin production. It was proposed that *M. citrifolia's* antidiabetic action was owing to a stimulatory impact on the remaining ß-cells, causing them to release more insulin [56, 57].

Wound Healing Activity

Morinda citrifolia leaf extract has been shown to have wound healing activity, which culminated in a decrease in lipid peroxide levels in the treated group owing to the existence of antioxidant carotene, flavonol, and iridoid glycosides [58]. The topical wound healing effects of fresh Tahitian *M. citrifolia* leaf juice at 1 mg/mL and its ethanol, hexane, and methanol fractions of leaves at 10–200 g/mL concentrations were examined. Due to the interaction of ligand with the PDGF/A2A receptor, methanol extract of leaves dramatically enhanced the closure of wound and decreased half of the closure time in treated animals compared to that of control [59].

Activity on Liver Protection

A study accompanied by Wang et al. [60] reported that drinking noni juice every day diminishes triglycerides, cholesterol, and hs-CRP. There was also a reduction in LDL and homocysteine, in addition to an escalation of HDL. Acute liver impairment caused by carbon tetrachloride (CCl_4) was reversed in Sprague-Dawley (SD) female rats by 20% of noni fruit juice in swallowing water [61].

Anti-arthritic Activity

Saraswathi et al. [62] observed a dosage-dependent noteworthy reduction in paw thickness, secondary lesions, arthritic index, mononuclear infiltration, and pannus formation owing to the existence of flavonoids and phenols in the fruit juice of *M. citrifolia* when administered orally to the rats with arthritis at the dose range of 1.8–3.6 mL/kg. A total of 82 osteoarthritis patients who were given 3 oz of TNJ daily stated substantial enhancements in symptoms including pain and movement [63].

Analgesic and Gastric Ulcer Healing Activity

Using the hotplate assay, Tahitian *M. citrifolia* fruit juice has been shown to be effective for pain relief in rats. Subsequently feeding with *M. citrifolia* juice at 10% or 20%, the rats tolerated more pain [3]. Mature Thai dried unripe *M. citrifolia* aqueous fruit extract demonstrated in vivo anti-inflammatory gastro-oesophageal actions in rats at doses ranging from 0.63 g/kg to 2.50 g/kg, with potential mechanisms for minimizing acute gastric lesions, prohibiting oesophageal reflux, and functioning as an antisecretory agent related to ranitidine in addition to lansoprazole [64]. *Morinda citrifolia* fruit aqueous extract worked as an absorber of ranitidine inducer, due to the presence of scopoletin, which activated the receptor 5-HT4 [65].

Activity Against Metabolic Disorders Associated with Obesity

The current emphasis is on *M. citrifolia* product development and application as a strategy for the management of obesity in animal models [9]. In an experiment, changes in abdominal weights in terms of fat were identified between treated and non-treated groups with noni leaf extract. It was similarly discovered that adding sucrose into the leaf extract of noni lowered the triacylglycerol levels in the treated population with the supplemented extract much more than the population with metabolic syndrome (MS) who did not receive the extract. The findings of Inada et al. revealed that consuming an aqueous leaf extract of noni lowered not only abdominal fats but also the quantities of triacylglycerols linked with an elevated adiponectin level in tissues of adipose. The use of ethanol extracts of dried leaves of noni produced positive results in the decrease of lipid and faecal fat content in plasma of blood, as well as the lowering the levels of insulin, leptin, and adiposity in rats fed an elevated fat diet for 9 weeks. The medication expressed a substantial metabolic abnormality improvement associated with obesity. Furthermore, the key pathways in the process of obesity reduction mostly included the breakdown and utilization of amino acids and glucose, predominantly in the citric acid cycle, creatine and choline metabolism, and the intestinal microbiota dynamics [66]. In humans, Wang et al. [60] used 188 mL of iridoids containing juice of noni plant as a medicine substitute to diminish the effects of oxidative stress-induced dyslipidaemia qualified to cigarette smoking, which resulted in a significant reduction of cholesterol, triglycerides, LDL, and homocysteine levels, in addition to an increase in high-density lipoprotein levels.

NFJ and Its Derivative Products

Morinda citrifolia contains many bioactive compounds as well as established therapeutic applications at laboratory level; however, conversion of these properties to products with an extraordinary standard of excellence is still limited [12]. Fruit juice is presently the most common creation of *M. citrifolia*, either through traditional fermentation techniques or by means of industrial manufacture with well-ordered processes and handling stages [24]. Pure fruit juice, whole fruit powder, dried fruit puree, and concentrated fruit juice are drinks produced from the noni fruit. In the sector, other preparations of noni fruits, such as fruit syrups, are also found [7]. The noni fruit powder is produced from the whole fruit deprived of kernels or juice by means of evaporation [67]. Noni puree is made by dripping and pounding the pulp juice with no added substances like water or sugar [68].

Products of various parts of *M. citrifolia*, such as leaves, fruits, roots, and barks, are marketed on the market encompassing noni juice, capsules, powder products, or tea made from different parts of this plant. Among them, the juice of noni is in use all over the world for its nutraceutical and medicinal properties. The noni fruit is a small, edible fruit with a bitter taste that smells like bad cheese and looks like a grenade. Despite its origins in tropical regions, the noni fruit juice has become popular all over the world. The first complete food product to be authorized was noni fruit juice when the EU's new Food Act took effect in 1997. For the preparation of noni juice, a variety of natural and industrial methods may be employed. In this way it is possible to produce 100% pure noni juice at home. Today, the juice of the noni fruit is considered to be a safe and functional food that can strengthen the immune system. In addition, noni fruit juice could prove to be an effective blend that may provide some health benefits [25, 69].

Health Benefits of Noni Juice

A fruit known as *M. citrifolia* is the source of the tropical beverage known as noni fruit juice. The noni fruit tree is native to Southeast Asia, primarily French Polynesia. The noni fruit is a tiny, edible fruit that resembles a grenade, has a bitter flavour, and smells like poor cheese. Noni fruit juice has gained popularity all across the world, despite its tropical origins. It is possible to make noni juice commercially or at home. A 100% pure noni juice is produced using the DIY approach [25, 69]. Noni fruit juice is now regarded as a healthy, functional food that could strengthen the immune system. Noni juice may also offer defence against systemic inflammation, blood lipid and homocysteine increase, and DNA damage brought on by tobacco smoke. Additionally, it may help manage weight, maintain bone health in women, raise immunity, boost physical endurance, prevent glycation of proteins, enhance gum health, function as an antioxidant, have anticancer properties, and reduce blood pressure.

Inferences and Outlook

Morinda citrifolia, popularly identified as noni, is a plant of remarkable high value in several grounds of application. The Polynesians have been using these structures for more than 2000 years, largely because of their therapeutic properties in the areas of skin diseases, tuberculosis, burns, and arthritis. In *M. Citrifolia*, a total of over

200 compounds have been shown to be biologically active and are highly beneficial for medical applications. Food, chemical, and pharmaceutical sectors are currently using *M. citrifolia*. In order to improve the quality of products already available such as NFJ, it is still necessary to pursue further standardization research in order to fully understand the true potential of this plant.

References

1. Ali M, Ansari SH, Ali S, Khan SA, Alam S, Hussain A. Phytochemical and pharmacological potential updates of *Morinda citrifolia* linn (noni). Res J Pharm Technol. 2013;6(11):1285–94.
2. Jahurul MH, Patricia M, Shihabul A, Norazlina MR, George MR, Noorakmar AW, Lee JS, Jumardi R, Jinap S, Zaidul IS. A review on functional and nutritional properties of noni fruit seed (*Morinda citrifolia* L.) and its oil. Food Biosci. 2021;41:101000.
3. Wang MY, West BJ, Jensen CJ, Nowicki D, Su C, Palu AK, Anderson G. *Morinda citrifolia* (noni): a literature review and recent advances in noni research. Acta Pharmacol Sin. 2002;23(12):1127–41.
4. Chan-Blanco Y, Vaillant F, Perez AM, Reynes M, Brillouet JM, Brat P. The noni fruit (*Morinda citrifolia* L.): a review of agricultural research, nutritional and therapeutic properties. J Food Compos Anal. 2006;19(6–7):645–54.
5. Patil P, Madkaikar H, Shah N. *Morinda citrifolia* L. (noni)–its ethnobotanical knowledge, phytochemical studies, pharmacological aspects, and future prospects. Int J Nov Res Dev. 2022;7(3):331–52.
6. Rojas-Sandoval J. *Morinda citrifolia* (Indian mulberry). Forestry Compendium. 2017;34854.
7. Ali M, Kenganora M, Manjula SN. Health benefits of *Morinda citrifolia* (noni): a review. Pharm J. 2016;8(4).
8. Honey J, Neha B, Ranjan B, Birendra S, Thakur A. Scientific basis of noni plant (*Morinda citrfolia*). Asian J Res Pharm Sci. 2012;2(2):45–7.
9. Inada AC, Figueiredo PS, Santos-Eichler RA, Freitas KD, Hiane PA, Castro AP, Guimarães RD. *Morinda citrifolia* Linn. (noni) and its potential in obesity-related metabolic dysfunction. Nutrients. 2017;9(6):540.
10. Shalan NA, Mustapha NM, Mohamed S. Noni leaf and black tea enhance bone regeneration in estrogen-deficient rats. Nutrition. 2017;33:42–51.
11. Potterat O, Hamburger M. *Morinda citrifolia* (noni) fruit-phytochemistry, pharmacology, safety. Planta Med. 2007;73(03):191–9.
12. Abou Assi R, Darwis Y, Abdulbaqi IM, Vuanghao L, Laghari MH. *Morinda citrifolia* (noni): a comprehensive review on its industrial uses, pharmacological activities, and clinical trials. Arab J Chem. 2017;10(5):691–707.
13. Saminathan M, Rai RB, Dhama K, Tiwari R, Chakraborty S, Amarpal RG, Kannan K. Systematic review on anticancer potential and other health beneficial pharmacological activities of novel medicinal plant *Morinda citrifolia* (noni). Int J Pharmacol. 2013;9(8):462–92.
14. Singh DR. *Morinda citrifolia* L. (noni): a review of the scientific validation for its nutritional and therapeutic properties. J Diabetes Endocrinol. 2012;3(6):77–91.
15. Meng C, Keo S, Chea S, San M, Trylim B, Seng S, Chhoun S, Chhan S. Phytochemical and thin layer chromatography analyses of *Morinda citrifolia* L. of Tboung Khmum Province, Cambodia. Asian J Pharmacogn. 2017;1(3):5–11.
16. Deng S, West BJ, Palu AK, Jensen CJ. Phytochemical, antioxidant and toxicological investigation of *Morinda citrifolia* L. blossoms. ISRN Anal Chem. 2012;2012:1–5.
17. Ranvir GD, Rajurkar SR, Ballurkar BV, Gangane GR. Physicochemical and phytochemical evaluations on noni (*Morinda citrifolia* linn.) Fruit powder and extracts. Adv Anim Vet Sci. 2017;5(4):160–6.

18. Motshakeri M, Ghazali HM. Nutritional, phytochemical and commercial quality of noni fruit: a multi-beneficial gift from nature. Trends Food Sci Technol. 2015;45(1):118–29.
19. Fontes RF, Andrade JK, Rajan M, Narain N. Chemical characterization of different parts of noni (*Morinda citrifolia*) fruit and its freeze-dried pulp powder with emphasis on its bioactive compounds and antioxidant activities. Food Sci Technol. 2023;43:e103722.
20. de Araújo Souza SK, Passos MP. Ethnobotanical use of *Morinda citrifolia*: a literature review. Res, Soc Dev. 2021;10(15):e352101523280.
21. Omar G, Abdallah L, Barakat A, Othman R, Bourinee H. In vitro haemostatic efficacy of aqueous, methanol and ethanol plant extracts of three medicinal plant species in Palestine. Braz J Biol. 2019;80:763–8.
22. Hirazumi A, Furusawa E, Chou SC, Hokama Y. Immunomodulation contributes to the anticancer activity of *Morinda citrifolia* (noni) fruit juice. In: Proceedings of the Western Pharmacology Society. 39;1996. p. 7–9.
23. Morton JF. The ocean-going noni, or Indian Mulberry (*Morinda citrifolia, Rubiaceae*) and some of its "colorful" relatives. Econ Bot. 1992;241–56.
24. Dixon AR, McMillen H, Etkin NL. Ferment this: the transformation of noni, a traditional Polynesian medicine (*Morinda citrifolia, Rubiaceae*). Econ Bot. 1999;51–68.
25. West BJ, Deng S, Isami F, Uwaya A, Jensen CJ. The potential health benefits of noni juice: a review of human intervention studies. Food Secur. 2018;7(4):58.
26. Usha R, Sashidharan S, Palaniswamy M. Antimicrobial activity of a rarely known species, *Morinda citrifolia* L. Ethnobot Leafl. 2010;2010(3):7.
27. Zhang WM, Wang W, Zhang JJ, Wang ZR, Wang Y, Hao WJ, Huang WY. Antibacterial constituents of Hainan *Morinda citrifolia* (noni) leaves. J Food Sci. 2016;81(5):M1192–6.
28. Srinivasahan V, Durairaj B. Antimicrobial activities of hydroethanolic extract of *Morinda citrifolia* fruit. Int J Curr Microbiol App Sci. 2014;3(9):26–33.
29. Sunder J, Singh DR, Jeyakumar S, Kundu A, Srivastava RC. Antimicrobial activity of *Morinda citrifolia* solvent extracts. Indian Vet J. 2012;89(4):9–11.
30. Manimala N, Britto SJ, Raj LJ, Prabhu S. Antibacterial activities and preliminary phytochemical analysis of fruit extracts of noni (*Morinda citrifolia* L.). Int J Noni Res. 2010;5(1–2):22–38.
31. Jayaraman SK, Manoharan MS, Illanchezian S. Antibacterial, antifungal and tumor cell suppression potential of *Morinda citrifolia* fruit extracts. Int J Integr Biol. 2008;3(1):44–9.
32. Jainkittivong A, Butsarakamruha T, Langlais RP. Antifungal activity of *Morinda citrifolia* fruit extract against *Candida albicans*. Oral Surg Oral Med Oral Pathol Oral Radiol Endod. 2009;108(3):394–8.
33. Almeida ÉS, de Oliveira D, Hotza D. Properties and applications of *Morinda citrifolia* (noni): a review. Compr Rev Food Sci Food Saf. 2019;18(4):883–909.
34. Kamata M, Wu RP, An DS, Saxe JP, Damoiseaux R, Phelps ME, Huang J, Chen IS. Cell-based chemical genetic screen identifies damnacanthal as an inhibitor of HIV-1 Vpr induced cell death. Biochem Biophys Res Commun. 2006;348(3):1101–6.
35. Fletcher HM, Dawkins J, Rattray C, Wharfe G, Reid M, Gordon-Strachan G. *Morinda citrifolia* (noni) as an anti-inflammatory treatment in women with primary dysmenorrhoea: a randomised double-blind placebo-controlled trial. Obstet Gynecol Int. 2013;2013:195454.
36. Palu AK, West BJ, Jensen CJ. Noni seed oil topical safety, efficacy, and potential mechanisms of action. J Cosmet Dermatol Sci Appl. 2012;2(2):74–8.
37. Dussossoy E, Brat P, Bony E, Boudard F, Poucheret P, Mertz C, Giaimis J, Michel A. Characterization, anti-oxidative and anti-inflammatory effects of Costa Rican noni juice (*Morinda citrifolia* L.). J Ethnopharmacol. 2011;133(1):108–15.
38. Serafini MR, Santos RC, Guimaraes AG, Dos Santos JP, da Conceicao Santos AD, Alves IA, Gelain DP, de Lima Nogueira PC, Quintans-Júnior LJ, Bonjardim LR, de Souza Araújo AA. *Morinda citrifolia* Linn leaf extract possesses antioxidant activities and reduces nociceptive behavior and leukocyte migration. J Med Food. 2011;14(10):1159–66.
39. Okusada K, Nakamoto K, Nishida M, Fujita-Hamabe W, Kamiya K, Mizushina Y, Satake T, Tokuyama S. The antinociceptive and anti-inflammatory action of the CHCl3-soluble phase

and its main active component, damnacanthal, isolated from the root of *Morinda citrifolia*. Biol Pharm Bull. 2011;34(1):103–7.
40. Ali M, Singh P, Singh L, Kumar S. Phytochemical constituents and pharmacological activities, profile of Morinda Citrifolia: a review. World J Pharm Res. 2020;9:421–33.
41. Arunachalam V, Morinda citrifolia L. (Rubiaceae): a multi-purpose tree for coastal ecosystems and its variability in Konkan region of India. Genet Resour Crop Evol. 2018;65(6):1751–65.
42. Krishnaiah D, Bono A, Sarbatly R, Anisuzzaman SM. Antioxidant activity and total phenolic content of an isolated *Morinda citrifolia* L. methanolic extract from Poly-ethersulphone (PES) membrane separator. J King Saud Univ Eng Sci. 2015;27(1):63–7.
43. Saminathan M, Rai RB, Dhama K, Jangir BL, Suresh S, Ranganath GJ, Sophia I, Karuppanasamy K, Barathiraja S, Gopalakrishnan A. Effect of *Morinda citrifolia* (noni) fruit juice on antioxidant, hematological and biochemical parameters in N-methyl-N-nitrosourea (NMU) induced mammary carcinogenesis in Sprague-dawley rats. Int J Pharmacol. 2014;10(2):109–19.
44. Matsuda H, Masuda M, Murata K, Abe Y, Uwaya A. Study of the anti-photoaging effect of noni (*Morinda citrifolia*). In: Melanoma-From Early Detection to Treatment. IntechOpen; 2013.
45. Liu JM, Haroun-Bouhedja FE, Boison-Vidal C. Analysis of the in vitro inhibition of mammary adenocarcinoma cell adhesion by sulphated polysaccharides. Anticancer Res. 2000;20(5/A):3265–72.
46. Nualsanit T, Rojanapanthu P, Gritsanapan W, Lee SH, Lawson D, Baek SJ. Damnacanthal, a noni component, exhibits antitumorigenic activity in human colorectal cancer cells. J Nutr Biochem. 2012;23(8):915–23.
47. Chanthira Kumar H, Lim XY, Mohkiar FH, Suhaimi SN, Mohammad Shafie N, Chin Tan TY. Efficacy and safety of *Morinda citrifolia* L.(noni) as a potential anticancer agent. Integr Cancer Ther. 2022;21:15347354221132848.
48. Abu N, Zamberi NR, Yeap SK, Nordin N, Mohamad NE, Romli MF, Rasol NE, Subramani T, Ismail NH, Alitheen NB. Subchronic toxicity, immunoregulation and anti-breast tumor effect of Nordamnacantal, an anthraquinone extracted from the stems of *Morinda citrifolia* L. BMC Complement Altern Med. 2018;18(1):1–10.
49. Lim SL, Goh YM, Noordin MM, Rahman HS, Othman HH, Bakar NA, Mohamed S. *Morinda citrifolia* edible leaf extract enhanced immune response against lung cancer. Food Funct. 2016;7(2):741–51.
50. Candida T, França JP, Chaves AL, Lopes FA, Gaiba S, Sacramento CK, Ferreira LM, França LP. Evaluation of antitumoral and antimicrobial activity of *Morinda citrifolia* L. grown in Southeast Brazil. Acta Cir Bras. 2014;29:10–4.
51. Gupta RK, Banerjee A, Pathak S, Sharma C, Singh N. Induction of mitochondrial-mediated apoptosis by *Morinda citrifolia* (noni) in human cervical cancer cells. Asian Pac J Cancer Prev. 2013;14(1):237–42.
52. Hornick CA, Myers A, Sadowska-Krowicka H, Anthony CT, Woltering EA. Inhibition of angiogenic initiation and disruption of newly established human vascular networks by juice from *Morinda citrifolia* (noni). Angiogenesis. 2003;6:143–9.
53. Lv L, Chen H, Ho CT, Sang S. Chemical components of the roots of noni (Morinda citrifolia) and their cytotoxic effects. Fitoterapia. 2011;82(4):704–8.
54. Puranik DS, Faraz M, Nagaraju B, Patan Fayaz NA, Shanthraj P, Bolouri A. Preclinical evaluation of antidiabetic activity of noni fruit juice. Int J Bioassays. 2013;2(02):475–82.
55. Nerurkar PV, Nishioka A, Eck PO, Johns LM, Volper E, Nerurkar VR. Regulation of glucose metabolism via hepatic forkhead transcription factor 1 (FoxO1) by *Morinda citrifolia* (noni) in high-fat diet-induced obese mice. Br J Nutr. 2012;108(2):218–28.
56. Mustaffa F, Indurkar J, Ali NI, Hanapi A, Shah M, Ismail S, Mansor SM. A review of Malaysian medicinal plants with potential antidiabetic activity. J Pharm Res. 2011;4(11):4217–24.
57. Nayak BS, Marshall JR, Isitor G, Adogwa A. Hypoglycemic and hepatoprotective activity of fermented fruit juice of *Morinda citrifolia* (noni) in diabetic rats. Evid Based Complement Alternat Med. 2010;2011:875293.
58. Rasal VP, Sinnathambi A, Ashok P, Yeshmaina S. Wound healing and antioxidant activities of Morinda citrifolia leaf extract in rats. Iran J Pharmacol Ther. 2008;7(1):49–52.

59. Palu A, Su C, Zhou BN, West B, Jensen J. Wound healing effects of noni (*Morinda citrifolia* L.) leaves: a mechanism involving its PDGF/A2A receptor ligand binding and promotion of wound closure. Phytother Res. 2010;24(10):1437–41.
60. Wang MY, Peng L, Weidenbacher-Hoper V, Deng S, Anderson G, West BJ. Noni juice improves serum lipid profiles and other risk markers in cigarette smokers. Sci World J. 2012;2012:594657.
61. Wang MY, Nowicki D, Anderson G, Jensen J, West B. Liver protective effects of *Morinda citrifolia* (noni). Plant Foods Hum Nutr. 2008;63:59–63.
62. Saraswathi CD, Prakash WS, Kunal PW. Antiarthritic activity of *Morinda citrifolia* L. fruit juice in Complete Freund's adjuvant induced arthritic rats. J Pharm Res. 2012;5(2):1236–9.
63. Wang MY, Lutfiyya MN, Weidenbacher-Hoper V, Peng L, Lipsky MS, Anderson G. *Morinda citrifolia* L.(noni) improves the quality of life in adults with osteoarthritis. Funct Foods Health Dis. 2011;1(2):75–90.
64. Mahattanadul S, Ridtitid W, Nima S, Phdoongsombut N, Ratanasuwon P, Kasiwong S. Effects of *Morinda citrifolia* aqueous fruit extract and its biomarker scopoletin on reflux esophagitis and gastric ulcer in rats. J Ethnopharmacol. 2011;134(2):243–50.
65. Nima S, Kasiwong S, Ridtitid W, Thaenmanee N, Mahattanadul S. Gastrokinetic activity of *Morinda citrifolia* aqueous fruit extract and its possible mechanism of action in human and rat models. J Ethnopharmacol. 2012;142(2):354–61.
66. Jambocus NG, Saari N, Ismail A, Khatib A, Mahomoodally MF, Hamid AA. An investigation into the antiobesity effects of *Morinda citrifolia* L. leaf extract in high fat diet induced obese rats using a 1H NMR metabolomics approach. J Diabetes Res. 2016;2016:2391592.
67. Nelson SC. Proceedings of the 2002 Hawai'i Noni Conference. In 2002 Hawai'i Noni Conference (2002: Hilo, Hawaii) 2003. Cooperative Extension Service, College of Tropical Agriculture and Human Resources, University of Hawai'i at Mānoa.
68. Nelson SC, Elevitch CR. Noni: the complete guide for consumers and growers. PAR; 2006.
69. Brown AC. Anticancer activity of *Morinda citrifolia* (noni) fruit: a review. Phytother Res. 2012;26(10):1427–40.

Under-Exploited Wild Edible Fruits of the Western Ghats: A Review on Antioxidant Potential and Phytochemistry

J. M. Sasikumar and Sibbala Subramanyam

Abstract

This article reports a review on findings related to the antioxidant potential and phytochemistry of under-exploited wild edible fruits consumed in the Western Ghats, India. It summarizes potent antioxidant properties and the phytochemical components of the fruits. The fruits reported by the researchers exhibited their noticeable activity against reactive oxygen species (ROS) and reactive nitrogen species (RNS) as well as lipid peroxides. The discoveries, highlighted on antioxidant properties of the fruits, are amassed to pursue further hunt for antioxidants leading to the revelation of their therapeutic values.

Keywords

Antioxidant activity · Phytochemical analysis · Therapeutics · Wild fruits

Introduction

Fruits have gained the grade of functional food since a long time [1] as they deliver health profits in addition to physiological requirements. Human kind gains health benefits from repetitive consumption of fruits [2]. Numerous epidemiological researches and intervention studies in the past depicted the worth of fruit consumption in the prevention of life-threatening chronic and degenerative ailments [3].

J. M. Sasikumar (✉)
Department of Microbiology, Karpagam Academy of Higher Education (KAHE), Coimbatore, India
e-mail: sasikumar.jmahalingam@kahedu.edu.in

S. Subramanyam
Vignans Foundation for Science, Technology and Research, Guntur, India

P. Thangaraj (ed.), *Pomotherapeutic Insights on Wild Edible Fruits*,
https://doi.org/10.1007/978-981-96-0543-9_15

These defensive potentials of the fruits are exerted by several phytochemical constituents, i.e. phenolic compounds and vitamins [4]. Their ability to prevent diseases is owing to the capability of the polyphenolics in scavenging reactive oxygen species (ROS), and hence the cell impairment caused by the ROS is effectually prevented. The ROS or free radicals encompassing singlet oxygen, superoxide radical (O), hydroxyl ion (OH-) and the reactive nitrogen species (NO*) are known to cause severe damage to cellular biomolecules, i.e. DNA, lipids, proteins and carbohydrates [5], which leads to causation of degenerative diseases such as cancer, atherosclerosis, diabetes and cardiovascular ailments [6]. The ROS scavenging properties of phenolic compounds rely on their redox potential that allows them to donate hydrogen atoms, to quench singlet oxygen and to chelate metal ions [7]. Thus, the purpose of exploring natural antioxidants from fruits that possess potential to shield mankind from impairment caused by ROS induced oxidative stress has been well addressed. Previous researches addressed antioxidant properties and phenolic components of the wild edible fruits. Thus, the current article extensively reviews the phytochemistry and antioxidant of under-exploited wild edible fruits from the Western Ghats.

Berberis tinctoria Lesch.

Botany and Therapeutic Value

Berberis tinctoria Lesch. (Berberidaceae), commonly identified as Nilgiri barberry, is known as '*jakkala hannu*' by the Badaga ethnic community or '*oosikala*' by the Kurumba tribal community of Nilgiri district, Western Ghats. The plant has been reported from Shola forests of Palni Hills and the Nilgiris, Western Ghats, India, at an altitude of 1800 m [8]. *Berberis tinctoria* raises up to 2–3 ft in height with hard stem bearing long branches with slim leafy twigs [9] (Fig. 1). The fruits of the plant are predominantly consumed by the Badaga and Toda community people of the district.

Phytochemistry

The phenolic and flavonoid composition of the fruits of *B. tinctoria* was examined and reported [10]. The content of phenolics was 410 ± 0.02 mg/GAE of 100 g of fresh weight, and the flavonoid concentration was 320 mg/quercetin equivalents (QE) of 100 g fresh weight. Fruits of the plant were analysed for mineral content, total polyphenolic and total saponins and physico-chemical parameters such as moisture content, crude fibre, crude ash, lipid and protein [11]. The fruits were found to be rich in potassium (177.898 ± 38.722 mg/100 g), aluminium (52.393 ± 15.986 mg/100 g) and magnesium (75.477 ± 12.845 mg/100 g), and they have high starch (206.4 ± 12.1 glucose equivalents mg/g) and protein content (471.8 ± 15.8 bovine serum albumin equivalents mg/g). High composition of total

Fig. 1 Fruits of *B. tinctoria* (created by the first author)

phenolics (498.6 ± 6.9 gallic acid equivalents mg/g) and flavonoids (389.7 ± 5.7 quercetin equivalents mg/g) was also assessed.

Antioxidant Potency

Despite exploitation of the root, bark and leaves of *B. tinctoria* for various biological activities, the bioactivity studies on the fruits are meagre. The antioxidative potential of *B. tinctoria* fruits was investigated using different assays [10]. The fruits scavenged DPPH free radical ion (EC_{50} 9.85 ± 1.33 μg/mL), ABTS radical, superoxide radical ($O_2^{\cdot-}$) (EC_{50} 64.65 ± 0.82 μg/mL), hydroxyl ion (OH^-) (EC_{50} 79.98 ± 1.02 μg/mL), nitric oxide (NO) (EC_{50} 75.21 ± 1.32 μg/mL) and chelated ferrous ion (EC_{50} 45.24 ± 1.42 μg/mL) effectively. Besides, anti-haemolytic activity of the fruits exhibited a concentration-dependent inhibition (EC_{50} 71.1 ± 0.22 μg/mL) of haemolysis of erythrocytes from cow. The above-said anti-haemolytic potency of this *Berberis* fruit may be owing to the capacity of phenolic antioxidants. The fruit extract's Trolox equivalent antioxidant capacity (TEAC) was observed in the range from 1.063 to 2.364 mM TE/g.

In another study, the ethyl acetate extract of the fruits was found to possess strong DPPH scavenging potential with an IC_{50} value of 38.2 ± 11.9 μg/mL, and the methanol extract exhibited strong reduction (20.1 ± 0.9 mg AAE/g) of Mo (VI) to Mo (V) [11]. They also reported high metal chelating property (115.5 ± 5.2 mg EDTA E/g) by the ethyl acetate extract of *B. tinctoria* fruits. Vignesh et al. [12] evaluated the antioxidant potential of zinc oxide nanoparticles (ZnONPs) synthesized using the fruits of *B. tinctoria*. The ZnONPs fabricated using fruits of this plant exhibited significant DPPH scavenging activity with an IC_{50} value of 15.96 μg/mL. The ABTS cation was also scavenged significantly by the fruit extract (76.03 μM Trolox equivalent/g extract. The fruit extract reduced TPTZ-Fe (III) complex to TPTZ-Fe

Fig. 2 The fruits of *E. kologa* (created by the first author)

(III) effectively with the value of 39.02 mM Fe (II)/mg extract). The total antioxidant activity using the phosphomolybdenum method displayed radical scavenging activity (34.72 mg/AAE/g) equivalent to ascorbic acid. The ZnONPs from the fruit extract were also found to chelate metal efficiently (118.21 mg EDTA equivalent/g).

Elaeagnus kologa Schlecht.

Botany and Therapeutic Value

Elaeagnus kologa Schlecht. (Elaeagnaceae), locally called '*kolangal hannu*', is endemic and distributed in the Nilgiri Hills of the Western Ghats. The plant is a large shrub with thorn bearing elliptic obovate leaves and projecting sliver scale (Fig. 2). Despite wide consumption by the local communities, the fruits are underutilized [13].

Phytochemistry

The methanol extract of *E. kologa* fruit was subjected to measure total phenolic composition and total flavonoid content [14]. The fruit extract of the species yielded phenolic composition of 212 ± 0.02 mg GAE/100 fresh weight (FW), and the flavonoid composition was found to be 220 ± 0.12 mg QE/100 g of FW.

Antioxidant Potency

The antioxidant capacity of methanol extract of *E. kologa* fruit was examined [14]. It exhibited potent quenching capacity against radicals, i.e. DPPH, $O_2^{\cdot-}$, OH^- and NO. An impressive scavenging activity against DPPH by the fruit extract was observed with an EC_{50} value of 10.2 ± 1.33 μg mL^{-1}. The OH^- scavenging ability of the fruit extract was found be outstanding (EC_{50} 51.1 ± 1.02 μg mL^{-1}). The extract was able to scavenge superoxide radicals in PMS/NADH-NBT system with an EC_{50} value of 102.4 ± 0.82 μg mL^{-1}. The methanol extract from *E. kologa* fruit was found to be capable of inhibition of NO generation from sodium nitroprusside (SNP), and it has an EC_{50} value of 25.2 ± 1.32 μg mL^{-1}. The iron-chelating ability of the fruit was dose dependent, and it was capable of chelating Fe^{3+} ions (EC_{50} 90.4 ± 1.42 μg mL^{-1}). A notable inhibition of LPO (EC_{50} 75.6 ± 0.22 μg mL^{-1}) which was induced by $FeCl_2$ in liver homogenate was also observed.

Elaeocarpus oblongus Gaestn. (Syn. *Elaeocarpus E. tectorius* (Lour.) Poir.

Botany and Therapeutic Value

Elaeocarpus oblongus Gaestn. (Elaeocarpaceae), a tree species, bears pearl-like edible fruits (Fig. 3). The tree is found in the Nilgiri district of the Western Ghats at the elevation of 6000 ft. The fruits are eaten by the Badaga ethnic community and other tribal communities, i.e. Todas, Kotas, Irulars, Kattunaickens and Kurumbas. The fruits are reported to be emetic towards rheumatism and used also used against pneumonia, ulcer, piles, leprosy and dropsy [8]. They are traditionally used to treat rheumatism and body pain [15].

Fig. 3 *E. oblongus* plant (created by the first author)

Phytochemistry

Physico-chemical parameters and pharmacognostical study *E. oblongus* fruit were conducted and reported [15]. Major composition of manganese (53.3 μg/g) and zinc (46.2 μg/g) was observed in the fruits. The fruit was reported to contain protein (59.3 mg/100 g), fibres (1.4 mg/100 g), phosphorous (1.6 mg/100 g), iron (26.00 mg/100 g), calcium (3.1 mg/100 g), carotenes 37 (mg/100 g), thiamine (190 mg/100 g), niacin (0.02 mg/100 g) and riboflavin (0.06 mg/100 g) [16]. A quantitative analysis of methanol extract exhibited phenolic composition of 10.96 ± 0.14 mg GAE/g dw and flavonoid composition of 1.25 ± 0.04 mg CE/g dw. Certain phenolic components, i.e. gallic acid, protocatechuic acid, chlorogenic acid, vanillic acid, caffeic acid, syringic acid, *p*-coumaric acid, ferulic acid and *trans*-cinnamic acid (Fig. 4), were detected in the fruits upon HPLC analysis of methanol extract [17].

Antioxidant Potency

Elaeocarpus oblongus fruit extracts were found to possess good antioxidant capacity [18]. They exhibited potent scavenging properties towards DPPH (EC_{50} 0.250 ± 0.02 mg/mL), $O_2^{\cdot-}$ (EC_{50} 0.5 ± 1.2 mg/mL), OH^- (EC_{50} 0.5 ± 0.48 mg/mL) and NO (EC_{50} 0.25 ± 0.55 mg/mL) radicals. Besides, Fe^{2+} chelation (EC_{50} 0.25 ± 1.42 mg/mL) and LPO inhibition (EC_{50} 0.506 ± 0.38 mg/mL) were well executed by the fruits. Another work showed the *E. oblongus* fruits displayed DPPH inhibitory ability with an IC_{50} value of 571.7 ± 2.04 μg/mL. The ABTS radicals were also effectively inhibited by the fruit extract (IC_{50} value 469.0 ± 1.82 μg/mL) [17].

Flueggea leucopyrus Willd. (Euphorbiaceae) Phyllanthaceae

Botany and Therapeutic Value

Flueggea leucopyrus, known as '*poolanchi*' in Tamil language, reaches to the height of 3–4 ft. with round and green colour leaves. The flowers are in pale green colour and seem to be similar to peppercorns. This plant bears a capsule fruit with one seed, globose and white in colour when ripened. It fruits during April and November. The seeds look like the seeds of mustard (Fig. 5).

Phytochemistry

Phytochemical investigation and antioxidant activity of methanol extract of *F. leucopyrus* fruits were studied and reported [19]. The phytochemical evaluation has shown the contents of anthocyanin (0.34 ± 0.09 CGE/100 g), ascorbic acid

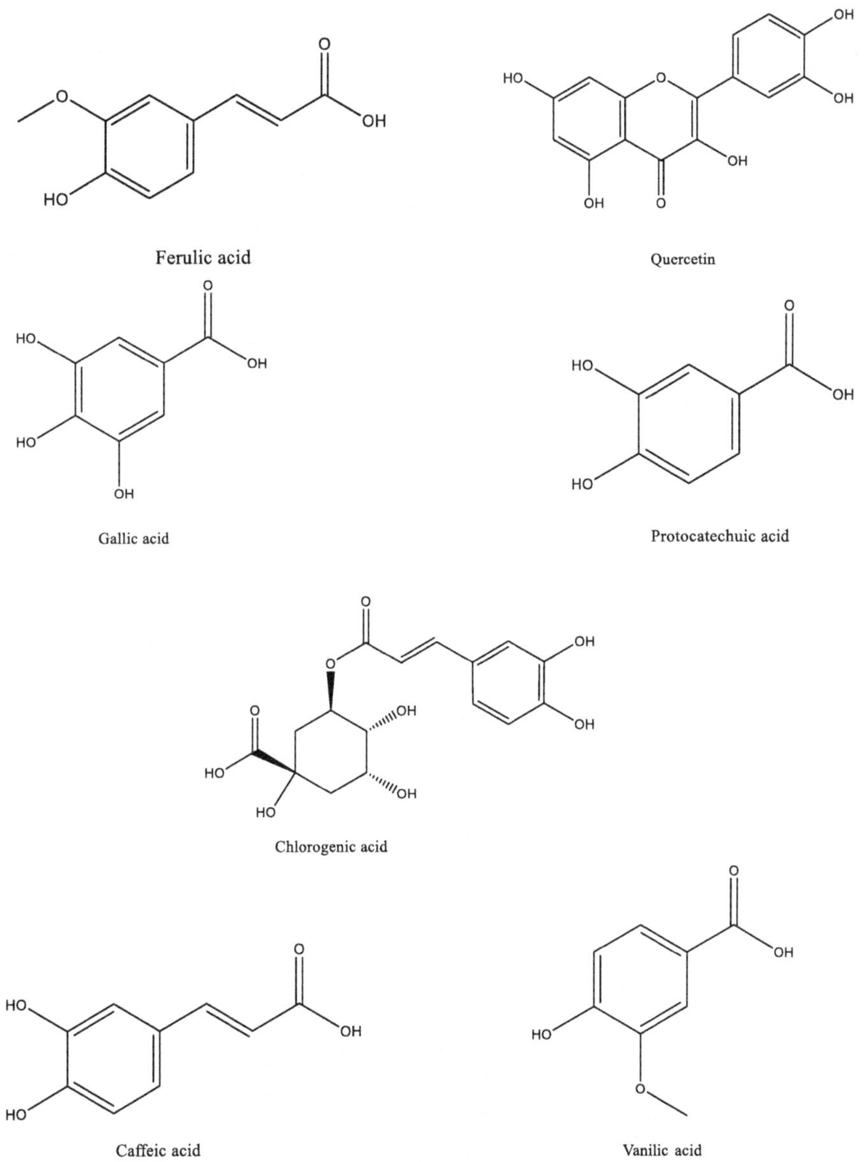

Fig. 4 Chemical structures of the compounds found in the wild fruits

(39.4 ± 4.93 AAE/100 g), total phenolics (21.7 ± 4.92 GAE/100 g) and total flavonoids (42.7 ± 0.46 QE/100 g). A study on qualitative phytochemical analysis revealed the presence of oil in chloroform extract of the fruits; flavones and oil in hexane extract; and oil, phenol, flavones, quinones, saponins and tannins in methanol extract [20].

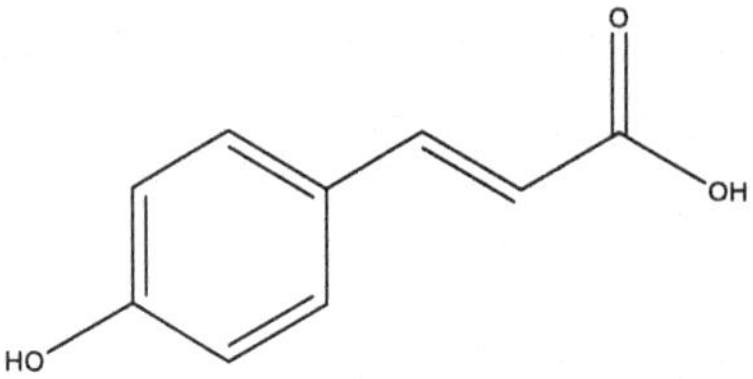

P- Coumaric acid

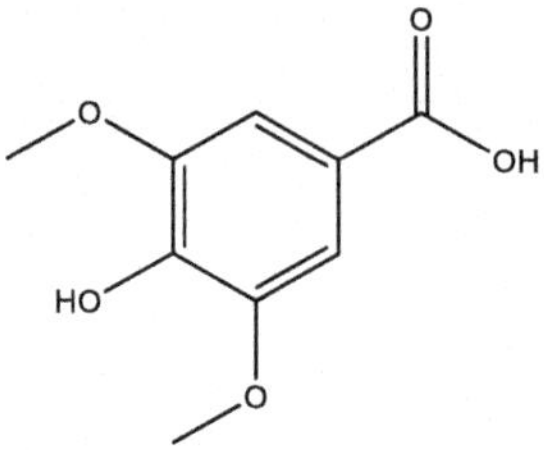

Syringic acid

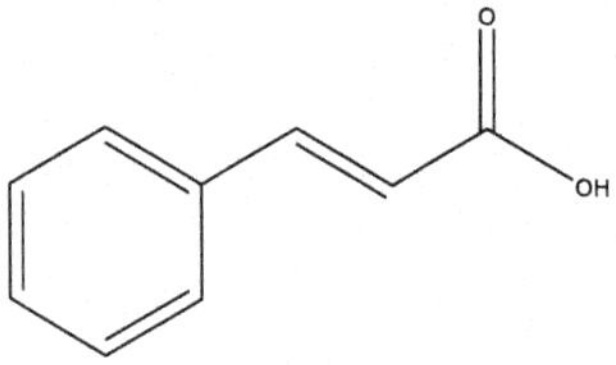

Trans- Cinnamic acid

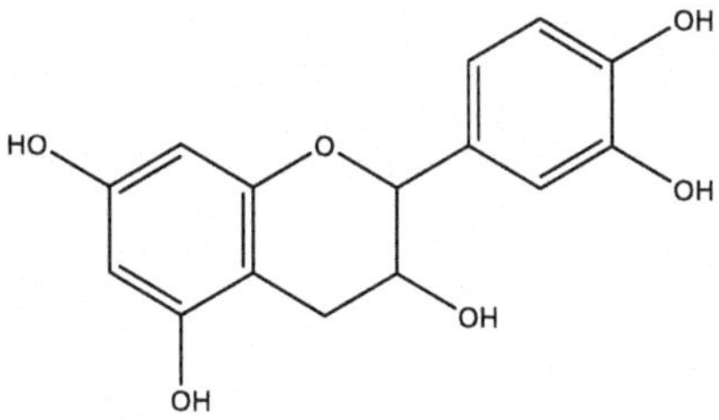

Catechin

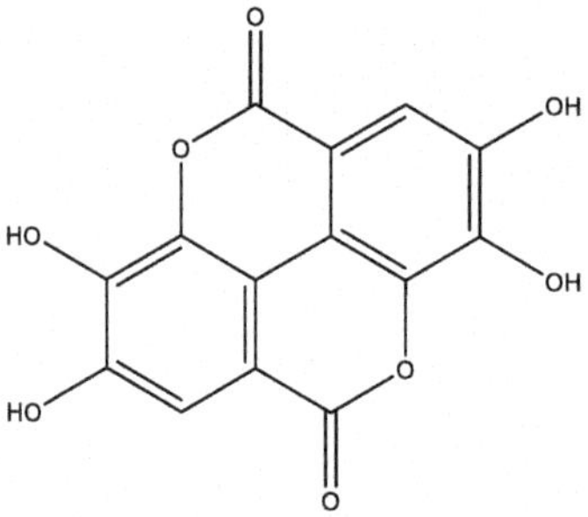

Ellagic acid

Fig. 15.4 (continued)

Antioxidant Potency

The methanol extract of the *F. leucopyrus* fruits was reported to quench DPPH free radicals efficiently with an IC_{50} 76.8 ± 1.22 mg/mL [19].

Fig. 5 Fruits of *F. leucopyrus* (India Biodiversity portal)

Fig. 6 *Gaultheria fragrantissima* fruits (India Biodiversity portal)

Gaultheria fragrantissima Wall.

Botany and Therapeutic Value

This species is an ever green shrub or a small tree and grows up to 100–300 cm tall. The plant is generally variable in size and shape of the leaves. If the species is found in forest, it grows a large shrub with large leaves. If it is growing on slopes, it is a small shrub with narrow leaves (Fig. 6).

Phytochemistry

Karuppusamy et al. [19] published phytochemical examination and antioxidant potential of methanol extract from *G. fragrantissima* fruits. The phytochemical investigation revealed the contents of anthocyanin (2.47 ± 0.09 CGE/100 g), ascorbic acid (67.6 ± 8.41 AAE/100 g), total phenolics (80.4 ± 3.18 GAE/100 g) and total flavonoids (94.3 ± 1.35 QE/100 g).

Antioxidant Potency

The methanol extract of the *G. fragrantissima* fruits was found to scavenge DPPH free radicals effectively with an IC_{50} 240.2 ± 1.28 mg/mL [19].

Grewia tiliifolia Vahl.

Botany and Therapeutic Value

The tree *G. tiliifolia* (family Tilliaceae) reaches a height of 20 m with 8-m length and 65-cm diameter trunk. The species bears yellow flowers on the thick axillary peduncles (Fig. 7). The fresh and ripe fruits are consumed, and its medicinal values have been documented in Ayurveda, Siddha and Unani medical systems [21, 22]. In ethnomedicine, this fruit is used to cure blood related disorders, and respiratory and cardiac ailments [23].

Fig. 7 *Grewia tiliifolia* (India Biodiversity portal)

Phytochemistry

Phytochemical examination and antioxidant potential of methanol extract from *G. tiliifolia* fruits were evaluated [19]. The outcome of phytochemical investigation showed the contents of anthocyanin (2.6 ± 0.09 CGE/100 g), ascorbic acid (70.5 ± 3.07 AAE/100 g), total phenolics (44.1 ± 1.81 GAE/ 100 g) and total flavonoids (47.1 ± 0.92 QE/100 g). Methanol extract of *G. tiliifolia* fruits was subjected for GC-MS analysis [24]. The major compounds detected in the analysis were 9,12-octadecadienoic acid, methyl ester and trans-13-octadecenoic acid [24]. The mineral analysis investigation showed that the fruits of *G. tiliifolia* owned zinc (5.4 ± 0.53 mg/100 g), copper (2.3 ± 0.05 mg/100 g), manganese (2.7 ± 0.015 mg/100 g), iron (31.3 ± 0.05 mg/100 g), sodium (239.3 ± 1.15 mg/100 g), potassium (1302 ± 2 mg/100 g), phosphorous (2.52 ± 0.04 mg/100 g), calcium (109.4 ± 0.2 mg/100 g), magnesium (239.3 ± 1 mg/100 g) and nitrogen (0.96 ± 0.02 mg/100 g) [25].

Antioxidant Potency

The methanol extract of the *G. tiliifolia* fruits was detected to quench DPPH free radicals capably with an IC_{50} 120.2 ± 1.86 mg/mL [19].

Mahonia leschenaultii (Wall. Wight & Arn.)

Botany and Therapeutic Value

Mahonia leschenaultii is a small tree or shrub of 4–7 m tall with compound, imparipinnate leaves. Leaves are opposite, sessile, ovate, glabrous, rigid, coriaceous with shiny top and pale beneath. Flowers are bisexual, yellow in colour, in recemes fascicled at the apex of the stem. The fruit is a berry, 8 mm across, ellipsoid-globose, purple in colour with bitter taste and three to five seeded (Fig. 8). The fruits are edible and used as diuretic and demulcent in dysentery [26].

Phytochemistry

Karuppusamy et al. [19] reported phytochemical screening and antioxidant capacity of methanol extract of *M. leschenaultii* fruits. The phytochemical analysis revealed the contents of anthocyanin (8.58 ± 0.02 CGE/100 g), ascorbic acid (69.9 ± 9.76 AAE/100 g), total phenolics (86.8 ± 0.3 GAE/ 100 g) and total flavonoids (95.5 ± 1.76 QE/100 g).

Fig. 8 *Mahonia leschenaultii* (India Biodiversity portal) licensed under CC-BY 4.0

Antioxidant Potency

The methanol extract of the *M. leschenaultii* fruits was observed to be active against the DPPH stable free radicals with an IC_{50} 361.2 ± 3.69 mg/mL [19, 27].

Rubus ellipticus Smith.

Botany and Therapeutic Value

Rubus ellipticus Smith., belonging to the family Rosaceae, is generally mentioned as Himalayan raspberry or Indian yellow raspberry. Locally it is referred to as '*Mulli hannu*'. Being native to tropical and subtropical India and Asia, the yellow aggregate fruit-bearing plant is found in the Nilgiris and Palani Hills, Western Ghats, India, at 1800-m altitude [8] (Fig. 9).

Phytochemistry

The fruits of *R. ellipticus* obtained from the Nilgiris were evaluated for their polyphenolic composition [28]. They were found to possess high content of total phenolics (610 mg gallic acid equivalents/100 g of fresh weight) and total flavonoids

Fig. 9 The fruits of *R. ellipticus* (created by the first author)

(92 mg quercetin equivalents/100 g of fresh weight). Besides, high-performance liquid chromatography (HPLC) analysis assisted to identify two polyphenolics, i.e. ellagic acid (retention time 4.5) and quercetin (retention time 5.45). The fruits of *R. ellipticus* from the Himalayan region yielded gallic acid, chlorogenic acid, caffeic acid, ellagic acid, m-coumaric acid, 3-hydroxybenzoic acid, 4-hydroxybenzoic acid, ferulic acid, vanillic acid, trans-cinnamic acid, catechin, kaempferol, phloridzin and cyanin [29–32] (Fig. 4).

Antioxidant Potency

Fruits of *R. ellipticus* from the Nilgiris, India, was examined for their antioxidant/free radical scavenging capacity [28]. The methanol extract of *R. ellipticus* was found capable of scavenging DPPH radical (EC_{50} 9.85 ± 1.33 μg mL^{-1}) which was more effective than the activity of ascorbic acid (EC_{50} 11.24 ± 0.02 μg mL^{-1}) and butyl hydroxy toluene (EC_{50} 26.12 ± 0.04 μg mL^{-1}). The extract of the fruit exhibited remarkable quenching potency against superoxide anion radical ($O_2^{\cdot-}$) with an EC_{50} value of 64.65 ± 0.82 μg mL^{-1}. The hydroxyl ion (OH^-) was effectively inhibited by the methanol extract of *R. ellipticus* fruit (EC_{50}79.98 1.02 μg mL^{-1}). The fruit extract also exerted a noticeable scavenging activity against the reactive nitrogen species (RNS), nitric oxide (NO), with an EC_{50} value of 75.21 ± 1.32 μg mL^{-1}. A potent Fe^{2+} chelation (EC_{50} 45.24 1.42 μg mL^{-1}) and ferrous reducing (Fe^{3+} to Fe^{2+}) capacity (OD value at 1000 μg mL^{-1} 1.435 ± 0.021) of the extract was observed. The fruit extract was also active against lipid peroxide (LPO) formation (EC_{50} 71.1 0.22 μg mL^{-1}). The DPPH radical scavenging potency (196.4 1.84 μg/ml AEAC/100 g) of *R. ellipticus* fruits from Palani Hills, Western Ghats, was also reported [19]. Several studies from other parts of the nation also vindicated the antioxidant potent of the *R. ellipticus* fruits. Thus, further investigations on this fruit are vital to develop commercial antioxidants.

Fig. 10 *Rubus niveus* (India Biodiversity portal) licensed under CC-BY 4.0

Rubus niveus Thunb.

Botany and Therapeutic Value

Rubus niveus Thunb, (Rosaceae), locally called '*belle mulli hannu*', is distributed in the Nilgiris, Western Ghats, India (Fig. 10). The plant also referred as Mysore raspberry or hill raspberry. The plant is used for the treatment of ailments like menorrhagia [33].

Phytochemistry

Phytochemicals like steroids, saponins, flavonoids, tannins and triterpenoids were extracted from the fruits [31]. The phenolics encompassed gallic acid, catechin and chlorogenic acid. The fruits were reported to contain crude protein (3.28 ± 0.87 g/100 g), crude fibre (5.9 ± 0.25 g/100 g), crude lipid (1.1 ± 0.07 g/100 g) and carbohydrates (85.35 ± 0.25 g/100 g) [32].

Antioxidant Potency

In vitro antioxidant property of *R. niveus* fruits was performed using standard assays [18]. A strong quenching activity against DPPH (EC_{50} 0.2 ± 0.002 mg/mL), $O_2^{\cdot -}$ (EC_{50} 0.5 ± 0.82 mg/mL), OH^- (0.25 ± 0.22 mg/mL) and NO (0.25 ± 1.3 mg/mL) was executed by the fruits of *R. niveus*. In addition, the fruits were able to chelate Fe^{2+} (0.25 ± 1.02 mg/mL) efficiently and reduced LPO (0.502 ± 0.24 mg/mL) excellently. The methanol extract of *R. niveus* fruit exhibited remarkable activity against ABTS radical (2.19 mM), DPPH free radical (22.84 mM) and by FRAP assay (2.06 mM AE/100 g).

Ziziphus rugosa Lam.

Botany and Therapeutic Value

Ziziphus rugosa Lam., belonging to family Rhamnaceae, is located in dry deciduous forests. The plant is large straggling thorny 3–6-m shrub or small tree with alternate, dark green, broadly ovate or elliptic, serrate leaves and subcordate or rounded or oblique at the base. The tree's wood is red in colour and hard. The sub-globose or obovoid globose fruits are orange or black in colour with 9–12 mm in length and 6–7 mm in width (Fig. 11).

In the Western Ghats, the Kodava community people consume the fruits as raw and ripe as a source of nutrition. This fruit is commonly consumed during famine time. The ripe fruit is ground to prepare a local dish called '*dosa*'. The fruit is emollient to treat irritation in throat and broncho-pulmonary disease. In addition, boils are treated by powder from fruits and leaves. The local people of villages of the upper Ghat and two villages belong to central Western Ghats of Karnataka state, India, use the fruits of *Z. rugosa* to keep body hydrated and as coolant. The fruit is also used to treat diarrhoea and wounds [33].

Fig. 11 Fruits of *Z. rugosa* (India Biodiversity Portal licensed under CC-BY 4.0)

Phytochemistry

The analysis of phytochemistry of the fruits of *Z. rugosa* was carried out by Karuppusamy et al. [19]. The work revealed the anthocyanin content (0.34 ± 0.02), ascorbic acid (35.0. ± 3.21 AAE/100 g), total phenolics (41.8 ± 0.20 GAE/100 g) and total flavonoids (41.8 ± 1.20 QE/100 g). The proximate composition and mineral analysis investigation revealed that the fruits of *Z.rugosa* possessed moisture content (62.2%), proteins (11%), fat (5%), total carbohydrates (26%), zinc (3.074 ppm), copper (62.4 ppm), manganese (609.1 ppm), iron (916.728 ppm), sodium (185%), potassium (16,5%), phosphorous (16.5%), calcium (17.1%), magnesium (22%) and nitrogen (19.4%) (35).

Antioxidant Potency

The antioxidant capacity of *Z. rugosa* fruits was measured using DPPH scavenging test [30]. The methanol extract of the fruits was capable of scavenging the DPPH free radicals with an IC_{50} value of 61.88 μg/mL. Karuppusamy et al. [19] reported the effective quenching of DPPH radicals (IC_{50} ± mg/mL) by the methanol extract of *Z. rugosa* fruits.

References

1. Hasler CM. Functional food: their role in disease prevention and health promotion. Food Technol. 1998;52:63–70.
2. Steinmetz KA, Potter JD. Vegetables, fruit and cancer prevention—review. J Am Dietic Assoc. 1996;325–51.
3. Scalbert A, Manach C, Morand C, Remesy C, Jimenez L. Dietary polyphenols and the prevention of diseases. Crit Rev Food Sci Nutr. 2005;45:287–306.
4. Prior RL. Fruits and vegetables in the prevention of cellular damage. Am J Clin Nutr. 2003;78:578S.
5. Sogut S, Zoroglu SS, Ozyurt H, Yilmaz HR, Ozugurlu F, Sivasli E. Changes in nitric oxide levels and antioxidant enzyme activities may have a role in the pathophysiological mechanisms involve in autism. Clin Chim Acta. 2003;331:111–7.
6. Aruoma OI. Nutrition and health aspects of free radicals and antioxidants. Food Chem Toxicol. 1994;32:671–85.
7. Rice-Evans C, Miller NJ, Paganga G. Antioxidant properties of phenolic compounds. Trends Plant Sci. 2001;2:152–9.
8. Wealth of India. *Berberis tinctoria* (Berberidaceae). Ambastha publication and Information Directorate, CSIR, New Delhi, India; 1988. p. 144–8.
9. Fyson PF. The flora of he Nilgiris and Palney Hill Tops, Vol - 1, Bishen Singh and Mahendra Pal Singh, Dehra Dun. 1974.
10. Sasikumar JM, Maheshu V, Smilin AG, Mm G, Joji C. Antioxidant and antihemolytic activities of common Nilgiri barberry berberis tinctoria Lesch. from South India. Int Food Res J. 2012;19(4):1601–7.
11. Vignesh A, Veerakumar KP, Selvakumar S, Rakkiappan R, Vasanth K. Nutritional assessment, antioxidant, anti-inflammatory and antidiabetic potential of traditionally used wild plant, *Berberis tinctoria* Lesch. Trends Phytochem Res. 2021;5(2):71–91.

12. Vignesh A, Selvakumar S, Vasanth K. Green synthesis and characterization of zinc oxide nanoparticles using *Berberis tinctoria* Lesch. leaves and fruits extract of multibiological applications. Nanomed Res J. 2021;6(2):128–47.
13. Paulsamy S, Kumar S, Anandhakumar AM, Satishkumar P. Elaeagnus kologa Schecht.—an underutilized edible and endemic fruit plant in Nilgiris, the western Ghats. Indian J Nat Prod Resour. 2008;1:258–60.
14. Sasikumar JM, Adithya ES, Christabel PH, Shamna R. Antioxidant capacity and phenolic content of *Elaeagnus kologa* Schlecht. An underexploited fruit from India. Free Radiac Antioxid. 2012;2(3):1–7.
15. Muthuswamy R, Senthamarai R. Pharmacognostical studies on the fruit of Elaeocarpus oblongus Gaertn. Pharmacogn J. 2014;6(3):72–8.
16. Nayagam MC, Pushparaj MS, Rajan S. Less known edible fruits yielding plants of Nilgiris, Ancient. Sci Life. 1993;12:363–76.
17. Mundaragi A, Thangadurai D, Sangeetha J, Bhat S. Phenolic composition, volatile constituents and antioxidant potential of wild edible fruit *Elaeeocarpus tectorius* (Lour). Poir (Elaeeocarpaceae). Farmacia. 2019;67(2):311–7.
18. Shamna R, Sasikumar JM, Adithya ES, Christabel PH, Lakshmi MS. Antioxidant capacity of *Rubus niveus* Thunb. and *Elaeocarpus oblongus* Gaestn. fruits. Food. 2011;5(1):96–101.
19. Karuppusamy S, Muthuraja G, Rajasekaran KM. Antioxidant activity of selected lesser known edible fruits from Western Ghats of India. Indian J Nat Prod Resour. 2011;2(2):174–8.
20. Ravi V, Thalavai PA, Renuga R, Sivapriya O, Vijayakanth P. Antifungal of fruit extracts of *Flueggea leucopyrus* Willd. against phytopathogenic fungi *Aspergillus* spp. IOSR J Pharm Biol Sci. 2019;141(1):1–7.
21. Ullah W, Uddin G, Siddique BS. Ethnic uses, pharmacological and phytochemical profile of Genus Grewia. J Asian Nat Prod Res. 2012;14:186–95.
22. Darmasoth RD, Rao BG. Phytochemical and pharmacological review of *Grewia tillaefolia* (Vahl.). Int Res J Pharm. 2019;123(1):39–42.
23. Poonam K, Singh GS. Ethnobotanical study of medicinal plants used by the Taunga community in Terai Arc Landscape, India. J Ethnopharmacol. 2009;123(1):167–76.
24. Singh S, Bharadwaj P, Sharma A, Bharadwaj G, Bhagat NR, Kumar P, Naryal A, Giri A. Phytochemical screening of Dhaman (Grewia tilifolia Vahl.) fruit in sub-tropical region of India. J Plant Dev. 2023;30:55–67.
25. Valvi SR, Rathod VS. Mineral composition of some wild edible fruits from Kolhapur district. Int J Appl Biol Pharm Technol. 2011;2:392–6.
26. Kirthikar KR, Basu BD. Indian medicinal plants, Allahabad. Lalit Mohan Basu Publisher; 1935.
27. Vishnu Kumar S, Jemima DK, Dharani S, Narayanan N. Evaluation of phytochemical profile, antibacterial and antioxidant activity of Mahonia leschenaultii (Wallich ex Wight & Arnott) Takeda. Plant Arch. 2023;23(1):285–92.
28. Sasikumar JM, Poulin RC, Meseret CE, Selvakumar P. *In vitro* analysis of antioxidant capacity of Indian yellow raspberry (*Rubus ellipticus* Smith.). Int Food Res J. 2015;22(4):1338–46.
29. Belwel T. Trends of polyphenolics and anthocyanins accumulation along ripening stages of wild edible fruits of Himalayan region. Sci Rep. 2019;9:1–11.
30. Saini R. Antioxidant and antiproliferative activities of phenolics isolated from fruits of Himalayan yellow Raspberry (*Rubus ellipticus*). J Food Sci Technol. 2014;51:3369–75.
31. Badhani A, Rawat B, Eawat RS. Variation in chemical constituents and antioxidant activity in yellow Himalayan (*Rubus ellipticus* Smith.) and hill raspberry (*Rubus niveus* Thunb.). J Food Biochem. 2015;39:663–72.
32. Muniyandi K. Phenolics, tannins, flavonoids and anthocyanins contents influenced antioxidant and anticancer activities of Rubus fruits from Western Ghats, India. Food Sci Hum Wellness. 2019;8:73–81.
33. Hegde VM, Vasudeva R, Kamatekar SL, Sthapit BR, Parthasarathy VA, Rao VR. Traditional knowledge associated with tropical fruit tree genetic resources: comparison of upper-ghat and coastal situation of Central Western Ghats, India. Indian J Plant Resour. 2015;28:95–105.

Edible Fruit-Yielding Wild Plants of Tamil Nadu: A Preliminary Checklist

G. Gnanasekaran, E. Angelin Felicia, S. P. Nithya, and W. Arisdason

Abstract

The present chapter is an enumeration of edible fruit-yielding wild plants of Tamil Nadu, the southernmost state of India, with 102 flowering plant species belonging to 37 families. For each species, the taxonomic details, such as botanical name, habit, Tamil name(s), distribution (world and Tamil Nadu) and note (uses) are provided. Among the different growth forms, tree is the dominant one with 51 species followed by shrubs (34), herbs (08), lianas (05) and climbers (04). Solanaceae, Rutaceae, Rhamnaceae, Rubiaceae and Rosaceae are the dominant families represented by a minimum of five species.

Keywords

Conservation · Edible fruits · Forest · Livelihood · Wild plants

Introduction

Wild plants that yield edible fruits play a vital role in human health, as wild fruits are the rich source of nutrients. They are associated with specific traditional knowledge and ethnobotanical practices. Wild edible fruits are denoted as fruits that are collected from non-cultivated wild plants from forest [1, 2]. The consuming of wild

G. Gnanasekaran (✉)
Department of Botany, Madras Christian College (Autonomous),
Tambaram East, Chennai, Tamil Nadu, India
e-mail: gnanasekaran@mcc.edu.in

E. Angelin Felicia · S. P. Nithya · W. Arisdason
Botanical Survey of India, Southern Regional Centre, TNAU Campus,
Coimbatore, Tamil Nadu, India

P. Thangaraj (ed.), *Pomotherapeutic Insights on Wild Edible Fruits*,
https://doi.org/10.1007/978-981-96-0543-9_16

edible plant parts especially the fruits is the inevitable part of the dietary habits of living beings worldwide. Wild edible fruits are an essential supplement in the diet of human beings, especially tribal people, and also can be sold in markets for generating daily income. The consumption of wild fruits is depending on the traditional knowledge on the edibility, to alleviate the nutrient deficiency.

In this modernized era, people have been attracted to highly processed foods or fast food culture, which affect the human health negatively. Due to modernization and industrialization, the reliance on these wild edible plants is declining in a fast rate, so that the nutritive values of such plants have been ignored by the people. The increase in malnutrition due to the modern food habits poses the urgency of creating awareness on the necessity of improved and sustainable food practices. Consequently, the wild edible plants that are enriched with vitamins, anthocyanins, flavonoids, micro- and macro-nutrients and fibre possess an integral part in maintaining the people's health in proper manner [3].

The wild edible plants possess various biologically active components such as antioxidant, antimicrobial, anti-inflammatory, free radical scavenging and anticancer activities, which can be applied to treat different chronic diseases, namely, asthma, diabetes, cancer, Alzheimer's and heart diseases [4–6]. Thus, it is of utmost need to identify the neglected and under-utilized species that show potential for national plant germplasm conservation and use with an aim to provide food security and to improve the livelihoods of rural communities [7]. The living beings are depending upon over 7000 wild edible plants, but still many useful plants are yet to be explored [8]. Therefore, it is essential to document the traditional knowledge and associated ethnobotanical practices related to the edible plants which are growing in wild; thus these plants will be utilized appropriately and also conserved. These plants are available from open sources and are prone to anthropogenic pressures and climatic changes, which will ultimately lead to the extinction of these species. The major threats to wild edible plants include deforestation, drought, habitat fragmentation and agricultural land degradation. Furthermore, the changes in land use and over-exploitation of plants also pose impact on these plants.

Globally, the studies related to the importance of wild edible plants and their products gained more attention in recent times. Several floristic documentations along with the ethnobotanical uses of plants have also been published in recent decades across the world. The studies carried out in the past decades revealed the value of wild edible plants, the key ingredient in human diet. Therefore, a preliminary attempt has been made to enumerate the edible fruit-yielding wild plants of Tamil Nadu based on personal observations made during the plant explorations conducted in different parts of the state and scrutiny of pertinent literature [9–16]. Though many inventories on ethnobotanical as well as various uses of the plants in specific regions have been made and recorded [17–22], there is no exclusive comprehensive data available on the wild edible fruit-yielding plants of Tamil Nadu, which are widely utilized by local people for their livelihood purposes, especially for consumption.

The present study is a preliminary attempt to enumerate the edible fruit-yielding wild plants of Tamil Nadu. The plants are arranged family-wise in alphabetical order, and the members of each family are arranged alphabetically with correct botanical name, habit, distribution (nativity, followed by district-wise distribution in Tamil Nadu alphabetically) and note (uses).

Enumeration of Edible Fruit-Yielding Wild Plants

Anacardiaceae

1. Botanical name: *Buchanania axillaris* (Desr.) Ramamoorthy (Fig. 1a)
   Habit: Tree
   Tamil name: Kalmaangai, Kaattu maangai and Mudai maangai
   Distribution: Native to India and Sri Lanka; Tamil Nadu: Most of the districts
   Note: Fruits edible, also used in traditional medicines
2. Botanical name: *Buchanania cochinchinensis* (Lour.) M.R. Ameida
   Habit: Tree
   Tamil name: Kaattu maangai, Saarai paruppu
   Distribution: Native to the Indian subcontinent, South Central China and Southeast Asia; Tamil Nadu: Coimbatore, Dharmapuri, Dindigul, Kanniyakumari, Nilgiris, Salem, Theni, Tirunelveli, Tiruvannamalai and Virudhunagar
   Note: Fruits edible, also used in traditional medicines.
3. Botanical name: *Semecarpus anacardium* L.f.
   Habit: Tree
   Tamil name: Seeran kottai
   Distribution: Native to the Indian subcontinent to Myanmar; Tamil Nadu: Coimbatore, Dharmapuri, Dindigul, Kanniyakumari, Nilgiris, Salem, Tiruvannamalai, Vellore and Viluppuram
   Note: Fruits edible
4. Botanical name: *Spondias pinnata* (L.f.) Kurz
   Habit: Tree
   Tamil names: Kaattu maamaram, Mambulichai, Puli ilai and Puli maa
   Distribution: Native to the Indian subcontinent to China and Malesia; Tamil Nadu: Coimbatore, Dharmapuri, Dindigul, Erode, Kanniyakumari, Namakkal, Nilgiris, Salem, Theni, Tiruvannamalai, Vellore, Viluppuram and Virudhunagar
   Note: Fruits edible

Fig. 1 **Edible fruit-yielding wild plants of Tamil Nadu**: (**a**) *Buchanania axillaris* (Desr.) Ramamoorthy; (**b**) *Carissa spinarum* L.; (**c**) *Ehretia microphylla* Lam.; (**d**) *Capparis brevispina* DC.; (**e**) *Diospyros chloroxylon* Roxb.; (**f**) *Diospyros ferrea* (Willd.) Bakh.; (**g**) *Scutia myrtina* (Burm.f.) Kurz; (**h**) *Manilkara hexandra* (Roxb.) Dubard

Annonaceae

5. Botanical name: *Huberantha korinti* (Dunal) Chaowasku
   Habit: Spreading shrub
   Tamil names: Sirunetti and Uluvindai
   Distribution: Native to southern India and Sri Lanka; Tamil Nadu: Coimbatore, Cuddalore, Dindigul, Kancheepuram, Madurai, Nilgiris, Tirunelveli, Tiruvallur and Viluppuram
   Note: Fruits edible

Apocynaceae

6. Botanical name: *Carissa carandas* L.
   Habit: Shrub
   Tamil names: Kala and Perungala
   Distribution: Native to India and Bangladesh; Tamil Nadu: Most of the districts
   Note: Fruits traded in local markets and used in traditional medicine. Fruits pickled. Fruits deseeded, dipped in sugar syrup, coloured and sold as an alternate to cherries, which are used as toppings in cakes and other bakery products.
7. Botanical name: *Carissa spinarum* L. (Fig. 1b)
   Habit: Shrub
   Tamil names: Sirukala
   Distribution: Native to Africa to Indo-China, Australia to New Caledonia; Tamil Nadu: Most of the districts
   Note: Fruits edible

Arecaceae

8. Botanical name: *Borassus flabellifer* L.
   Habit: Tree
   Tamil name: Panai maram
   Distribution: Native to India to Indochina and Indonesia; Tamil Nadu: Most of the districts
   Note: Tender kernels and fruits are edible. All parts of the species are useful. An old Tamil literature, 'Thaalai Vilasam', lists nearly 1000 uses for this single species.
9. Botanical name: *Phoenix loureiroi* Kunth
   Habit: Tree
   Tamil name: Malai eecham
   Distribution: Native to Indian subcontinent to southern China and the Philippines; Tamil Nadu: Most of the districts
   Note: Fruits edible

10. Botanical name: *Phoenix loureiroi* var. *pedunculata* (Griff.) Govaerts
    Habit: Tree
    Tamil name: Malai eecham
    Distribution: Native to the Indian subcontinent; Tamil Nadu: Dindigul
    Note: Fruits edible
11. Botanical name: *Phoenix sylvestris* (L.) Roxb.
    Habit: Tree
    Tamil name: Periya eecham
    Distribution: Native to the Indian subcontinent to western Myanmar; Tamil Nadu: Cuddalore, Dharmapuri, Dindigul, Kancheepuram, Thanjavur, Tiruvallur, Tiruchirappalli, Tiruvannamalai, Vellore and Viluppuram
    Note: Fruits edible

Asteraceae

12. Botanical name: *Guizotia abyssinica* (L.f.) Cass.
    Habit: Herb
    Tamil name: Payellu
    Distribution: Native to northeastern tropical Africa; Tamil Nadu: Dharmapuri, Namakkal, Salem, Tiruchirappalli, Tiruvannamalai and Vellore
    Note: Seeds edible and source of oil

Berberidaceae

13. Botanical name: *Berberis lycium* Royle
    Habit: Herb
    Tamil name: Himalaya maramanjal
    Distribution: Native to eastern Afghanistan to Himalayas and India; Tamil Nadu: Nilgiris
    Note: Fruits edible and also used in traditional medicine
14. Botanical name: *Berberis tinctoria* Lesch.
    Habit: Shrub
    Tamil name: Saaya maramanjal
    Distribution: Native to southern India and Sri Lanka; Tamil Nadu: Coimbatore, Dindigul and Nilgiris
    Note: Fruits edible

Boraginaceae

15. Botanical name: *Cordia dichotoma* G. Forst.
    Habit: Tree
    Tamil names: Mooku chali pazham and Narivizhi
    Distribution: Native to the Indian subcontinent to Nansei-shoto and southwestern Pacific; Tamil Nadu: Most of the districts
    Note: Fruits edible

16. Botanical name: *Cordia diffusa* K.C. Jacob
    Habit: Tree
    Tamil name: Nila narivizhi
    Distribution: Native to India; Tamil Nadu: Coimbatore and Thoothukudi
    Note: Fruits edible
17. Botanical name: *Ehretia acuminata* R. Br.
    Habit: Tree
    Tamil names: Kalvirasu and Kuruvichai
    Distribution: Native to tropical and subtropical Asia to eastern Australia; Tamil Nadu: Most of the districts
    Note: Fruits edible
18. Botanical name: *Ehretia aspera* Willd.
    Habit: Tree
    Tamil name: Kalvirasu
    Distribution: Native to Pakistan to Hainan and Peninsular Malaysia; Tamil Nadu: Most of the districts
    Note: Fruits edible
19. Botanical name: *Ehretia microphylla* Lam. (Fig. 1c)
    Habit: Shrub
    Tamil name: Kuruvi pazham
    Distribution: Native from India to Nansei-shoto and Solomon Islands; Tamil Nadu: Most of the districts
    Note: Fruits edible

Burseraceae

20. Botanical name: *Garuga pinnata* Roxb.
    Habit: Tree
    Tamil names: Arunelli, Enni kaarai, Karuvaembu and Karuvilaangan
    Distribution: Native to India to southeastern China and Indochina; Tamil Nadu: Coimbatore, Dharmapuri, Dindigul, Nilgiris, Salem, Theni, Tiruchirappalli, Tirunelveli, Tiruvannamalai and Viluppuram
    Note: Fruits edible

Cactaceae

21. Botanical name: *Acanthocereus tetragonus* (L.) Hummelinck
    Habit: Shrub
    Tamil name: Mukkonam kalli
    Distribution: Native to southern Florida (USA) to the Caribbean, southern Texas (USA) to Venezuela; Tamil Nadu: Most of the districts
    Note: Fruits edible

22. Botanical name: *Brasiliopuntia brasiliensis* (Willd.) A. Berger
    Habit: Shrub
    Tamil name: Perikaai kalli
    Distribution: Native to Brazil to eastern Peru and northern Argentina; Tamil Nadu: Most of the districts
    Note: Fruits edible
23. Botanical name: *Cylindropuntia ramosissima* (Engelm.) F.M. Knuth
    Habit: Shrub
    Tamil name: Uruttai chappathi kalli
    Distribution: Native to southern Nevada (USA) to northwestern Mexico; Tamil Nadu: Coimbatore
    Note: Fruits edible
24. Botanical name: *Opuntia monacantha* (Willd.) Haw.
    Habit: Shrub
    Tamil name: Chappathikalli
    Distribution: Native to eastern and southern Brazil to Uruguay; Tamil Nadu: Most of the districts
    Note: Fruits edible

Calophyllaceae

25. Botanical name: *Mesua ferrea* L.
    Habit: Tree
    Tamil names: Nangu, Pongu and Siru naga poo
    Distribution: Native to the Indian subcontinent to Indochina and western and central Malesia; Tamil Nadu: Coimbatore, Dindigul, Kanniyakumari, Madurai, Nilgiris, Theni and Tirunelveli
    Note: Fruits edible

Capparaceae

26. Botanical name: *Cadaba fruticosa* (L.) Druce
    Habit: Straggling shrub
    Tamil name: Vizhuthi
    Distribution: Native to the Indian subcontinent; Tamil Nadu: Most of the districts
    Note: Fruits pickled and also used in traditional medicine
27. Botanical name: *Capparis brevispina* DC. (Fig. 1d)
    Habit: Shrub
    Tamil names: Kaattu aathondai and Pudhar aathondai
    Distribution: Native to India and Sri Lanka; Tamil Nadu: Cuddalore, Kancheepuram, Pudukkottai, Ramanathapuram, Salem, Tiruchirappalli, Tirunelveli, Tiruvallur and Viluppuram
    Note: Fruits edible

28. Botanical name: *Capparis decidua* (Forssk.) Edgew.
    Habit: Shrub
    Tamil name: Ilai aathondai
    Distribution: Native to Mauritania to Andaman and Nicobar Islands; Tamil Nadu: Tirunelveli
    Note: Fruits edible and used in traditional medicine
29. Botanical name: *Capparis zeylanica* L.
    Habit: Liana
    Tamil name: Aathondai
    Distribution: Native to southeastern China to tropical Asia; Tamil Nadu: Most of the districts
    Note: Fruits edible

Celastraceae

30. Botanical name: *Celastrus paniculatus* Willd.
    Habit: Liana
    Tamil names: Kodiyam, Thippadigam and Vaaluluvai
    Distribution: Native to the Indian subcontinent to southern China and Indochina; Tamil Nadu: Most of the districts
    Note: Fruits edible
31. Botanical name: *Elaeodendron glaucum* (Rottb.) Pers.
    Habit: Tree
    Tamil names: Karuvaali and Khiri
    Distribution: Native to Sri Lanka; Tamil Nadu: Most of the districts
    Note: Fruits edible

Cornaceae

32. Botanical name: *Alangium salviifolium* (L.f.) Wangerin
    Habit: Tree
    Tamil names: Azhinjil and Yerazhinjil
    Distribution: Native to eastern tropical Africa, Comoros and the Indian subcontinent; Tamil Nadu: Most of the districts
    Note: Fruits edible

Cucurbitaceae

33. Botanical name: *Coccinia grandis* (L.) Voigt
    Habit: Climber
    Tamil name: Kovai
    Distribution: Native to tropical Africa, western Arabian Peninsula and tropical and subtropical Asia
    Note: Unripe fruits used in preparation of curries, and ripe fruits are eaten raw.

34. Botanical name: *Cucumis maderaspatanus* L.
    Habit: Climber
    Tamil name: Musumusukkai
    Distribution: Native to India to Nansei-shoto and Australia; Tamil Nadu: Most of the districts
    Note: Fruits edible
35. Botanical name: *Diplocyclos palmatus* (L.) C. Jeffrey
    Habit: Climber
    Tamil name: Aivirali
    Distribution: Native to tropical and southern Africa, southern Arabian Peninsula and tropical and subtropical Asia; Tamil Nadu: Most of the districts
    Note: Fruits edible

Dilleniaceae

36. Botanical name: *Dillenia indica* L.
    Habit: Tree
    Tamil names: Ugaa kai, Uvaa, Uvaa thekku
    Distribution: Native to India to China and western and central Malesia; Tamil Nadu: Chennai, Kancheepuram and Nilgiris
    Note: Fruits used as vegetable; used in traditional medicine

Dioscoreaceae

37. Botanical name: *Trichopus zeylanicus* Gaertn.
    Habit: Herb
    Tamil name: Arogyapacha
    Distribution: Native to southwestern India to peninsular Malaysia; Tamil Nadu: Kanniyakumari and Tirunelveli
    Note: Fruits are eaten by Kani tribes for its anti-fatigue properties.

Ebenaceae

38. Botanical name: *Diospyros chloroxylon* Roxb. (Fig. 1e)
    Habit: Tree
    Tamil names: Karu vakkanai and Vaganai
    Distribution: Native to central and southern India to Bangladesh; Tamil Nadu: Chengalpattu, Coimbatore, Cuddalore, Kancheepuram, Pudukkottai, Tiruvallur, Tiruvannamalai, Vellore and Viluppuram
    Note: Fruits edible
39. Botanical name: *Diospyros ferrea* (Willd.) Bakh. (Fig. 1f)
    Habit: Shrub
    Tamil names: Irumbili and Irumbuli
    Distribution: Native to tropical Africa to southwestern Pacific; Tamil Nadu: Most of the districts
    Note: Fruits edible

40. Botanical name: *Diospyros malabarica* (Desr.) Kostel.
    Habit: Tree
    Tamil names: Ettu kottai maram and Thumbika
    Distribution: Native to tropical Asia; Tamil Nadu: Chennai, Coimbatore, Cuddalore, Dharmapuri, Kancheepuram, Nilgiris, Salem, Tirunelveli, Tiruvallur, Tiruvannamalai and Viluppuram
    Note: Fruits edible
41. Botanical name: *Diospyros melanoxylon* Roxb.
    Habit: Tree
    Tamil names: Beedi ilai, Kaarai and Karumthumbi
    Distribution: Native to tropical Asia; Tamil Nadu: Chengalpattu, Coimbatore, Dharmapuri, Kancheepuram, Nilgiris, Salem, Tirunelveli, Tiruvannamalai and Viluppuram
    Note: Fruits edible
42. Botanical name: *Diospyros montana* Roxb.
    Habit: Tree
    Tamil names: Thuvarai, Vakkanai and Vellai
    Distribution: Native to tropical Asia; Tamil Nadu: Coimbatore, Dharmapuri, Kancheepuram, Nilgiris, Salem, Tiruchirappalli, Tirunelveli, Tiruvallur, Tiruvannamalai, Vellore and Viluppuram
    Note: Fruits edible

Elaeagnaceae

43. Botanical name: *Elaeagnus conferta* Roxb.
    Habit: Straggling shrub
    Tamil names: Kolungai and Kulari
    Distribution: Native to Himalayas to China and western Malesia; Tamil Nadu: Coimbatore, Dindigul, Erode, Kanniyakumari, Nilgiris, Salem, Theni, Tirunelveli and Virudhunagar
    Note: Fruits edible
44. Botanical name: *Elaeagnus indica* Servett.
    Habit: Straggling shrub
    Tamil name: Kulil chedi
    Distribution: Native to southern India; Tamil Nadu: Coimbatore, Dharmapuri, Dindigul, Erode, Kanniyakumari, Namakkal, Nilgiris, Salem, Theni, Tirunelveli, Tiruvannamalai, Vellore, Viluppuram and Virudhunagar
    Note: Fruits edible
45. Botanical name: *Gaultheria fragrantissima* Wall.
    Habit: Shrub
    Distribution: Native to the Indian subcontinent to China (Yunnan) and Peninsula Malaysia; Tamil Nadu: Dindigul and Nilgiris
    Note: Fruits edible

Erythroxylaceae

46. Botanical name: *Erythroxylum monogynum* Roxb.
    Habit: Tree
    Tamil names: Devadhaaru and Sembulichan
    Distribution: Native to India to Myanmar and Sri Lanka; Tamil Nadu: Most of the districts
    Note: Fruits edible

Fabaceae

47. Botanical name: *Mucuna pruriens* (L.) DC.
    Habit: Climber
    Tamil name: Poonaikaali
    Distribution: Native to tropical and subtropical Old World; Tamil Nadu: Most of the districts
    Note: Seeds edible

Lamiaceae

48. Botanical name: *Gmelina arborea* Roxb. ex Sm.
    Habit: Tree
    Tamil names: Kumizhamaram, Perungumizh and Umithekku
    Distribution: Native to Pakistan to China and northern Indochina; Tamil Nadu: Coimbatore, Dharmapuri, Dindigul, Kancheepuram, Madurai, Namakkal, Nilgiris, Salem, Tiruchirappalli, Tirunelveli, Tiruvannamalai and Virudhunagar
    Note: Fruits edible

Moraceae

49. Botanical name: *Streblus asper* Lour.
    Habit: Tree
    Tamil names: Kuruvipala and Pirasu
    Distribution: Native to southern China to tropical Asia; Tamil Nadu: Most of the districts.
    Note: Fruits edible

Myrtaceae

50. Botanical name: *Myrtus communis* L.
    Habit: Shrub
    Tamil name: Kulinaval
    Distribution: Native to Macaronesia to Pakistan; Tamil Nadu: Dindigul and Nilgiris
    Note: Fruits edible

51. Botanical name: *Rhodomyrtus tomentosa* (Aiton) Hassk.
    Habit: Shrub
    Tamil name: Malai koyya
    Distribution: Native to tropical and subtropical Asia; Tamil Nadu: Coimbatore, Dindigul, Madurai, Nilgiris and Theni
    Note: Fruits edible
52. Botanical name: *Syzygium jambos* (L.) Alston
    Habit: Tree
    Tamil names: Jambunaaval, Perunaaval and Samba naaval
    Distribution: Native to China (southern Yunnan) and western Malesia; Tamil Nadu: Coimbatore, Dindigul, Kanniyakumari, Namakkal, Nilgiris, Salem, Theni and Tirunelveli
    Note: It is also cultivated for its edible fruits.

Olacaceae

53. Botanical name: *Olax scandens* Roxb.
    Habit: Liana
    Tamil names: Kadal azhinji and Kadal ranchi
    Distribution: Native to India, Sri Lanka, Bangladesh to Peninsula Malaysia, Indonesia (Java to Lesser Sunda Islands, Bali); Tamil Nadu: Chengalpattu, Coimbatore, Cuddalore, Dharmapuri, Kancheepuram, Namakkal, Nilgiris, Salem, Tiruchirappalli, Tiruvallur, Tiruvannamalai, Vellore and Viluppuram
    Note: Fruits edible
54. Botanical name: *Ximenia americana* L.
    Habit: Shrub
    Tamil names: Kadal azhinji and Siru illanthai
    Distribution: Native to tropics and subtropics; Tamil Nadu: Coimbatore, Salem, Tiruvannamalai and Viluppuram
    Note: Fruits edible

Phyllanthaceae

55. Botanical name: *Baccaurea courtallensis* (Wight) Müll.Arg.
    Habit: Tree
    Tamil names: Mootu pazham and Pulichan pazham
    Distribution: Native to India; Tamil Nadu: Coimbatore, Kanniyakumari, Nilgiris and Tirunelveli
    Note: Fruits edible
56. Botanical name: *Bridelia retusa* (L.) A. Juss.
    Habit: Tree
    Tamil names: Mullumarathu and Mulvaengai
    Distribution: Native to the Indian subcontinent to southern China and western Malesia; Tamil Nadu: Most of the districts
    Note: Fruits edible

57. Botanical name: *Phyllanthus indofischeri* Bennet
    Habit: Tree
    Tamil name: Karunelli
    Distribution: Native to southern India; Tamil Nadu: Coimbatore, Dharmapuri, Erode, Nilgiris and Tiruvannamalai
    Note: Fruits edible

Primulaceae

58. Botanical name: *Embelia drupacea* (Dennst.) M.R. Almeida & S.M. Almeida
    Habit: Shrub
    Distribution: Native to southern India, Sri Lanka and western Malesia; Tamil Nadu: Nilgiris
    Note: Fruits edible

Putranjivaceae

59. Botanical name: *Drypetes sepiaria* (Wight & Arn.) Pax & K. Hoffm.
    Habit: Tree
    Tamil name: Veerai
    Distribution: Native to India and Sri Lanka; Tamil Nadu: Most of the districts
    Note: Fruits edible

Rhamnaceae

60. Botanical name: *Sageretia filiformis* (Roth) G. Don
    Habit: Tree
    Tamil name: Siru sillaa
    Distribution: Native to Himalayas to northwestern India and northern Thailand, Andaman Islands; Tamil Nadu: Coimbatore, Dharmapuri, Nilgiris, Salem, Tiruchirappalli, Tiruvannamalai and Vellore
    Note: Fruits edible
61. Botanical name: *Scutia myrtina* (Burm.f.) Kurz (Fig. 1g)
    Habit: Tree
    Tamil names: Indu, Kokku mullu and Thutthvi
    Distribution: Native to Ethiopia to southern Africa, India to China and Indochina; Tamil Nadu: Most of the districts
    Note: Fruits edible
62. Botanical name: *Ziziphus mauritiana* Lam. var. *fruticosa* (Haines) Sebastine & A.N. Henry
    Habit: Shrub
    Distribution: Native to India; Tamil Nadu: Tiruchirappalli, Tiruvannamalai and Vellore
    Note: Fruits edible

63. Botanical name: *Ziziphus nummularia* (Burm.f.) Wight & Arn.
    Habit: Shrub
    Tamil name: Korgodi
    Distribution: Native to eastern Mediterranean region to India and Arabian Peninsula; Tamil Nadu: Coimbatore, Dharmapuri, Dindigul, Kanniyakumari, Madurai, Pudukkottai, Tirunelveli, Tiruvannamalai and Virudhunagar
    Note: Fruits edible
64. Botanical name: *Ziziphus oenopolia* (L.) Mill.
    Habit: Shrub
    Tamil names: Surai ilanthai, Soorai mullu and Soorai pazham
    Distribution: Native to China (southern Yunnan, Guangxi) to tropical Asia and northern Australia; Tamil Nadu: Most of the districts
    Note: Fruits edible
65. Botanical name: *Ziziphus rugosa* Lam.
    Habit: Shrub
    Tamil names: Kaattilandai, Soorai and Suduthoraitti
    Distribution: Native to Indian subcontinent to China and Indochina; Tamil Nadu: Coimbatore, Dharmapuri, Dindigul, Erode, Kanniyakumari, Namakkal, Nilgiris, Salem, Theni, Tirunelveli, Tiruvannamalai, Vellore, Viluppuram and Virudhunagar
    Note: Fruits edible

Rosaceae

66. Botanical name: *Eriobotrya japonica* (Thunb.) Lindl.
    Habit: Tree
    Tamil names: Ilakotta and Nokkatta maram
    Distribution: Native to China (Chongqing, Hubei); Tamil Nadu: Dharmapuri, Dindigul, Namakkal, Nilgiri, Salem and Tirunelveli
    Note: Fruits edible
67. Botanical name: *Rubus ellipticus* Sm.
    Habit: Shrub
    Tamil name: Veyli pazham
    Distribution: Native to Indian subcontinent to China and Indochina and the Philippines; Tamil Nadu: Coimbatore, Dindigul, Namakkal, Nilgiris, Salem, Theni and Tirunelveli
    Note: Fruits edible
68. Botanical name: *Rubus fairholmianus* Gardner
    Habit: Shrub
    Distribution: Native to southwestern India and Sri Lanka; Tamil Nadu: Dindigul
    Note: Fruits edible
69. Botanical name: *Rubus niveus* Thunb.
    Habit: Shrub
    Tamil name: Kolalindu

Distribution: Native to Afghanistan to central China, Indochina and Indonesia (Java); Tamil Nadu: Coimbatore, Dindigul, Namakkal, Nilgiris, Salem, Theni, Tirunelveli and Virudhunagar
Note: Fruits edible

70. Botanical name: *Rubus racemosus* Roxb.
Habit: Shrub
Distribution: Native to the Western Himalayas and southern India; Tamil Nadu: Dindigul and Nilgiris
Note: Fruits edible

Rubiaceae

71. Botanical name: *Benkara malabarica* (Lam.) Tirveng.
Habit: Shrub
Tamil names: Mul pavattai, Pidavu and Pida
Distribution: Native to southern India and Sri Lanka; Tamil Nadu: Most of the districts
Note: Fruits edible

72. Botanical name: *Canthium coromandelicum* (Burm.f.) Alston
Habit: Shrub
Tamil names: Kaaraichedi and Nalla kaarai
Distribution: Native to southern India, Sri Lanka, Nepal and Bangladesh; Tamil Nadu: Most of the districts
Note: Fruits edible

73. Botanical name: *Meyna spinosa* Roxb. ex Link
Habit: Tree
Tamil name: Manakkarai
Distribution: Native to Indian subcontinent to Indochina; Tamil Nadu: Chengalpattu
Note: Fruits edible

74. Botanical name: *Pavetta tomentosa* Roxb. ex Sm.
Habit: Tree
Distribution: Native to the Indian subcontinent to southern China and the Philippines; Tamil Nadu: Coimbatore, Dharmapuri, Kanniyakumari, Namakkal, Nilgiris, Perambalur, Salem, Theni, Tiruchirappalli, Tiruvannamalai, Viluppuram and Virudhunagar
Note: Fruits edible

75. Botanical name: *Tamilnadia uliginosa* (Retz.) Tirveng. & Sastre
Habit: Tree
Tamil names: Vara kaarai and Wagatta
Distribution: Native to the Indian subcontinent to Indochina; Tamil Nadu: Chengalpattu, Kanniyakumari, Nilgiris, Tirunelveli, Tiruvannamalai and Vellore
Note: Fruits edible

Rutaceae

76. Botanical name: *Aegle marmelos* (L.) Corrêa
    Habit: Tree
    Tamil name: Vilvam
    Distribution: Native to the Indian subcontinent; Tamil Nadu: Most of the districts
    Note: Fruits edible
77. Botanical name: *Clausena anisata* (Willd.) Hook.f. ex Benth.
    Habit: Tree
    Tamil name: Kattu karivaeppilai
    Distribution: Native to tropical and southern Africa, the Indian subcontinent to southern central China; Tamil Nadu: Most of the districts
    Note: Fruits edible
78. Botanical name: *Glycosmis mauritiana* (Lam.) Tanaka
    Habit: Shrub
    Tamil name: Konji
    Distribution: Native to the Indian subcontinent to Hainan and western Malesia; Tamil Nadu: Most of the districts
    Note: Fruits edible
79. Botanical name: *Glycosmis pentaphylla* (Retz.) DC.
    Habit: Tree
    Tamil names: Anam and Kula pannai
    Distribution: Native to the Indian subcontinent to China and western and central Malesia; Tamil Nadu: Dharmapuri, Dindigul, Madurai, Namakkal, Salem, Theni, Tirunelveli, Tiruvannamalai, Vellore and Virudhunagar
    Note: Fruits edible
80. Botanical name: *Limonia acidissima* L.
    Habit: Tree
    Tamil name: Vila maram
    Distribution: Native to the Indian subcontinent to Andaman Islands; Tamil Nadu: Most of the districts
    Note: Fruits edible
81. Botanical name: *Pamburus missionis* (Wall. ex Wight) Swingle
    Habit: Tree
    Tamil names: Kaattu narinji and Kurundhu
    Distribution: Native to southeastern India and Sri Lanka; Tamil Nadu: Chennai, Coimbatore, Cuddalore, Dharmapuri, Kancheepuram, Nagapattinam, Nilgiris, Pudukkottai, Ramanathapuram, Salem, Tiruchirappalli, Tiruvallur, Tiruvannamalai and Viluppuram
    Note: Fruits edible
82. Botanical name: *Zanthoxylum armatum* DC.
    Habit: Liana
    Distribution: Native to the Indian subcontinent to temperate eastern Asia and Malesia; Tamil Nadu: Nilgiris
    Note: Fruits valued as spice

83. Botanical name: *Zanthoxylum asiaticum* DC.
Habit: Liana
Tamil names: Kindu mullu, Milagaranai and Sirusoori
Distribution: Native to Ethiopia to Eswatini, western Indian Ocean and tropical and subtropical Asia; Tamil Nadu: Most of the districts
Note: Fruits edible

Salicaceae

84. Botanical name: *Flacourtia indica* (Burm.f.) Merr.
Habit: Shrub
Tamil names: Kaatukala and Sothaikala
Distribution: Native to Ethiopia to southern Africa, southeastern China to tropical Asia; Tamil Nadu: Most of the districts
Note: Fruits edible
85. Botanical name: *Flacourtia inermis* Roxb.
Habit: Tree
Tamil name: Loveloikkaai
Distribution: Native to Malesia to Papuasia, Caroline Islands (Palau); Tamil Nadu: Kanniyakumari and Tirupathur
Note: Fruits edible
86. Botanical name: *Flacourtia jangomas* (Lour.) Raeusch.
Habit: Tree
Tamil name: Loveloikkaai
Distribution: Native to Nepal to China (southern Yunnan, western Guangxi) and northern Myanmar; Tamil Nadu: Coimbatore, Kanniyakumari and Tirunelveli
Note: Fruits edible
87. Botanical name: *Flacourtia montana* J. Graham
Habit: Tree
Tamil name: Malai sothaikala
Distribution: Native to western and southern India; Tamil Nadu: Coimbatore
Note: Fruits edible

Sapindaceae

88. Botanical name: *Allophylus cobbe* (L.) Forsyth f.
Habit: Tree
Tamil names: Amalai and Sirusalle
Distribution: Native to India to the Philippines; Tamil Nadu: Most of the districts
Note: Fruits edible
89. Botanical name: *Lepisanthes tetraphylla* (Vahl) Radlk.
Habit: Small tree
Tamil names: Neikottan and Koogaimutti

Distribution: Native to southern China to tropical Asia; Tamil Nadu: Most of the districts
Note: Fruits edible

90. Botanical name: *Schleichera oleosa* (Lour.) Oken
Habit: Large Tree
Tamil names: Chakkattai, Kolamaa, Konji-vanji, Miruga kottai
Distribution: Native to the Indian subcontinent to Indochina and central and southern Malesia; Tamil Nadu: Chennai, Coimbatore, Dharmapuri, Dindigul, Erode, Kanniyakumari, Namakkal, Nilgiris, Salem, Theni, Tirunelveli, Tiruvannamalai, Vellore, Viluppuram and Virudhunagar
Note: Fruits edible

Sapotaceae

91. Botanical name: *Manilkara hexandra* (Roxb.) Dubard (Fig. 1h)
Habit: Tree
Tamil name: Kannupalai, Paala maram and Ulakkai palai
Distribution: Native to India to China (S. Guangxi) and Peninsula Malaysia; Tamil Nadu: Most of the districts
Note: Fruits edible
92. Botanical name: *Manilkara kauki* (L.) Dubard
Habit: Tree
Distribution: Native to Indochina to northern Queensland (Australia); Tamil Nadu: Coimbatore, Kanniyakumari and Nilgiris
Note: Fruits edible
93. Botanical name: *Xantolis tomentosa* (Roxb.) Raf.
Habit: Tree
Tamil name: Paalai
Distribution: Native to India, Sri Lanka and Indochina; Tamil Nadu: Coimbatore, Nilgiris and Theni
Note: Fruits are made into pickles and curries.

Solanaceae

94. Botanical name: *Physalis angulata* L.
Habit: Herb
Tamil name: Sodakku thakkali
Distribution: Native to tropical and subtropical America; Tamil Nadu: Most of the districts
Note: Fruits edible
95. Botanical name: *Physalis lagascae* Roem. & Schult.
Habit: Herb
Tamil name: Sodakku thakkali
Distribution: Native to Mexico to Venezuela and Argentina, Cuba; Tamil Nadu: Most of the districts
Note: Fruits edible

96. Botanical name: *Physalis peruviana* L.
    Habit: Herb
    Tamil name: Pottipazham
    Distribution: Native to Bolivia to western Brazil; Tamil Nadu: Coimbatore, Dharmapuri, Dindigul, Namakkal, Nilgiris, Salem, Theni, Tirunelveli, Vellore and Viluppuram
    Note: Fruits edible
97. Botanical name: *Physalis philadelphica* Lam.
    Habit: Herb
    Distribution: Native to Mexico to Central America, Cuba to Haiti; Tamil Nadu: Most of the districts
    Note: Fruits edible
98. Botanical name: *Physalis pruinosa* L.
    Habit: Herb
    Tamil name: Sodakku thakkali
    Distribution: Native to Mexico to western Venezuela; Tamil Nadu: Most of the districts
    Note: Fruits edible
99. Botanical name: *Solanum melongena* L. subsp. *cumingii* (Dunal) J. Samuel
    Habit: Shrub
    Tamil name: Mullu kathiri
    Distribution: Native to western Indian Ocean and tropical and subtropical Asia; Tamil Nadu: Most of the districts
    Note: Fruits used as a vegetable
100. Botanical name: *Solanum pubescens* Willd.
    Habit: Shrub
    Tamil name: Kaattu sundai
    Distribution: Native to Arabian Peninsula and India; Tamil Nadu: Coimbatore, Dharmapuri, Dindigul, Erode, Kanniyakumari, Namakkal, Nilgiris, Salem, Theni, Tirunelveli, Tiruvannamalai, Vellore, Viluppuram and Virudhunagar
    Note: Fruits pickled and eaten

Vitaceae

101. Botanical name: *Ampelocissus latifolia* (Roxb.) Planch.
    Habit: Shrub
    Tamil name: Kaattu kodi mundhiri
    Distribution: Native to the Indian subcontinent to northern Thailand; Tamil Nadu: Chengalpattu, Coimbatore, Dindigul, Kancheepuram, Nilgiris, Tirunelveli, Tiruvannamalai and Virudhunagar
    Note: Fruits edible

Zygophyllaceae

102. Botanical name: *Balanites aegyptiaca* (L.) Delile
Habit: Tree
Tamil names: Nanjundan and Koolathi
Distribution: Native to Africa, Israel to Arabian Peninsula; Tamil Nadu: Coimbatore, Namakkal, Tiruchirappalli, Tiruvannamalai and Virudhunagar
Notes: Fruits edible

Conclusion

Out of the 102 edible fruit-yielding wild plants enumerated here, trees (51) are the dominant growth form, followed by shrubs (34), herbs (8), lianas (5) and climbers (4). Solanaceae and Rutaceae are the two dominant families with seven fruit-yielding wild plants each, followed by Rhamnaceae with six and Rubiaceae and Rosaceae with five each. This preliminary checklist of edible fruit-yielding wild plants of Tamil Nadu will serve as the baseline data for effective protection, conservation and sustainable utilization of these wild plants of the state for posterity. Perhaps, these fruit-yielding plants, especially the native elements, may be introduced as an intercrop in cultivated lands on trial basis, and on success they may be suggested for mass cultivation.

References

1. Shackleton CM. The prevalence of use and value of wild edible herbs in South Africa. S Afr J Sci. 2003;99:23–5.
2. Harisha RP, Padmavathy S, Nagaraja BC. Traditional Ecological Knowledge (TEK) and its importance in South India: perspective from local community. Appl Ecol Environ Res. 2015;14(1):311–26.
3. Motti R. Wild edible plants: a challenge for future diet and health. Plan Theory. 2022;11:344.
4. Gescher A, Pastorino U, Plummer SM, Manson MM. Suppression of tumour development by substances derived from the diet-mechanisms and clinical implications. Br J Clin Pharmacol. 1998;45:1–12.
5. Lock K, Pamerleau J, Causer L, Altmann DR, McKee M. The global burden of diseases attributable to low consumption of fruit and vegetables: implications for the global strategy on diet. Bull World Health Organ. 2005;83:100–8.
6. Li Y, Zhang J, Xu D, Zhou T, Zhou Y, Li S, Li., H. Bioactivities and health benefits of wild fruits. Int J Mol Sci. 2016;17(8):1258.
7. Mahapatra AK, Panda PC. Wild edible fruit plants of Eastern India. Bhubaneshwar: Regional Plant Resource Centre; 2009.
8. Grivetti LF, Ogle BM. Value of traditional foods in meeting macro- and micro nutrients needs: the wild plant connection. Nutr Res Rev. 2000;13(1):31–46.
9. Gamble JS. Flora of the Presidency of Madras. 11 Parts. (Parts 1–7 by Gamble & 8–11 by C.E.C. Fischer). London: Adlard & Son Ltd.; 1915–1936.
10. Nair NC, Henry AN. Flora of Tamil Nadu, India. Ser. 1: analysis. Volume 1. Coimbatore: Botanical Survey of India; 1983.

11. Henry AN, Kumari GR, Chithra V. Flora of Tamil Nadu, India. Ser. 1: analysis. Volume 2. Coimbatore: Botanical Survey of India; 1987.
12. Matthew KM. *Flora of Palni Hills*. Part I. Ranunculaceae—Alangiaceae. Tiruchirappalli: The Rapinat Herbarium, St. Joseph's College (Autonomous); 1999.
13. Manickam VS, Jeya Jothi G, Ignacimuthu S. Flora of Tirunelveli Hills (Southern Western Ghats). Volume III. Monochlamydeae. Dehra Dun: Bishen Singh Mahendra Pal Singh; 2022.
14. Manickam VS, Murugan C, Jeya Jothi G. Flora of Tirunelveli Hills (Southern Western Ghats). Volume I. Polypetalae. Dehra Dun: Bishen Singh Mahendra Pal Singh; 2008.
15. Vijayasankar R, Ravikumar K, Ravichandran P. Plant resources of Tiruvannamalai District, Tamil Nadu, India. Dehra Dun: Bishen Singh Mahendra Pal Singh; 2012.
16. Narasimhan D, Irwin SJ. Flowering plants of Tamil Nadu: a compendium. Chennai: Care Earth Trust; 2021.
17. Alagesaboopathi C, Balu S, Dwarakan P. Edible fruit yielding plants of Shevaroy Hills in Tamil Nadu. Anc Sci Life. 1996;16:148–51.
18. Arinathan V, Mohan VR, Britto JD, Murugan C. Wild edibles used by Palliyars of the Western Ghats, Tamil Nadu. Indian J Tradit Knowl. 2007;6(1):163–8.
19. Ramachandran VS. Wild edible plants of the Anamalais, Coimbatore District, Western Ghats, Tamil Nadu. Indian J Tradit Knowl. 2007;6(1):173–6.
20. Sasi R, Rajendran A, Mahrajan M. Wild edible plant diversity of Kotagiri hills- a part of Nilgiri Biosphere Reserve, Southern India. J Res Biol. 2011;2:80–7.
21. Anand SP, Deborah S. Enumeration of wild edible fruits from Boda hills and Kolli hills. Int J Appl Biol Pharm Technol. 2016;7(2):96–102.
22. Sanu C, Jeevith S, Sheeba TS. Ethnomedicinal knowledge used by Mullukurumbas of Nilgiris, Western Ghats, India. Asian J Ethnobiol. 2023;6(2):115–26.

Application of Herbal-Based Edible Coating on Fruits: A Promising Technology to Counter Postharvest Losses

Murugan Rajan, Maria Terezinha Santos Leite Neta, and Narendra Narain

Abstract

Consumption of fruits can reduce the risk of development of chronic degenerative diseases because they are rich in antioxidants especially polyphenols; hence the demand for fruits has increased of late in the market. Fruits have a short shelf life, and about 30% fruits are damaged by insects or microbes during transport and preservation. In India, postharvest loss of fresh produces is a serious problem during pre- and postharvest conditions. In order to decrease the postharvest losses of fruits, there is a need for research on novel techniques to prolong the shelf life of food products. On the other hand, consumer's expectation is low consumption of chemicals on minimally processed fruits, and hence more attention is been paid to the search for naturally occurring antimicrobial and antioxidant substances. Recently, the herbal compounds are being increasingly studied for their suitability as an additive in the edible coating of fruits which would be an interesting and innovative technology for commercial application and an alternative to the use of postharvest treatments leading to the increment of shelf-life of fruits without using any chemical preservatives. In this present review, an attempt is made to summarize the postharvest losses of fruits in India and the current state of research findings on the nutraceutical properties of edible coating of fruits using herbal compounds.

M. Rajan (✉)
Centre for Research and Postgraduate Studies in Botany, Ayya Nadar Janaki Ammal College (Autonomous), Sivakasi, Tamil Nadu, India

Maria Terezinha Santos Leite Neta · N. Narain
Laboratory of Flavor and Chromatographic Analysis, Federal University of Sergipe, Sao Cristovao - SE, Brazil

P. Thangaraj (ed.), *Pomotherapeutic Insights on Wild Edible Fruits*,
https://doi.org/10.1007/978-981-96-0543-9_17

Keywords

Antioxidant · Antimicrobial · Edible coating · Herbs · Polyphenols · Shelf life

Introduction

In developing countries, more than 70% people depend on fruits which are commonly consumed by humans in the diet. Furthermore, fruits play a vital role as antioxidants to prevent the development of chronic diseases due to the presence of active principles in them [1]. Thus, sustainable agriculture has to achieve food security in combination with economic viability and social responsibility and have as little effect on biodiversity and natural ecosystems as possible. Generally, consumers prefer to eat fresh foods and usually prefer the fresh fruits on the basis of their texture, color, and quality. Thus, before marketing, most of the fruits are harvested in bulk and preserved in a suitable environment. The harvested fruits are metabolically active, undergoing ripening and senescence processes that must be controlled to prolong postharvest quality and their shelf life. However, during the storage and transportation times, the food materials may deteriorate by respiration which affects the nutrition, flavor, color, sensory, and texture qualities and cause microbial contaminations [2, 3]. Since every commodity shows different response during storage, it is difficult to preserve the shelf life and nutritional quality of all fruits. Thus, fresh fruits require careful storage to prevent deterioration as they are very susceptible [4].

The postharvest and marketing procedures need to be developed in order to maintain the fruits at their best quality. An example of the importance on the field to postharvest handling is the discovery that ripening of fruit can be delayed and thus their storage prolonged by preventing or decreasing the fruit tissue respiration. However, they are perishable living products that require coordinated activity by growers, storage operators, processors, and retailers to maintain quality and hence reduce food loss and waste. Aular (2006) reported that 40% losses of crops occurred during the handling of postharvest plant products, differing among products, production areas, and time of the year [5].

India with diverse agroclimatic zones produces a very wide variety of fruits, and the country is the world's second largest producer of fruits, having production of about 75.27 million tons which contribute to 12.49% of the total world production [6]. However, the Central Institute of Post-Harvest Engineering and Technology (CIPHET) reported the extended loss of harvest and postharvested fruits is about 6%. In addition, nutritional loss (loss of vitamins, antioxidants, and health-promoting substances) and decreased market value are another important loss that occurs in fresh produce. Thus, it is extremely important to develop sustainable technologies to maintain the quality and shelf life of fruits [7]. The main objective of this present review is to focus on developing ways, including coating of fruits so as to reduce the postharvest losses of fruits in India and study the nutritional quality of edible-coated fruits.

Postharvest Losses of Fruits

Srinivas et al. [8] conducted a survey in Karnataka to assess postharvest losses of "Totapuri" (Bangalora) and "Alphonso" (Badami) mangos in Karnataka. Total postharvest losses of "Totapuri" occurring in the field were 3.5% (orchard/field), 4.9% (during transportation), 4.1% (storage), and 5.4% (at retail level). However, 1.9% orchard/field, 3.7% transportation, 3.5% storage, and 5.3% at retail level losses were observed in the "Alphonso" variety. The major causes of losses may occur by mechanical injuries, spoilage, either over mature/shriveling, or immature/unmarketable sizes, pilferage, and damage by birds/hailstorms.

Murthy et al. [9] reported the postharvest losses in "Banganapalli" mangoes at different stages of marketing in Andhra Pradesh. The average postharvest loss at the farm level was 15.6%, which was due to the harvest of immature and small fruits. The postharvest loss of mangoes during storage and retail level was estimated at 8.8% and 5.25%, respectively. The major cause for this loss was pressing injury, which caused about 51% of the fruit damage. The other factors for retail level losses were a black spot (31%) and injury due to mechanical and physical causes. The total postharvest loss in "Banganapalli" mangoes from production to consumption was estimated to be 29.70%.

Gajanana et al. [10] conducted a survey to estimate the postharvest loss of banana cv. Poovan in the local market of Tamil Nadu. It was observed that 3.9% of loss occurred at farm level sorting. During the transportation, the loss was estimated to be in the range of 2.19–2.52% due to the long distances of transportation. At wholesale and retail market storage, the losses were 2.52% and 7.50%, respectively.

Murthy et al. [11] conducted a survey in Bijapur district of Karnataka on grapes. The loss in sorting, transportation, and local and distant retail markets was observed to be at 7.31%, 4.24%, 2.85%, and 3.27%, respectively. In Himachal Pradesh, postharvest losses in apple [12], mango, peach, and Kinnow/orange were found to be 14.48%, 24.45%, 18.31%, and 24.51%, respectively. The losses were more in the retailer's level in all the selected fruits except apple.

Sreenivasa Murthy et al. [13] studied the marketing losses of banana in Karnataka. This study reported 5.53% at the field level, 6.65% at the wholesale level, and 16.66% at the retail level of postharvest losses in wholesale marketing cycle systems, whereas in the cooperative marketing systems, the loss was 7.82%, 1.77%, and 8.72%, respectively, of the corresponding stages.

The fruits of apple, banana, citrus, grapes, guava, mango, papaya, and sapota constituted about 84% of produce in the country. The losses of different fruits in farm level operations were estimated to vary between 4.2% and 13.9%. The losses of different fruits in storage at various levels were observed to be in the range of 1.2–4.1%. Total losses in farm operations were the highest in the case of guava (13.9%) followed by those in apple (11.1%) and mango (10.6%). However, the losses in farm operations for other fruits were in the relatively lower range of 4–7%. Guava also exhibited the highest storage loss (4.1%), whereas the storage losses for other fruits were in the range of 1.5–2.4%. Transportation losses were observed to be generally high in all the selected fruits but were more pronounced in the case of guava, mango, and grapes. The overall total losses were the highest (18.0%) for guava and the lowest (5.8%) for sapota.

In light of the above incidence of the huge postharvest losses of fruit in India and new challenges faced under trade liberalization and globalization, serious efforts are being made to reduce postharvest losses. New technologies have been introduced to prolong horticultural products, including physical treatments (ultraviolet light, heat treatment, low temperature storage), hydrothermal treatment application, chemical treatments (chlorine-based solutions, calcium chloride, salicylic acid, gibberellic acid (GA_3), naphthalene acetic acid (NAA) peroxyacetic acid, organic acids, hydrogen peroxide, and electrolyzed water), gaseous (ozone, ethylene, and 1-methylcyclopropene) treatments, and plastic packaging (packaging films using polyethylene, polyethylene terephthalate, polypropylene, and polyvinyl chloride) usage aiming to create modified or controlled atmospheres to maintain and extend shelf life of agricultural produces [14].

Generally, these treatments have been used commercially to slow down the ripening of fruits and maintain the fruit cell wall integrity, peel color, aroma, and flavor. However, prolonged duration affects the shelf life and nutrient quality of fruits and reduces the dietary substances like sugars and ascorbic acid, and it can increase susceptibility to CO_2 injury and chilling disorders. In addition, inadequate management of these processes can result in outbreaks of foodborne pathogens and financial loss for all players along the supply chain, from growers to consumers [15]. On the other hand, synthetic polymers used as food packages tend to react with food materials releasing naphtha-based compounds which may be carcinogenic on ingestion by humans [16] and contribute to environmental pollution [17, 18]. Thus, there is a need to discover advance techniques for the preservation of processed food products as well as fresh food products like fruits, which is of direct use to food industry. Nowadays food industries are focusing on the eco-friendly biodegradable packaging such as edible films and coatings [19]. The edible coating forms a protective barrier around the material commodity and can be consumed along with the coated product [20]. Moreover, edible coatings are used to create a modified atmosphere which reduces also the weight loss during transport and storage [21]. Thus, this review focuses on the recent findings regarding the use of a protective edible coating to maintain or enhance the shelf life of fresh fruits.

Edible Coating

The thin layer of coating material can be applied to the fruit surfaces which provide barrier to oxygen, microbes of external source, moisture, solute movement, gas exchange, oxidative reaction rates, and respiration as well as to reduce physiological disorders which are known as "edible coatings" [22, 23]. The main application of the edible coating is to increase the shelf life of fresh cut fruits, and it is protected from postharvest damages and environmental damages [24]. The edible coatings serve as a carrier of texture enhancer and antioxidants, and it is used as a nutraceutical. Moreover, the edible coating should be stable, generally recognized as safe

(GRAS) under high relative humidity [25, 26]. Edible coating or edible films are mostly tasteless, colorless, and odorless, but they should have good mechanical properties.

According to Pavlath and Orts [27], a different type of material was used for coating and wrapping for various fruits to extend their shelf life, which can be eaten together with foods and is considered an edible coating. They are applied directly on the food surface by dipping, spraying, or brushing to create a modified atmosphere [18, 28]. An edible coating is defined as one that can extend the storage life of fresh fruits without causing anaerobiosis and which reduces decay without affecting their quality. Earlier studies on edible coatings have been used to reduce water loss, but nowadays formulated edible coatings with a wider range of permeability characteristics have extended the potential for fresh produce application which increases further their market value [29, 30]. Edible coatings are also known as biodegradable coatings which have special functional requirements, for example, moisture barrier, solute or gas barrier, water/lipid solubility, color appearance, mechanical characteristics, nontoxicity, etc. The effect of edible coatings on fruits mainly depends on temperature, alkalinity, thickness, and type of coating, associated with the variety and condition of fruits [30].

Furthermore, edible coatings may control adhesion, cohesion, and durability and improve the appearance of coated foods. Antimicrobials and antioxidants containing edible coating may inhibit microbial growth and extend product shelf life. In the preservation of paneer, edible coating with antimicrobial agents could be the best alternative since such techniques have been successfully applied in some dairy products, e.g., cheese, and other food products like fruits, confectionary products, chocolate, etc. The edible coating is a thin continuous layer of edible material formed on or placed between food and food components [31]. The coating varies from transparent to translucent depending on formulation, purity, sources, and composition. The plasticizers are low molecular weight high boiling point molecules that impart flexibility and extensibility to the edible polymeric film, e.g., whey proteins. They facilitate the processing of edible coating films by reduction of the intermolecular forces between adjacent polymer chains, resulting in a reduction of the polymer glass transition temperature.

History of Edible Coating

A historical view of edible coatings was wax which was the first edible coating used on fruits. The Chinese applied wax coatings to oranges and lemons in the twelfth and thirteenth centuries to slow down the respiratory gas exchange. They found that wax-coated fruits could be stored longer than non-waxed fruits [32]. In the 1930s, hot-melt paraffin waxes were applied in the apples and pears which were commercially available as edible coatings. Lowings and Cuts [33] reported non-phytotoxic, tasteless, and odorless sucrose fatty acid esters (SFAE), sodium carboxymethyl cellulose, and mono- and diglyceride edible coating material which were effective in preserving fruits. Since the 1980s, SFAE mixtures have been commercially available for coating fruits under the trade names "TAL Pro-long" and Semper Fresh. Erbil and MuFtuGil [34] reported that wax emulsion coating of peach surfaces decreased water vapor and oxygen transmission, thus diminishing respiration rate and increasing shelf life. Nisperos-Carriedo et al. [35] and Baldwin et al. [36] observed that oils or waxes and cellulose had similar effects of preventing spoilage and retaining fresh-picked quality of tropical fruits. Several attempts have been made to develop other materials that could be used to coat, produce, and modify internal gas composition for short-term storage. El-Ghaouth et al. [37] and Zhang and Quantick [38] suggested that chitin and chitosan (deacetylated chitin) from marine invertebrates could be used as a transparent film for application as an edible coating on fruits. Park et al. [30] applied zein coating on the surface of tomatoes and reported that the film coating delayed color change and weight loss and maintained firmness during storage.

Recently, edible polymeric packaging materials have been made from polysaccharides, proteins, and lipids. They are generally used to minimize respiration in fruits, limit the movement of moisture and other gases, provide antimicrobial or antioxidant capabilities to the product, and extend the shelf life of the product [39, 40]. The edible films can also be classified into three types based on their origin: (a) hydrocolloids such as proteins, polysaccharides, and alginate, (b) lipids such as fatty acids, acylglycerol, and waxes, and (c) composites [41]. Thus, the application of edible coatings on fruits is an effective method to improve their quality and shelf-life, and it can be safely eaten as a part of the food. Presently there has been increased interest in using the herbal extract as an edible coating material for fruits driven by its beneficial properties.

Herbal-Based Edible Coating

Herbal-based edible coatings have shown to prevent water loss, control ripening process and respiration rate, delay oxidative browning, and reduce microbial growth without any toxicity. These herbal edible coatings increased the shelf life of fresh produce, and it also increased storage time of 15–35 days at low temperature. Consequently, herbal extracts are being increasingly studied as additives in the edible coating on fruits which would be an interesting and innovative option for commercial application and an alternative to the use of postharvest chemical treatment leading to the increment of shelf life of fruits. Recently, edible coatings are prepared by the mixing extracts of herbs such as thyme (*Thymus vulgaris*), cinnamon (*Cinnamon cassia* and *Cinnamon zeylanicum* Blume), oregano (*Origanum vulgare*), lemongrass (*Cymbopogon citratus*), neem (*Azadirachta indica*), *Aloe vera*, clove (*Syzygium aromaticum* L.), mint (*Mentha viridis* L.), and peppermint (*Mentha piperita* L.) [42–44]. The herbal extracts contain antimicrobial, antioxidant, and therapeutic properties; it also acts as a nutraceutical. Lemongrass, cinnamon oil, and oregano film-coated fruits showed lower oxygen permeability and contained antibacterial compounds of citral, cinnamaldehyde, and carvacrol, respectively [45, 46]. Thus, herbal edible coatings are a promising concept because of their nontoxic, eco-friendly nature while increasing the market value of fruits. Table 1 summarizes the beneficial postharvest quality effects of the edible coating of fruits using natural compounds. Edible-coated fruits could act as antioxidants and antimicrobial barrier, which also enhances the nutritional quality of fruits. On the other hand, some publications have revealed that extract incorporation in films may generate a heterogeneous surface with numerous small pores [47], which could account for the high water vapor permeability of the incorporated films. In the same way that natural extracts can be successfully incorporated into biodegradable films, the use of edible coatings in fruits could improve food quality and shelf life. The recent studies reported that natural compounds/herbal extract incorporated with edible films (chitosan, alginate, pectin, gelatin, etc.) also extends the shelf life, enhances the postharvest quality, and improves the nutrient quality (Table 2). Nevertheless, some publications have shown that this technology can better control weight loss and respiration rates, allowing for longer storage time as compared with samples without coating [48–50].

Application of Antioxidant Films

Recently, synthetic antioxidants (butylated hydroxy toluene, butylated hydroxy anisole, and tert-butyl hydroxyquinone) have raised some safety concerns, and regulatory agencies have restricted their use as food additives. Thus, researchers have targeted films containing antioxidant agents from natural sources from herbal extracts [51, 52]. These extracts should also contribute to nutritional quality aspects without affecting the food product integrity [28]. Some of the plant extracts like tea extracts [50, 53], fruit extracts [49, 54], ginseng extract [47], propolis [48], and various plant extracts possess excellent antioxidant activity, which can retard lipid oxidation and improve the quality and shelf life of various food model systems in different ways. The antioxidant activity of these extracts results mainly from phenolic compounds and their synergistic, antagonistic, and additive effects [18]. Recently, fruit extracts have been considered for application as natural bioactive additives for their coloring potential, pharmaceutical activities, and bioactivity, regarding aspects of hygiene, nutrition, and environmental consciousness [54]. Several studies on antioxidant extracts that confer color to films have been reported [47, 53–57]. In addition, it has been well documented that these extracts exhibit coloring and antioxidant properties that, in some cases, enable good control against photooxidation through reduced light transmission, especially UV radiation [47, 48, 53]. In general, the physical properties of the film, like moisture content and water solubility, remain unaltered upon the addition of extracts, because the extract and the film matrix interact well. Li et al. [53] analyzed gelatin-based film incorporated with tea extracts through FTIR and verified that the extracts establish hydrogen bonds with gelatin, to reduce free hydrogen. Research has indicated that applying an edible coating containing antioxidants to fresh-cut fruits effectively reduces browning while increasing the antioxidant capacity of the coated or packed food [58].

Table 1 Nutraceutical property of herbal extracts based on edible-coated fruits

S. No.	Commodity	Name of the edible coating	Texture quality	Nutrient quality	Sources
1	Pineapple	*Aloe vera*	Reduced the firmness and weight losses	Increased ascorbic acid content	Adetunji et al. [69]
2	Papaya		Extending the shelf life up to 15 days	Increased ascorbic acid content	Sharmin et al. [70] Marpudi et al. [71]
3	Grapes		Increasing storage life and maintaining their quality	Increased total phenolic content and antioxidant activity	Asghari et al. [72]
4	Sweet cherry		Increasing storage life	Improved total phenolic content, vitamin C, and catalase enzyme activity	Asghari et al. [72]
5	Kiwifruit		Maintained the firmness and yellowing due to ripening than chitosan and alginate coating	Prevented the loss of ascorbic acid content	Benitez et al. [73]
6	Apple	Neem extract	Extend the storage life	Antimicrobial property	Chauhan et al. [42]
7	Fruits	Linalool, eugenol, carvacrol, camphor, methyl cinnamate, and β-caryophyllene	Extend the storage life	Antioxidant and antimicrobial properties	Reuveni et al. [74], Kayastha [75]
8	Fresh-cut apples	The cinnamon bark ethanol extract	Increased the storage life for 12 days at 6 °C	Reduced the gram-negative and gram-positive bacteria	Muthuswamy et al. [76]
9	Kiwifruit and fresh-cut melon	Cinnamic acid and carvacrol	Extend the storage life	Reduced the microbial spoilage; reduced viable counts	Roller and Seedhar [77]
10	Fruits	Mint, peppermint oil, lemongrass, turmeric, green tea extract, marigold, thyme, clove oil, basil extract, and garlic extract	Extend the storage life and protect the texture qualities	Antioxidant and antimicrobial properties	Baranauskien et al. [78], Chauhan et al. [42], Chiu and Lai [79], Toro-Sanchez et al. [80]

11	Kinnow fruit	Corn starch, stearic acid, jojoba oil, monoglycerides, and emulsion of Jojoba oil and water	Superior effect on shelf life and stability	–	Din et al. [81]
12	Chilgoza (*Pinus gerardiana*)	Gum Cordia	Highest increase in shelf life (oxidation value, aroma, sensory) than carboxymethyl cellulose	Improve the peroxide and TBARS value	Abdul Haq et al. [82]
13	Mangaba fruits	Cassava starch (*Myrcia ovata Cambessedes*) and essential oil	Prolong the shelf life	Controlling foodborne bacteria and natural microorganism growth	Frazaao et al. [83]
14	Guava	Tulsi extract	Significant effect on the carbon dioxide evolution rate than others	–	Murmu et al. [84]
15	Beluga sturgeon fillets	Jujube gum and nanoemulsions containing nettle essential oil	Lowered weight and cooking losses, pH changes, textural and color deterioration, lipid oxidation, and microbial growth	Good antimicrobial and antioxidant barrier	Gharibzahedi and Mohammadnabi [85]
16	Fresh-cut Fuji apples	Lemongrass essential oil	Reduced respiration rates and ethylene production No browning and ethylene production	–	Salvia-Trujillo et al. [86]
17	Hayward kiwifruit slices; 'Dottato' fig (*Ficus carica* L.)	*Opuntia ficus-indica* mucilage	Improved quality of stored fresh-cut kiwifruits	Maintain the firmness and ascorbic acid and pectin contents	Allegra et al. [87]
18	Cut apricots	Basil-seed gum containing *Origanum vulgare* subsp. *viride* essential oil	Reduced total plate count, yeasts, and mold populations	Enhanced total soluble phenolic contents and antioxidant activity	Hashemi et al. [88]
19	'Dottato' fig (*Ficus carica* L.)	*Opuntia ficus*-Indica mucilage	Maintained fruit fresh weight, visual score values, and fruit firmness	Improved total carotenoid content; lower development of *Enterobacteriaceae*	Allegra et al. [87]

Table 2 Nutraceutical properties of mixtures of chemical and herbal extracts based on edible-coated fruits

S. No.	Commodity	Name of the edible coating	Texture quality	Nutrient quality	Sources
1	Blueberry (*Vaccinium corymbosum*)	Chitosan-*Aloe vera* coating	Expanding the shelf life	Reduced the postharvest contamination by fungi	Jorge et al. [89]
2	Fresh-cut melon, apple, date palm, papaya, and mango	Lemongrass oil with alginate, gelatin, apple puree, and chitosan	Improves the quality of fruits and extends their shelf life for 3 weeks	–	Chiu and Lai [79]
3	Grape berries (*Vitis vinifera* L. *Vitis labruscana* Bailey)	Lemongrass oil-chitosan	Greater retention of color and total soluble solid content	Greater efficiency in increasing microbial safety against *Salmonella* and preserving grape berries than larger droplets and antioxidant activity	Oh et al. [90]
5	Formosa plum	Essential oils (oregano and bergamot)-hydroxypropyl methylcellulose	Effective in reducing the respiration rate, ethylene production, total weight loss, and total cell count	–	Choi et al. [91]
6	Chinese winter jujube	Tea polyphenols-alginate	Reduced red indices, respiration rate, and electrolyte leakage	Reduced total chlorophylls and malonaldehyde contents and maintaining the ascorbic acid, total phenol contents, and antioxidant enzymes	Zhang et al. [92]
7	Nanfeng mandarin	*Ficus hirta* fruits-sodium alginate	Enhanced long-term quality and storage	Improved activities of antioxidant enzymes such as superoxide dismutase, catalase, chitinase, -1,3-glucanase, and phenylalanine ammonia lyase and stimulated the accumulation of phenolic compounds	Chen et al. [93]
8	Guavas (*Psidium guajava* L.)	Chitosan-cassava starch coatings enriched with *Lippia gracilis* Schauer	Extend the shelf life at room temperature; maintained the total soluble solids, pH, and titratable acidity	Good antimicrobial activity	Aquino et al. [94]

9	Butternut squash and *Listeria monocytogenes*	Chitosan-oleoresins (rosemary and olive)	Prevented the browning reactions	Improved antioxidant protection	Ponce et al. [95]
10	Fresh-cut pineapple	Lemongrass essential oil-alginate	Lower counts of yeast, mold, and total plate; decreased firmness and sensory scores (taste, texture, and overall acceptability); extend the shelf life and maintain quality	–	Azarakhsh et al. [96]
12	*Arbutus unedo*	Eugenol and citral-alginate	Improved postharvest quality; preserved sensory and storage life	Better nutritional attributes and reduced microbial spoilage	Guerreiro et al. [97]
13	Raspberry fresh fruit	Essential oils of citral and eugenol and pectin and alginate	Color parameter and Trolox equivalent antioxidant capacity were not significantly affected; citral reduced hue and firmness and increased weight loss; effective at reducing microbial spoilage and sensory attributes	/–	Guerreiro et al. [97]
14	Fresh-cut pineapple (*Ananas comosus*)	Chitosan-mucilages and pullulan	Preserve the quality properties (color, odor, flavor, texture); improved the quality and prolonged the shelf life	–	Trevino-Garza et al. [98]
15	Fresh and cut guavas	Cashew gum-carboxymethyl cellulose	Reduced water loss and changes in color	–	Forato et al. [99]
16	Raspberry	Essential oils (citral and eugenol) and alginate and pectin	–	Not cytotoxic; did not significantly affect nutritional quality; controlled microbial food spoilage	Guerreiro et al. [100]
17	Avocado (*Persea americana* Mill.) fruit	Leaf extract, chitosan, and carboxymethyl cellulose	Improve fruit quality and shelf life	–	Tesfay and Magwaza [101]

Application of Antimicrobial Films

The use of antimicrobial compounds containing edible coatings is an alternative to enhance the safety of fresh produces. The physical and chemical barrier provided by the epidermis, which prevents the development of microbes on the fruit surface, is removed during postharvest handling [59]. Dipping of aqueous solutions containing antimicrobials is the most practical way to extend the microbial stability of fruits. However, application of antimicrobial agents directly on the food surface may have limited benefits because the active substances are rapidly neutralized or diffused from the surface into the food product, thus limiting the effect of the antimicrobial compound [60]. In this sense, antimicrobial coatings may provide increased inhibitory effects against spoilage and pathogenic bacteria by maintaining effective concentrations of the active compounds on the food surfaces [61]. The antibacterial effectiveness of organic acids is thought to appear from the fact that protonated acids are membrane soluble and can enter the cytoplasm by simple diffusion [62]. Essential oils (EOs) are also alternatives to chemical preservatives, and their use in food meets the demands of consumers for natural products [63]. The activity of EOs and their active constituents has been widely studied against many microorganisms, including several pathogens [64–66]. The application of EOs on foods is yet limited due to their impact on organoleptic food properties and their variable activity in foods due to interactions with food components [67]. Nevertheless, the use of EOs to control microbial growth in foods has been proposed for several fruits, and the data are presented in Tables 1 and 2. Many factors must be considered in developing an antimicrobial edible coating, which includes the properties of the food and coating and the effectiveness of the antimicrobial agents incorporated into the coating. Despite the good results achieved so far with the incorporation of EOs into edible coatings, the major drawback is their strong flavor, which could change the original taste of foods [68].

Conclusion

In conclusion, all the research findings indicated edible coatings of fruits using natural compounds from herbs or natural compounds incorporated with chitosan, alginate, and pectin act as antioxidant and antimicrobial barriers, enhancing the external and internal quality characteristics and shelf life of highly perishable fruits. It is also an environment-friendly method of preservation of food for a long time and acts further as nutraceuticals. The development of new technologies to improve the delivery properties of edible coatings is a major issue for future research. Despite these extensive uses of edible coating, the potential use of this application of fruits in India is rather limited. The introduction of edible coating using herbs is a novel technique to overcome the postharvest losses of fruits as well as nutraceutical functions while improving the economic value of fruits.

Acknowledgments This study was funded in part by the *Conselho Nacional de Desenvolvimento Científico e Tecnológico* (CNPq) vide research Project *Instituto Nacional de Ciência e Tecnologia de Frutos Tropicais* (Process 465335/2014-4). Authors gratefully acknowledge the financial support of *Coordenação de Aperfeiçoamento de Pessoal de Nível Superior*—Brazil under *Programa Nacional de Pós-doutorado* (PNPD/CAPES- 086/2013—Financial code 001).

Conflict of Interest All authors declare that there is no conflict of interest.

Data Availability Statement The data supporting the findings of this study are available from the corresponding authors upon request.

References

1. Nahak G, Suar M, Sahu RK. Antioxidant potential and nutritional values of vegetables: a review. Res J Med Plant. 2014;8(2):50–81.
2. Gatto MA, Ippolito A, Linsalata V, Cascarano NA, Nigro F, Vanadia S, Venere DD. Activity of extracts from wild edible herbs against postharvest fungal diseases of fruit and vegetables. Postharvest Biol Technol. 2011;61(1):72–82.
3. Terry LA, Joyce DC. Elicitors of induced disease resistance in postharvest horticultural crops: a brief review. Postharvest Biol Technol. 2004;32(1):1–13.
4. Kader AA, Zagory D, Kerbel EL, Wang CY. Modified atmosphere packaging of fruits and vegetables. Crit Rev Food Sci Nutr. 1989;28(1):1–30.
5. Aular J. Consideraciones sobre el manejo postcosecha de frutas en Venezuela. In: Salamanca G, editor. Anales del Seminario Hortofrutícola Colombiano y I Congreso Iberoamericano sobre Sistemas de Procesado. Ibague-Tolima; 2006. p. 59–62.
6. NHB. Indian Horticulture Database-2011, National Horticulture Board, Ministry of Agriculture, Government of India, 85, Institutional Area, Sector-18, Gurgaon-122 015 India.
7. Sharma A, Yadav BS. Resistant starch: physiological roles and food applications. Food Rev Int. 2008;24(2):193–234.
8. Srinivas RN, Venkatesh Reddy T, Ravi PC, Lalitha A, Chinnappa Reddy BV. Post-harvest loss assessment of Totapuri and Alphonoso mangoes. J Food Sci Technol. 1997;34(1):70–2.
9. Murthy SD, Gajanana TM, Sudha M, Subrahmanyam KV. Post-harvest loss estimation in mango at different stages of marketing – a methodological perspective. Agric Econ Res Rev. 2002;15(2):188–200.
10. Gajanana TM. Marketing practices and post harvest loss assessment of banana var. Poovan in Tamil Nadu. Agric Econ Res Rev. 2002;15(1):56–65.
11. Murthy SD, Gajanana TM, Sudha M. Post harvest losses and its impact on marketing cost, margin and efficiency: a study on grapes in Karnataka. Indian J Agric Econ. 2004;59(4):773–86.
12. Singh RV. Evaluation of post harvest losses in apple in Himachal Pradesh. Agro-economic Research centre, H.P. University, Shimla. (Memo). 2002.
13. Sreenivasa Murthy D, Gajanana TM, Sudha M, Dakshinamoorthy V. Marketing losses and their impact on marketing margins: a case study of banana in Karnataka. Agric Econ Res Rev. 2007;20(7):47–60.
14. Gonzalez-Aguilar G, Ayala-Zavala J, Olivas G, de la Rosa L, Alvarez-Parrilla E. Preserving quality of fresh-cut products using safe technologies. J für Verb Leb. 2010;5:65–72.
15. Mahajan PV, Caleb OJ, Singh Z, Watkins CB, Geyer M. Postharvest treatments of fresh produce. Phil Trans: Math Phy Eng Sci. 2017;372:2013–309.
16. Parra DF, Tadini CC, Ponce P, Lugao AB. Mechanical properties and water vapour transmission in some blends of cassava starch edible films. Carbohydr Poly. 2004;58(4):475–81.
17. Arvanitoyannis I, Psomiadou E, Nakayama A. Edible films made from sodium caseinate, starches, sugars or glycerol; part 1. Carbohydr Poly. 1996;31(4):179–92.

18. Krochta JM, Mulder-Johnston CD. Edible and biodegradable polymer films: challenges and opportunities. Food Technol. 1997;51(2):61–74.
19. Galgano F, Condelli N, Favati F, Dibianco V, Perretti G, Caruso MC. Biodegradable packaging and edible coating for fresh-cut fruits and vegetables. Ital J Food Sci. 2015;27(1):1–20.
20. Guilbert S. Technology and application of edible protective films. In: Mathlouthi M, editor. Food packaging and preservation. London: Elsevier Applied Science; 1986. p. 371–94.
21. Baldwin EA. Edible coatings for fresh fruits and vegetables: past, present, and future. In: Krochta JM, Baldwin EA, Nisperos-Carriedo M, editors. Edible coatings and films to improve food quality. Lancaster: Technomic Publishers Co.; 1994. p. 25–63.
22. Baldwin EA, Nesperos-Carriedo MO, Chen X, Hagenmaier RD. Improving storage life of cut apple and potato with edible coating. Postharvest Biol Technol. 1996;9(2):151–63.
23. Park HJ. Development of advanced edible coating of fruits. Trends Food Sci Technol. 1999;10(8):250–60.
24. Mohamed AYE, Aboul AHE, Hassan AM. Utilisation of edible coating in extending the shelf life of minimally processed prickly pear. J App Sci Res. 2013;9(2):1202–8.
25. Undurraga PL, Oleta JA, Taito M. Effect of N, O-carboxymethyl chitosan, and nutrisaveon avocado fruits has during cool storage. In: Proceeding of the World Avocado Congress 3rd; 1995. p. 362–5.
26. Rojas-Grau MA, Montero-Calderon M, Martin-Belloso O. Effect of packaging condition on quality and shelf life of fresh-cut pineapple *(Ananas comoves)*. Postharvest Biol Technol. 2008;50(2–3):182–9.
27. Pavlath AE, Orts W. Edible films and coating: why, what and how? In: Embuscado ME, Hurber KC, editors. Edible coating and film for food application. Springer Sc. Business Media, LIC; 2009. p. 1–24.
28. Guilbert S, Gontard N, Gorris LGM. Prolongation of the shelf-life of perishable food products using biodegradable films and coatings. LWT – Food Sci Technol. 1996;29(1–2):10–7.
29. Avena-Bustillos RJ, Cisneros-Zevallos LA, Krochta JM, Saltveit ME. Application of casein-lipid edible film emulsions to reduce white blush on minimally processed carrots. Postharvest Biol Technol. 1994;4(4):319–29.
30. Park HJ, Chinnan MS, Shewfelt RL. Edible corn-zein film coatings to extend storage life of tomatoes. J Food Proc Pres. 1994;18(4):317–31.
31. Torres JA. Edible films and coatings from proteins. In: Hettiarachchy NS, Ziegler GR, editors. Protein functionality in food systems. New York., ISSN 0065–2415: Marcel Dekker; 1994. p. 467–507.
32. Hardenburg RE. Wax and related coatings for horticulture products. A bibliography. Agric Res Bull. 1967;5:1–15.
33. Lowings PH, Cutts DF. The preservation of fresh fruits and vegetables. In: Proceedings of the Institute of Food Science and Technology Annual Symposium. Nottingham, U.K. ; 1982. p. 52.
34. Erbil HY, MuFtuGil N. Lengthening the postharvest life of peaches by coating with hydrophobic emulsions. J Food Proc Pres. 1986;10:269–79.
35. Nisperos-Carriedo MO, Show PE, Baldwin EA. Changes in volatile flavor components of pineapple and orange juice as influenced by the application of lipid and composite films. J Agric Food Chem. 1990;38(6):1382–7.
36. Baldwin EA, Nisperos MO, Baker RA. Use of edible coatings to preserve quality of lightly (and slightly) processed products. Crit Rev Food Sci Nutr. 1995;35(6):509–52.
37. El-Ghaouth A, Ponnampalam R, Castaigne F, Arul J. Chitosan coating to extend the storage life of tomatoes. Hort Sci. 1992;27(9):1016–8.
38. Zhang DL, Quantick PC. Effects of chitosan coating on enzymatic browning and decay during post-harvest storage of litchi fruit. Postharvest Biol Technol. 1997;12(2):195–202.
39. Pascall MA, Lin SJ. The application of edible polymeric films and coatings in the food industry. J Food Process Technol. 2013;4(2):1–2.
40. Marsh K, Bugusu B. Food packaging-roles, materials and environmental issues. J Food Sci. 2007;72(3):39–54.

41. Donhowe IG, Fennema OR. The effects of plasticizers on crystallinity, permeability, and mechanical properties of methylcellulose films. J Food Process Preserv. 1993;17(4):247–57.
42. Chauhan S, Gupta KC, Agrawal M. Efficacy of natural extracts on the storage quality of Apple. Int J Curr Microbiol App Sci. 2014;3(3):1–6.
43. Padmaja N, John Don Bosco S. Preservation of Jujube fruits by edible *Aloe vera* gel coating to maintain quality and safety. Indian J Sci Res Technol. 2014;2(3):79–88.
44. Zinoviadou KG, Koutsoumanis KP, Biliaderis CG. Physico-chemical properties of whey protein is late films containing oregano oil and their antimicrobial action against spoilage flora of fresh beef. Meat Sci. 2009 Jul;82(3):338–45.
45. Rojas-Grau MA, Avena-Bustillos RJ, Friedman M, Henika PR, Martin-Belloso O, Mchugh TH. Mechanical, barrier, and antimicrobial properties of apple puree edible films containing plant essential oils. J Agric Food Chem. 2006;54(24):9262–7.
46. Rojas-Grau MA, Avena-Bustillos RJ, Olsen C, Friedman M, Henika PR, Martin-Belloso O, Pan Z, Mchugh TH. Effects of plant essential oils and oil compounds on mechanical, barrier and antimicrobial properties of alginate-apple puree edible films. J Food Eng. 2007;81(3):634–41.
47. Norajit K, Kim KM, Ryu GH. Comparative studies on the characterization and antioxidant properties of biodegradable alginate films containing ginseng extract. J Food Eng. 2010;98(3):377–84.
48. Pastor C, Sanchez-Gonzalez L, Chiralt A, Chafer M, Gonzalez-Martínez C. Physical and antioxidant properties of chitosan and methylcellulose based films containing resveratrol. Food Hydrocol. 2013 Jan;30(1):272–80.
49. Supapvanich S, Prathaan P, Tepsorn R. Browning inhibition in fresh-cut rose apple fruit cv. taaptimjaan using konjac glucomannan coating incorporated with pineapple fruit extract. Postharvest Biol Technol. 2012;73:46–9.
50. Das DK, Dutta H, Mahanta CL. Development of a rice starch-based coating with antioxidant and microbe-barrier properties and study of its effect on tomatoes stored at room temperature. LWT – Food Sci Technol. 2013;50(1):272–8.
51. Murcia MA, Martinez-Tome M. Antioxidant activity of resveratrol compared with common food additives. J Food Prot. 2001;64(3):379–84.
52. De'nobili MD, Perez CD, Navarro DA, Stortz CA, Rojas AM. Hydrolytic stability of l-(+)-ascorbic acid in low methoxyl pectin films with potential antioxidant activity at food interfaces. Food Bioproc Tech. 2013;6(1):186–97.
53. Li JH, Miao J, Wu JL, Chen SF, Zhang QQ. Preparation and characterization of active gelatin-based films incorporated with natural antioxidants. Food Hydrocoll. 2014;37:166–73.
54. Akhtar MJ, Jacquot M, Jasniewski J, Jacquot C, Imran M, Jamshidian M, Paris C, Desobry S. Antioxidant capacity and light-aging study of HPMC films functionalized with natural plant extract. Carbohydr Polym. 2012;89(4):1150–8.
55. Gomez-Estaca J, Bravo L, Gomez-Guillen MC, Aleman A, Montero P. Antioxidant properties of tuna-skin and bovine-hide gelatin films induced by the addition of oregano and rosemary extracts. Food Chem. 2009;112(1):18–25.
56. Gomez-Estaca J, Gimenez B, Montero P, Gomezguillen MC. Incorporation of antioxidant borage extract into edible films based on sole skin gelatin or a commercial fish gelatin. J Food Eng. 2009;92(1):78–85.
57. Bitencourt CM. Development and application of gelatin-based films added with ethanolic extract of Curcuma (*Curcuma longa* L.). In: Dissertation (Master's degree in science). Pirassununga: Faculty of Animal Husbandry and Food Engineering, University of São Paulo; 2013. p. 107 f.
58. Kaliana Sitonio ECA, Sartori T. Films and edible coatings containing antioxidants—a review. Braz J Food Technol. 2014;17(2):98–112.
59. Martin-Belloso O, Soliva-Fortuny R, Oms-Oliu G. Fresh-cut fruits. In: Hui YH, editor. Handbook of fruits and fruit processing. Ames, Iowa: Blackwell Publishing; 2006. p. 129–44.

60. Min S, Krochta JM. Inhibition of *Penicillium commune* by edible whey protein films incorporating lactoferrin, lactoferrin hydrolysate, and lactoperoxidase systems. J Food Sci. 2005;70(2):M87–94.
61. Gennadios A, Kurth LB. Application of edible coatings on meats, poultry and sea foods: a review. Leb-Wiss und-Technol. 1997 Jun;30(4):337–50.
62. Ricke SC. Perspectives on the use of organic acids and short chain fatty acids as antimicrobials. Poul Sci. 2003;82(4):632–9.
63. Burt S. Essential oils: their antibacterial properties and potential applications in foods: a review. Int J Food Microbiol. 2004;94(3):223–53.
64. Karatzas AK, Bennik MHJ, Smid EJ, Kets EPW. Combined action of S-carvone and mild heat treatment on *Listeria monocytogenes* Scott A. J Appl Microbiol. 2000;89(2):296–301.
65. Vazquez BI, Fente C, Franco CM, Vazquez MJ, Cepeda A. Inhibitory effects of eugenol and thymol on *Penicillium citrinum* strains in culture media and cheese. Int J Food Microbiol. 2001;67(1–2):157–63.
66. Delaquis PJ, Stanich K, Girard B, Mazza G. Antimicrobial activity of individual and mixed fractions of dill, cilantro, coriander and eucalyptus essential oils. Int J Food Microbiol. 2002;74(1–2):101–9.
67. Gutierrez J, Barry-Ryan C, Bourke P. The antimicrobial efficacy of plant essential oil combinations and interactions with food ingredients. Int J Food Microbiol. 2008;124(1):91–7.
68. Dhall RK. Advances in edible coatings for fresh fruits and vegetables: a review. Crit Rev Food Sci Nutr. 2013;53(5):435–50.
69. Adetunji CO, Fawole OB, Arowora KA, Nwaubani SI, Ajayi ES, Oloke JK, Majolagbe OM, Ogundele BA, Aina JA, Adetunji JB. Effects of edible coatings from *Aloe vera* gel on quality and postharvest physiology of *Ananas Comosus* (L.) fruit during ambient storage. Glob J Sci Front Res Bio-Tech Gen. 2012;12(5):39–43.
70. Sharmin MR, Islam MN, Alim MA. Shelf-life enhancement of papaya with aloe vera gel coating at ambient temperature. J Bangladesh Agric Univ. 2015;13(1):131–6.
71. Marpudi SL, Abirami LSS, Pushkala R, Srividya N. Enhancement of storage life and quality maintenance of papaya fruits using *Aloe vera* based antimicrobial coating. Indian J Biotechnol. 2011;10:83–9.
72. Asghari M, Khalili H, Rasmi Y, Mohammadzadeh A. Influence of postharvest nitric oxide and *Aloe vera* gel application on sweet cheery quality indices and storage life. Int J Agron Plant Prod. 2013;4(9):2393–8.
73. Benitez S, Achaerandio I, Pujola M, Sepulcre F. *Aloe vera* as an alternative to traditional edible coatings used in fresh cut fruits: a case of study with kiwifruit slices. LWT – Food Sci Technol. 2015;61(1):184–93.
74. Reuveni R, Fleisher A, Putieusky E. Fungistatic activity of essential oils from *Ocimum basilicum* chemo-types. Phytopathol Z. 1984 May;110(1):20–2.
75. Kayastha BL. Queen of herbs tulsi (*Ocimum sanctum*) removes impurities from water and plays disinfectant role. J Med Plant Stud. 2014;2(2):1–8.
76. Muthuswamy S, Rupasinghe HPV, Stratton GW. Antimicrobial effect of cinnamon bark extract on *Escherichia coli* O157:H7, *Listeria innocua* and fresh-cut apple slices. J Food Saf. 2008;28(4):534–49.
77. Roller S, Seedhar P. Carvacrol and cinnamic acid inhibit microbial growth in fresh- cut melon and kiwifruit at 4° and 8 °C. Lett App Microbiol. 2002;35(5):390–4.
78. Baranauskien R, Bylait E, Zukauskait J, Venskutonis RP. Flavor retention of pepper mint (*Mentha piperita* L.) essential oil spray-dried in modified starches during encapsulation and storage. J Agric Food Chem. 2007;55(8):3027–36.
79. Chiu P, Lai L. Antimicrobial activities of tapioca starch decolorized hsian-tsao leaf gum coatings containing green tea extracts in fruit-based salads, romaine heart sand pork slices. Int J Food Microbiol. 2010;139(1–2):23–30.
80. Toro-Sanchez DC, Ayala-Zavala J, Machi L, Santacruz H, VillegasOchoa M, Alvarez-Parrilla E, Gonzalez-Aguilar G. Controlled release of antifungal volatiles of thyme essential oil from Cyclodextrein capsules. J Incl Phenom Macrocycl Chem. 2010;67(3):431–41.

81. Din A, Nadeem M, Rafique M, Shabbir MA. Development and application of edible skin coatings to improve the quality of kinnow during storage. Acta Sci Technol. 2015;37(1):111–6.
82. Abdul Haq M, Junaid Alam M, Hasnain A. Gum Cordia: a novel edible coating to increase the shelf life of Chilgoza (*Pinus gerardiana*). LWT – Food Sci Technol. 2013;50(1):306–31.
83. Frazaao GGS, Blank AF, Santana LCLA. Optimisation of edible chitosan coatings formulations incorporating *Myrcia ovata* Cambessedes essential oil with antimicrobial potential against foodborne bacteria and natural microflora of mangaba fruits. LWT – Food Sci Technol. 2017;79:1–10.
84. Murmu SB, Mishra HN. Optimization of the Arabic gum based edible coating formulations with sodium caseinate and tulsi extract for guava. LWT – Food Sci Technol. 2017;80:271–9.
85. Gharibzahedi SMT, Mohammadnabi S. Effect of novel bioactive edible coatings based on jujube gum and nettle oil-loaded nanoemulsions on the shelf-life of Beluga sturgeon fillets. Int J Biol Macromol. 2017;95:769–77.
86. Salvia-Trujillo L, Rojas-Grau MA, Soliva-Fortuny R, Martín-Belloso O. Use of antimicrobial nanoemulsions as edible coatings: impact on safety and quality attributes of fresh-cut Fuji apples. Postharvest Biol Technol. 2015;105:8–16.
87. Allegra A, Inglese P, Sortino G, Settanni L, Todaro A, Liguori G. The influence of *Opuntia ficus-indica* mucilage edible coating on the quality of 'Hayward' kiwifruit slices. Postharvest Biol Technol. 2016;120:45–51.
88. Hashemi SMB, Khaneghah AM, Ghahfarrokhic MG, Eşd I. Basil-seed gum containing *Origanum vulgare* subsp. viride essential oil as edible coating for fresh cut apricots. Postharvest Biol Technol. 2017;125:26–34.
89. Jorge M, Vieiraa F-LML, de Rodríguezc DJ, Sousa MC, Vicenteb AA, Martinsb JT. Effect of chitosan–*Aloe vera* coating on postharvest quality of blueberry (*Vaccinium corymbosum*) fruit. Postharvest Biol Technol. 2016;116:88–97.
90. Oh YA, Oh YJ, Song AY, Won JS, Song KB, Min SC. Comparison of effectiveness of edible coatings using emulsions containing lemongrass oil of different size droplets on grape berry safety and preservation. LWT – Food Sci Technol. 2017;75:742–50.
91. Choi WS, Singh S, Lee YS. Characterization of edible film containing essential oils in hydroxypropyl methylcellulose and its effect on quality attributes of 'Formosa' plum (*Prunus salicina* L.). LWT – Food Sci Technol. 2016;70:213–22.
92. Zhang L, Li S, Dong Y, Zhi H, Zong W. Tea polyphenols incorporated into alginate-based edible coating for quality maintenance of Chinese winter jujube under ambient temperature. LWT – Food Sci Technol. 2016;70:155–61.
93. Chen C, Peng X, Zeng R, Chen M, Wan C, Chen J. *Ficus hirta* fruits extract incorporated into an alginate-based edible coating for Nanfeng mandarin preservation. Sci Hort. 2016;202:41–8.
94. Aquino DAB, Blank AF, Santana LCLA. Impact of edible chitosan–cassava starch coatings enriched with *Lippia gracilis* Schauer genotype mixtures on the shelf life of guavas (*Psidium guajava* L.) during storage at room temperature. Food Chem. 2015;171:108–16.
95. Ponce AG, Rour SI, Del Valle CE, Moreir MR. Antimicrobial and antioxidant activities of edible coatings enriched with natural plant extracts: *in vitro* and *in vivo* studies. Postharvest Biol Technol. 2008;49(2):294–300.
96. Azarakhsha N, Osmana A, Ghazalia HM, Tanb CP, Adzahanb NM. Lemongrass essential oil incorporated into alginate-based edible coating for shelf-life extension and quality retention of fresh-cut pineapple. Postharvest Biol Technol. 2014;88:1–7.
97. Guerreiroa AC, Gagoa CML, Faleiro ML, Miguela MGC, Antunesa MDC. The effect of alginate-based edible coatings enriched with essential oils constituents on *Arbutus unedo* L. fresh fruit storage. Postharvest Biol Technol. 2015;100:226–33.
98. Trevino-Garzaa MZ, Garciab S, Herediab N, Alanis-Guzmanc MG, Arevalo-Ninoa K. Layer-by-layer edible coatings based on mucilages, pullulan and chitosan and its effect on quality and preservation of fresh-cut pineapple (*Ananas comosus*). Postharvest Biol Technol. 2017;128:63–75.

99. Forato LA, de Britto D, de Rizzo JS, Gastaldi TA, Assis OBG. Effect of cashew gum-carboxymethylcellulose edible coatings in extending the shelf-life of fresh and cut guavas. Food Pack Shelf Life. 2015 Sep;5:68–74.
100. Guerreiro AC, Gagoa CML, Miguela MGC, Faleirob ML, Antunes MDC. The influence of edible coatings enriched with citral and eugenol on the raspberry storage ability, nutritional and sensory quality. Food Pack Shelf Life. 2016;9:20–8.
101. Tesfaya SZ, Magwaza LS. Evaluating the efficacy of moringa leaf extract, chitosan and carboxymethyl cellulose as edible coatings for enhancing quality and extending postharvest life of avocado (*Persea americana* Mill.) fruit. Food Pack Shelf Life. 2017;11:40–8.

Index

A
Agriculture, 3, 4, 59, 71, 134, 136, 216, 217, 238, 242, 243, 245, 328
Agroforestry, 219, 257
Alkaloids, 40, 43, 48, 83, 91, 94, 96, 105, 144, 146, 151, 192, 239, 275
Alzheimer's disease (AD), 47, 50, 92
Anthocyanins, 69, 78, 151, 198, 242, 292, 296, 297, 306
Anti-aging, 264, 276

B
Bael, 5, 254
Beverages, 2, 45, 56, 154, 209, 281
Bioactive molecules, 2, 193
Bioavailability studies, 128
Biotic interactions, 2, 12
Botanical investigations, 14

C
Cardioprotective, 55, 140
Cardiovascular diseases, 2, 13, 55, 68, 82, 118, 122, 123, 125, 137, 193
Cash crops, 4
Catechins, 44, 202, 203, 207, 255, 299, 300
Chlorogenic acid, 46, 292, 299, 300
Chronic diseases, 2, 50, 55, 59–61, 68, 69, 118, 121, 306, 328
Chronic illnesses, 2, 239
Climate change, 58, 70, 240, 242
Clinical trials, 41, 126–127, 129, 181
Conservation efforts, 57, 69–73, 109, 218
Cosmetics, 41, 136, 148
Culinary uses, 70, 116, 134
Cultural significance, 56, 70, 72, 105, 116, 241, 257

D
Deciduous trees, 223, 226, 228, 232
Dietary equilibrium, 20, 28
Dietary fibers, 2, 54, 69, 116, 117, 207
Dietary requirements, 14, 19–35, 238
Dietary supplements, 41, 158, 207, 257
Drug discovery, 79, 85, 90

E
Eastern Ghats, 2–15, 218, 225
Ecosystem services, 72, 216
Elaeocarpus serratus, 4, 5, 7–10
Ellagic acid, 239, 246, 248, 254, 299
Ethnic communities, 3, 14, 20, 288, 291
Ethnic people, 2–4, 28
Ethnobotanical surveys, 144
Ethnobotany, 40, 238
Exotic fruits, 54, 216

F
Fermentation, 281
Fertile, 3, 153
Folk knowledge, 4
Folk medicines, 45, 101, 136, 138, 140
Folk remedies, 144, 245
Food crops, 3, 238
Food habits, 20, 247, 306
Food insecurity, 238
Food safety, 14
Food security, 14, 55–58, 70, 135, 137, 158, 216, 233, 245, 306, 328
Food systems, 3, 59, 60, 71, 158
Forest produces, 2
Fruit morphology, 167

P. Thangaraj (ed.), *Pomotherapeutic Insights on Wild Edible Fruits*,
https://doi.org/10.1007/978-981-96-0543-9

G
Gas chromatography, 178
Genetic modification, 3
Genomics, 169
Gut microbiota, 55

H
Habitat destruction, 69
Harvesting techniques, 247
Hepatoprotective, 105, 156
Hypertension, 13, 27, 28, 90, 193, 275

I
Immune system, 14, 55, 63, 120–124, 240, 281
Indigenous communities, 56, 57, 71, 136, 241, 243
Indigenous fruits, 58, 59
Indigenous knowledge, 14, 55, 71, 173, 216, 233
Inflammatory bowel diseases, 122, 123, 129

J
Jamun, 254

L
Livelihood development, 14
Local inhabitants, 4, 20, 21, 26, 28

M
Malayali tribes, 3–7, 9–11, 221
Malnutrition, 2, 3, 12, 14, 70, 119, 134, 135, 193, 238, 306
Medicinal uses, 4, 116, 144, 146, 150, 223
Micronutrients, 2, 10, 12–14, 55, 70, 135, 197, 217, 233, 238, 242, 257
Microorganisms, 49, 121, 123, 335, 338
Morphological descriptions, 1

N
Neuroprotective, 55, 95, 96, 175, 239
Non-timber forest products (NTFPs), 2–4, 164, 168, 243–245, 247
Novel medication, 41
Nutraceuticals, 60, 68, 95, 135, 178, 181, 207, 238, 242, 281, 330, 332, 334–338
Nutritional analyses, 12, 14

O
Obesity, 55, 96, 280
Overharvesting, 240

P
Packaging, 267, 330, 332
Pharmacists, 39
Phenolic acids, 44, 48, 100, 146, 154, 197
Phoenix loureiroi, 4, 6–8, 10, 13, 14
Phytochemical profiling, 173
Phytomedicines, 39
Phytopharmacology, 180
Phytosterols, 157
Phytotherapeutic medicines, 40
Pollination, 42, 274
Polyphenols, 46, 54, 55, 59–61, 68, 116, 124, 125, 137, 158, 197, 239
Potherbs, 20

Q
Quercetin, 155, 239, 254, 288, 289, 299

R
Rainfall patterns, 70
Regional cuisines, 241
Resilience, 2, 58, 109, 244
Resource management, 72
Resveratrol, 239

S
Saponins, 43, 44, 46, 48, 91, 105, 197, 207, 254, 275, 279, 288, 293, 300
Seed dispersal, 145
Shelf life, 70, 78, 117, 134, 172, 265, 328, 330–338
Solanum torvum, 4, 6–8, 11, 13, 14, 25, 31, 219
Sustainable conservation, 15
Sustainable development, 56–58, 134, 264
Sustainable use, 134, 216
Synthetic medicines, 40–41

T
Taxonomic classification, 1
Therapeutics, 2–4, 38–40, 44, 50, 55, 62, 71, 78, 79, 82–85, 93, 94, 96, 97, 99, 116–129, 140, 142, 144, 146, 149, 154, 164, 168, 173–175, 192, 193,

225, 232, 239, 241, 244, 245, 257, 264, 276, 280, 281, 288, 290–292, 295–298, 300, 301, 332
Traditional healing practices, 78
Tribal markets, 3
Tropical fruits, 3, 164–181, 238, 332

U
Underutilized fruits, 54, 238, 257

V
Value addition, 58

W
Weeds, 20, 29
Wild berries, 57, 58, 241
World Health Organization (WHO), 39, 40, 79, 192, 265
Wound healing, 62–64, 99, 149, 150, 155, 205, 279

X
Xanthine oxidase pathway, 40

Z
Ziziphus rugosa, 4, 6–8, 12, 13, 301–302

The manufacturer's authorised representative in the EU is Springer Nature Customer Service Centre GmbH, Europaplatz 3, 69115 Heidelberg, Germany. If you have any concerns regarding our products, please contact ProductSafety@springernature.com

Printed and bound by CPI Group (UK) Ltd, Croydon, CR0 4YY
01/10/2026
02225939-0004